161 Topics in Current Chemistry

W0107403

Macrocycles

Editor: E. Weber and F. Vögtle

With contributions by
A. K. Burrell, J. Dohm, M. Hesse, P. Knops,
E. Koepp, Q. Meng, H.-B. Mekelburger,
A. Ostrowicki, N. Sendhoff, J. L. Sessler, F. Vögtle

With 46 Figures and 10 Tables

Springer-Verlag Berlin
Heidelberg GmbH

This series presents critical reviews of the present position and future trends in modern chemical research. It is addressed to all research and industrial chemists who wish to keep abreast of advances in their subject.

As a rule, contributions are specially commissioned. The editors and publishers will, however, always pleased to be receive suggestions and supplementary information. Papers are accepted for "Topics in Current Chemistry" in English.

ISBN 978-3-662-14965-2
DOI 10.1007/978-3-540-47581-1

ISBN 978-3-540-47581-1 (eBook)

© Springer-Verlag Berlin Heidelberg 1992
Originally published by Springer-Verlag Berlin Heidelberg New York in 1992
Softcover reprint of the hardcover 1st edition 1992

Typesetting: Th. Müntzer, Bad Langensalza; Printing: Heenemann, Berlin;

51/3020-543210 — Printed on acid-freepaper

Preface

Macrocycles are a highly topical subject. They constitute a large spectrum of compounds involving both artifical substances and natural products such as crowns, cryptands, cyclophanes, porphyrins, or macrolides. The former initiated the exiting area of host-guest supramolecular chemistry, which was highlighted by the award of the Nobel Prize for Chemistry to *D. J. Cram, J.-M. Lehn*, and *C. J. Pedersen* in 1987 but is still developing enormously. Porphyrins and macrolides are important active substances. No wonder that macrocycles are of immediate interest and everyone wants to know how they can be synthesized efficiently.

The present volume is intended to provide this knowledge, showing synthetic principles and creative strategies for the above-mentioned classes of macrocycles with stress on crowns, strained and cavity-shaped cyclophanes, expanded porphyrins and macrolide antibiotics.

General effects supporting macroring formation (high-dilution reaction, caesium salt assistance), also discussed, are now gaining wide use in macrocyclic chemistry. Although the synthetic aspect is placed in the forefront, startling properties of novel macrocycles are specified at several places in the volume as well. For instance, some of the expanded porphyrines are highly promising as far-red absorbing photosensitizers for use in photodynamic human therapy, others show potential applications in magnetic resonance imaging.

It is hoped that the book will be of particular value to those whose interest is the design and synthesis of macrocycles. The volume contains five chapters written by contributors who are eminently capable of meeting this expectation.

Bonn, Oktober 1991   Edwin Weber

Table of Contents

High Dilution Reactions —
New Synthetic Applications

Peter Knops, Norbert Sendhoff**, Hans-Bernhard Mekelburger,
and Fritz Vögtle*

Institut für Organische Chemie und Biochemie der Universität Bonn,
Gerhard-Domagk-Straße 1, W-5300 Bonn 1, FRG

Table of Contents

---

* To whom correspondence should be addressed.
** Present address: BASF AG, W-6700 Ludwigshafen, FRG.

Topics in Current Chemistry, Vol. 161
© Springer-Verlag Berlin Heidelberg 1991

The chemistry of medio- and macrocycles experienced an impetuous development during the past years, especially through the increasing significance of *supramolecular chemistry* [1] which was valued in 1987 by awarding the Nobel Prize to Pedersen, Cram, and Lehn [2].

The following contribution is intended to explain recent syntheses of macrocycles using the dilution principle after giving a short introduction into the basics of dilution principle reactions [3].

1 Introduction and Historical Review

The synthesis of alicyclic ring systems makes use of open-chained starting materials which are cyclized by a ring closure reaction. These are usually standard reactions, e.g., Dieckmann condensation of diesters, Glaser coupling of acetylenes, intramolecular Wurtz coupling, Acyloin condensation, etc.

However, the following problem is encountered when cyclizations are carried out in practice: yields of *carbo*cycles are the largest for common rings (5–7 carbon atoms), lower for small rings (3–4 carbon atoms), and poor for medium rings (8–12 carbon atoms). This is due to *strain effects* on one hand and *entropic effects* on the other hand: In small rings the strain opposes ring formation, but the probability of ring closure is higher than in the case of longer chains. In common rings the negative entropy of activation, i.e., the lower probability for a coincidence of the reactive centers of the molecule and a subsequent cyclization, is more than compensated by the considerably decreasing ring strain, whereas in medium rings, in addition to the low probability for a coincidence of the two reactive ends of the open-chained starting material, transannular strains further diminish the yield of the cyclization. Large rings do not have any ring strain. However, the probability for a coincidence of the reactive centers is extremely low, i.e., the yield of the cyclization is likewise very poor [4].

If, however, ring formation is carried out at low concentrations of the reactants, ring closure is favored compared to oligomerization because the reacting molecules are "isolated", therefore more time is available for the intramolecular reaction. This observation was made as early as 1912 by Ruggli in the reaction of 2,2'-diaminotolane (1) with succinyl chloride (2) which yields the lactam 3:

"... However, the expectation was that such a chloride can also react with the amino groups of different molecules and that in this way long chains and rings ... can develop. To reduce their formation a diluted solution was used; since after a reaction on one side has taken place the probability of polymerization at the expense of simple ring formation decreased with increasing dilution ..." [5]

Subsequently, experimental procedures of dilution principle reactions were further improved and standardized, so that today it is possible to synthesize many desired compounds by suitable choice of components and dilution conditions, often with the assistance of other effects, e.g., the *"rigid group principle"* [6], the *"cesium effect"* [7a, 8], and the *"template effect"* [9].

In the following chapters a progress report on recent syntheses using the dilution principle shall be given; some examples of reactions which work without dilution conditions, but nevertheless use the basic ideas of the dilution principle, are presented, as well.

2 Theoretical Approaches [10–15]

The practical aspects of the dilution principle, i.e., the choice between different possible reactants and reaction parameters (solvent, rate of addition, amount of solvent), are determined as empirically today as in those days when the basic ideas of the principle were developed. Quantitative approaches were attempted by several authors but the pretension of a complete mathematical description considering the multitude of possible types of reactions is rather illusory. One can only hope to obtain general rules for the experimental procedure. Some of these approaches are outlined briefly:

The *"theory of the effective molarity"* (EM) [10]: Galli and Mandolini defined the EM as the reactant concentration at which the *intra*molecular and the *inter*molecular processes occur at the same rate ($k_{intra}/k_{inter} = 1$). If the concentration of the reactants is small enough, the intramolecular ring formation is favored. For the method normally used, where the reactants are introduced slowly into a large volume of solvent, the rate of addition is decisive.

The *"Monte Carlo method"* [11]: This purely statistical method allows — on the assumption that the rate of cyclization is not dependent on the size of the macrocycle being formed — the following conclusions:

- with increasing dilution the portion of cyclic compounds increases at the expense of the formation of linear oligomers.
- the dilution principle is not confined to certain ring sizes.
- not the absolute rate constants are decisive for the success of a cyclization, but the ratio k/k_c (k = rate constant for the formation of linear compounds; k_c = rate constant for the cyclization). If the ratio k/k_c is small, the influence of dilution is minor.
- the best yields are obtained by using equivalent amounts of reactants (provided that the remaining parameters are constant).

The method of Fastrez [12]: It considers the reaction of two symmetrical, bifunctional monomers (A-A and B-B with A functions reactive toward B) which are added to an extremely large volume of solvent at a constant rate. The theoretical yield of the reaction is calculated — on the assumption that the probability of the formation of a macrocycle is not dependent on its size, i.e., with neglect of the entropic term — by setting up differential equations for the rate of formation of the different possible products (Fig. 1).

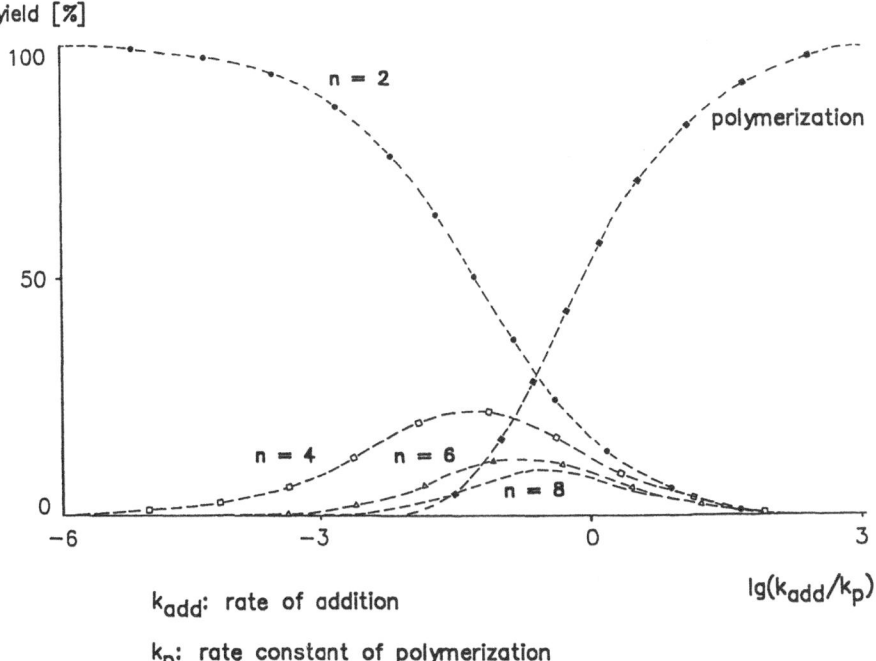

Fig. 1. Calculated yields of cyclic oligomers as a function of the rate of addition of the monomers [12]

From the calculations the following conclusions have been derived by Fastrez:

— at slow rates of addition, dimers are formed.
— at higher rates of addition, more and more larger cyclic oligomers are formed.
— to diminish the yield of the dimer (n = 2) from 90 to 10%, the rate of addition has to be enhanced by a factor of 10^4.
— the yields of cyclic tetramers, hexamers, etc. (n = 4, 6 ...) as a function of the rate of addition resemble a bell-shaped curve and are not very dependent on the rate of addition.
— under consideration of the entropic term, even lower yields of cyclic tetramers, hexamers, etc. (n = 4, 6 ...) are obtained.

It has to be emphasized, however, that conformational effects are not included in this theoretical consideration.

Other authors [10, 11b, 13] have derived that the maximum concentration which favors the intramolecular at the expense of the intermolecular reaction is about 10^{-2}–10^{-3} mol/l. High temperatures usually favor the intramolecular reaction, whereas low temperatures ($-70\,°C$) favor oligomer formation [14].

5

3 Examples of Recent Cyclizations Using the Dilution Principle

The following contribution compiles recently published cyclizations. Most of these examples have not been included in our former survey on the dilution principle [3a] and are thought as a continuation and update of that former survey up to 1990. As shown there, some characteristic experimental standard methods are specified here as well. (For a standard apparatus and a modified motor-driven syringe-type version see [3d, e].)

3.1 Formation of C−S Single Bonds [6b, 7b, 16−26]

Cyclizations by formation of carbon-sulfur-bonds belong to the best studied reactions which use the dilution principle. Usually bromides react with thiols [7b, 17, 18], thiuronium salts [19], or with sodium sulfide [20] and thioacetamide, respectively. The formation of C−S bonds represents one of the most valuable methods to synthesize medium-membered cyclophanes such as **4−8** [7b, 18b, 20d, 21].

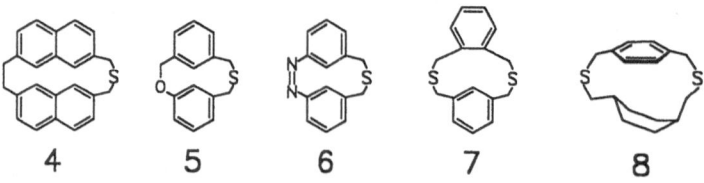

| 4 | 5 | 6 | 7 | 8 |

In general they can be transformed into strained [2.2]cyclophanes by subsequent extrusion of sulfur. (In case of the [3.2]cyclophane **6**, all attempts of desulfurization have been unsuccessful up to now [21]; no report on attempts to desulfurize hexahydrodithia[3.3]paracyclophane **8** is known so far [18h].)

Even severely strained ring systems such as the bridged biphenylene **9** [20c] and the helical chiral [2.2]metacyclophane **10** [22, 23] could be synthesized by formation of C−S bonds.

9   10

The cyclophanes **11–13** [18 a, g, 24] represent examples for the formation of three C−S bonds in one step; even the formation of six bonds at the same time to **14** [18 c, d] and **15** [17c] succeeded − although in low yields only:

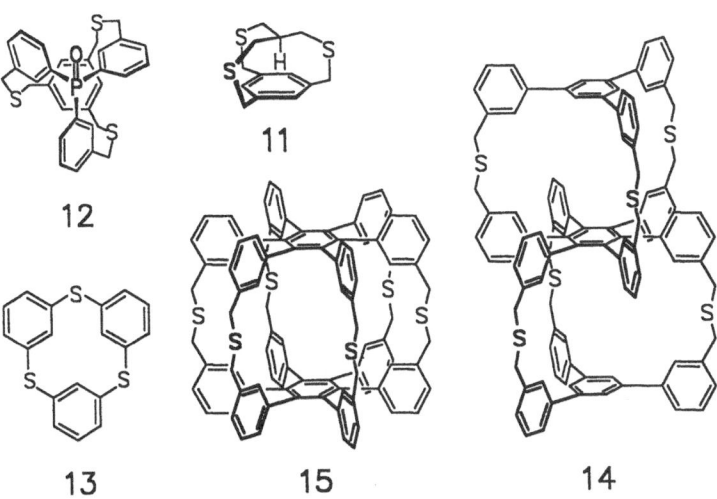

Experimental procedure for **14** [18c]:

Reaction of 1.63 g (3.60 mmol) 1,3,5-tris[3-(mercaptomethyl)phenyl]benzene, 2.00 g (1.80 mmol) hexakis[3-(bromomethyl)phenyl]benzene and 3.70 g (11.3 mmol) Cs_2CO_3 in 400 ml benzene and 200 ml ethanol
reaction temperature: boiling solvent
time of addition: −
reaction time: 8 h
yield: 2%

3.2 Formation of C−Se Single Bonds [27, 28]

Cyclizations by formation of carbon−selenium bonds represent a modern method with a high synthetic potential in the chemistry of cyclophanes. Selenocyanates such as **16** are accessible usually in excellent yields through the reaction of bromides with KSeCN [27]. The reaction with benzylic bromides under reductive conditions using the dilution principle results in good to excellent yields of [3.3]di-selenacyclophanes which can be deselenized photochemically, pyrolytically (without previous oxidation), or by reaction with arynes, Stevens rearrangement and subsequent reaction with Raney nickel. [2.2]Metacyclophane (**18**), for example, is accessible in 47% total yield by using this sequence of reactions starting with 1,3-bis(bromomethyl)benzene [28]. In this case, the method seems to be superior to the customary synthetic routes, e.g., the desulfurization of dithia[3.3]metacyclophane [29a] or the reaction of 1,3-bis(bromomethyl)benzene with phenyl lithium (yield 29%) [29b].

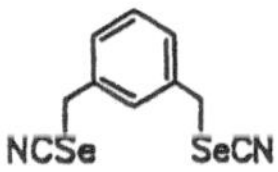

16 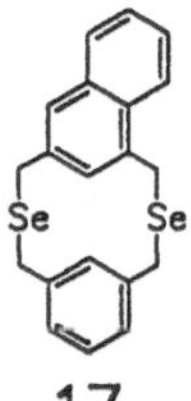   **17** 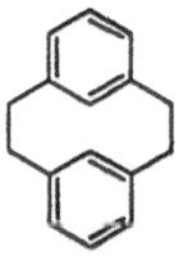   **18**

Experimental procedure for **17** [28]:

starting components: a) 1.32 g (5.0 mmol) 1,3-bis(bromomethyl)benzene in 100 ml ethanol/THF 1 : 1; b) 1.60 g (5.1 mmol) 1,3-bis(selenocyanatomethyl)naphthalene in 100 ml ethanol/THF 1 : 1
reaction medium: 1 l ethanol/THF 1 : 19 and 2 g NaBH$_4$
reaction temperature: 40–50 °C
time of addition: 13 h
additional reaction time: −
yield: 81%

The deselenization of **17** succeeded photochemically in 52% yield [28] (for comparison: The direct synthesis of this [2.2]cyclophane from the corresponding benzylic bromides and phenyllithium yielded only 5.7% [30]).

3.3 Formation of C−O Ether Bonds [9i, 31–37]

Oxygen atoms usually are less strong nucleophiles compared to sulfur atoms. Thus the formation of ethers belongs to the group of reactions using weakly reactive starting materials [12]. Many of the compounds which are important in supramolecular chemistry were cyclized by formation of an ether bond, e.g., several crown ethers [31], hemispherands [32], host molecules of host/guest chemistry [9i, 33], catenanes [34], and several natural products.

The synthesis of the cyclophane **22** from the dibromide **19** and the bis(phenol) **20** was carried out in various ways [9i]; on the one hand, stepwise via the open-chained intermediate **21**; on the other hand, in a one-step cyclization of **19** and **20**:

Here a rare template effect was noticed: On cyclizing the open-chained compound **21** under dilution conditions the macrocycle **22** was obtained in 28% yield, whereas the direct cyclization under these conditions yielded only 4%. If, however, benzene was added in the direct cyclization the yield increased to

23%. This increase of yield is attributed to a template effect of the added hydrocarbon; even without application of the dilution principle, **22** could be obtained in a one-step reaction from **19** and **20** in the presence of benzene in 21% yield.

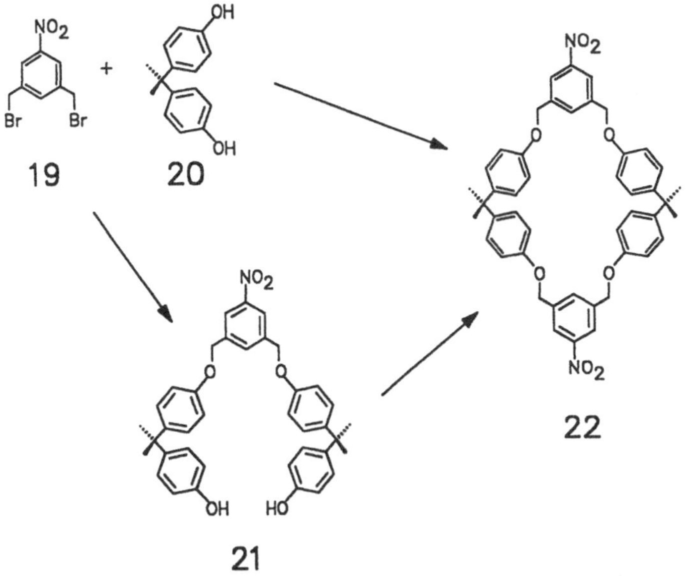

With only moderate dilution and without a template effect, the cyclic ether **24** which is an important precursor in the total synthesis of the racemic germacranolide-aristolactone (**25**) was obtained from the hydroxychloride **23** [35].

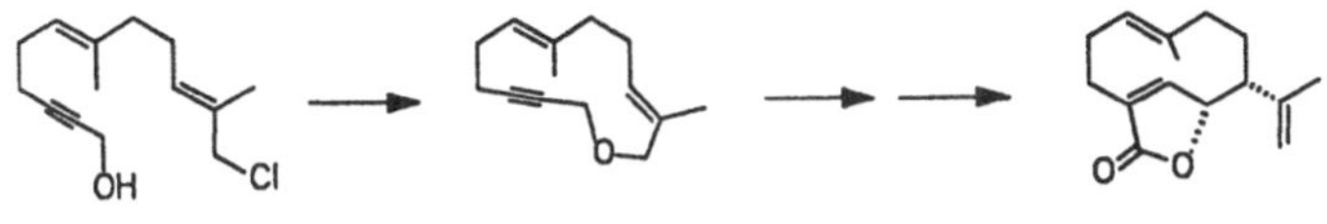

Experimental procedure for **24** [35]:

reaction of the hydroxychloride **23** with one equivalent of ethyl magnesiumbromide in THF/HMPT
concentration: 0.02 M
reaction temperature: 0 °C to boiling solvent
reaction time: 4 h
yield: 70%

Two further compounds which are attractive because of their topological aspects were cyclized by formation of ether bonds: Walba [31a] succeeded in realizing the Möbius belt in molecular dimensions: Starting with the crown ether ditosylate **26** he produced the molecular belt **27** and the Möbius belt **28** by cyclization. Here, **28** results from the twisting of **26** and subsequent "crosswise" ring closure which becomes possible at sufficient length of the ditosylate. The difference between **27** and **28** is the fact that **27** has two surfaces, an inner and an outer one, whereas **28** has only one single continuous surface because of the twisting.

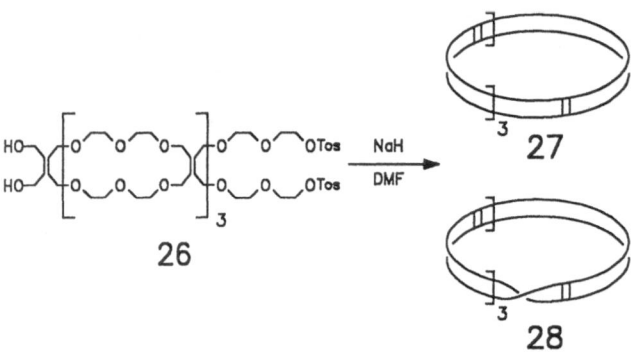

The existence of **28** was proved by Walba in an obvious and elegant manner: Ozonolysis of the $C=C$ double bonds in **27** leads to two 30-membered cyclic ketones, whereas in the ozonolysis of **28** one 60-membered cyclic ketone is produced.

Experimental procedure for **27** *and* **28** [31a]:

starting component: 32 mg (0.02 mmol) dioltosylate **26** in 5 ml DMF
reaction medium: 15 mg (0.63 mmol) NaH in 11.3 ml DMF
reaction temperature: room temperature
time of addition: 15 h
additional reaction time: 2 h
yield: 55% mixture of the polyethers **27** and **28**

Sauvage [36] succeeded in synthesizing the "molecular knot" **30** by cyclization of the binuclear helical phenanthroline-copper(I)-complex **29** with hexaethylene glycol diiodide in DMF under extreme dilution and addition of cesium salt. However, the yield amounted to less than 3%.

Both molecules, **28** and **30**, are topologically chiral; the chirality of **30** was proven by ^{1}H-NMR spectroscopy with the addition of KPF_6 and Pirkles reagent. This resulted in different chemical shifts for the hydrogen atoms of the "clover-leaf knot" **30** due to formation of diastereomeric complexes.

29   **30**

3.4 Formation of Amines and Amides

3.4.1 Synthesis of Amines [38–47]

Ring formations by nucleophilic substitution at saturated carbon atoms with primary amines as nucleophiles have rarely been carried out because the resulting secondary amines as a rule are more nucleophilic than the primary ones, and therefore competition reactions are favored. The synthesis of secondary amines often starts from toluene sulfonamides which can easily be deprotonated and alkylated. A large number of methods for detosylation exists; especially the acidic cleavage with H_2SO_4 or with HBr/phenol have proved to be reliable.

Most of the cycles which have been formed in this way belong to the field of host/guest chemistry, e.g., the azacrown ethers **31** [38] and **32** [39], tropocoronands such as **33** [40], or even cyclophanes such as the host compound **34** synthesized by Koga [41]. In the cavity formed by the two diphenylmethane units, **34** encloses naphthalene as a guest [42].

In the field of azacrowns the representative **35**, synthesized by Lehn and Hosseini [43], while is able to catalyze the reaction of phosphate to pyrophosphate, has to be emphasized.

By reaction of the tritosylamide **36** with the trimesylether **37** in DMF and subsequent detosylation, Lehn [44] succeeded to synthesize the bicyclic cavity **38** in 65% yield. Complexation experiments in acidic solution point to the inclusion of nitrate ions into the cavity of the host. However, no such inclusion could be found in the crystal.

The macrocycle **41** represents another example of large molecular cavities. The size of the cavity can be changed by switching the azobenzene units photochemically [(E)/(Z)-isomerism].

11

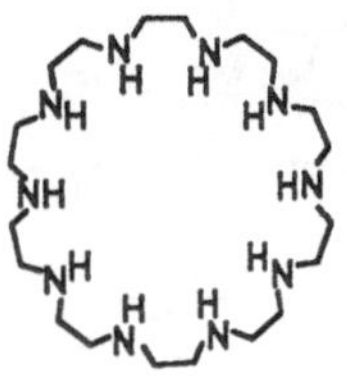

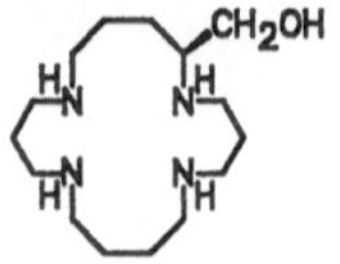

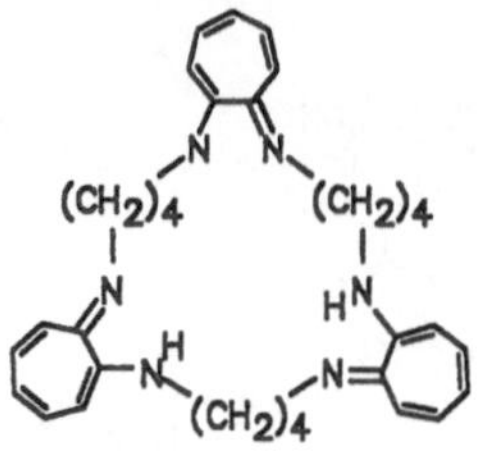

Fritz Vögtle et al.

31

32

33

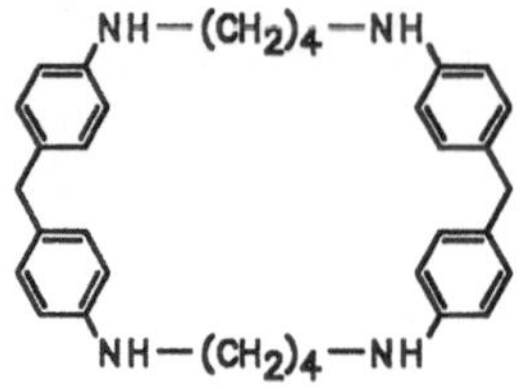

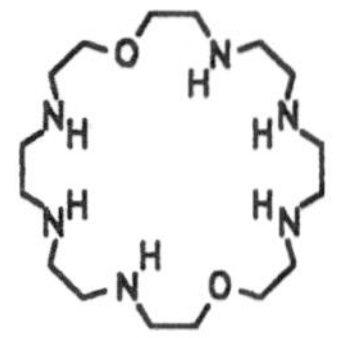

34

35

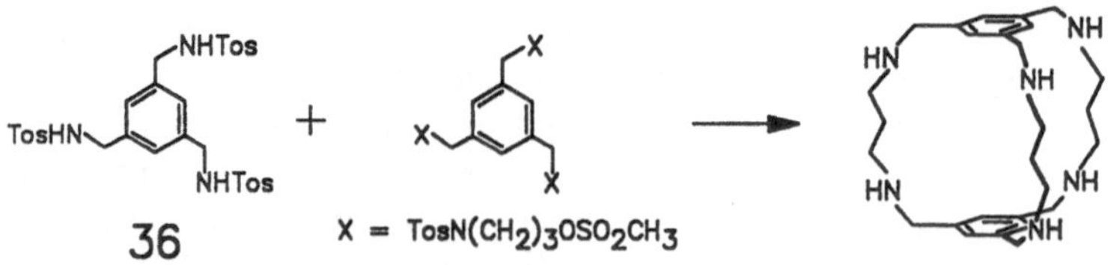

36

X = TosN(CH$_2$)$_3$OSO$_2$CH$_3$

37

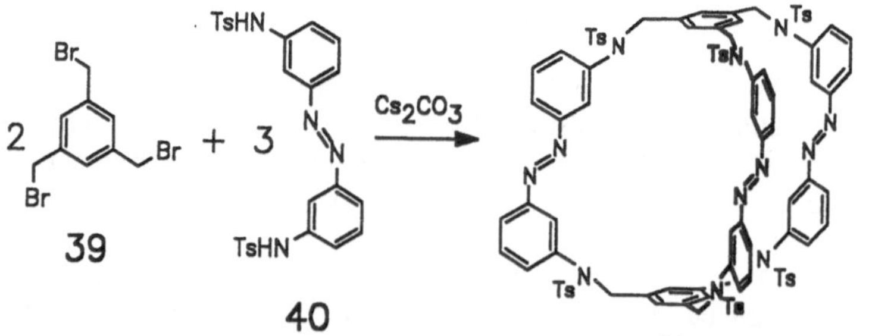

38

2

39

+ 3

40

Cs$_2$CO$_3$

41

The synthesis of **41** succeeded in 5% yield by reaction of 1,3,5-tris(bromome-thyl)benzene (**39**) with *N,N'*-ditosyl-3,3'-diaminoazobenzene (**40**) in DMF using potassium carbonate as base and without dilution conditions (see Sect. 4); if, on the other hand, the reaction was carried out in DMF under high dilution and in the presence of cesium carbonate, the yield of **41** could be increased to 12% [45].

By reaction of 5,5'-bis(bromomethyl)-2,2'-bipyridine (**42**) with 1,3,5-tris[*N*-(benzyl)aminomethyl]benzene (**43a**) and 1,3,5-tris[*N*-(1-naphthylmethyl)amino-methyl]benzene (**43b**), respectively, and sodium carbonate as auxiliary base the bipyridine units containing macrocycles **44 a, b** could be prepared as its sodium complexes in 4 and 6% yield, respectively. The sodium ion exerts a template effect in this reaction; however, if cesium carbonate was used as base, only traces of the macrocycles could be detected [46].

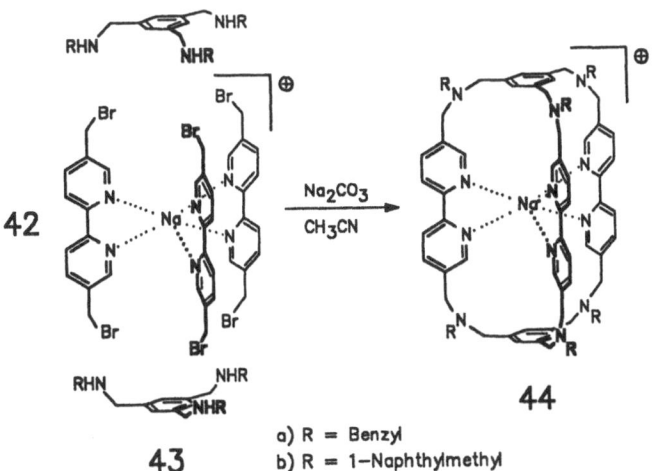

a) R = Benzyl
b) R = 1–Naphthylmethyl

Experimental procedure for **44a** · $Na^{\oplus}X^{\ominus}$ [46]:

starting component: 4.74 g (10.88 mmol) **43a** in 400 ml acetonitrile
mixture put into the reaction flask: 5.58 g (16.32 mmol) **42** and 30.0 g (280 mmol) sodium carbonate in 1500 ml acetonitrile
reaction temperature: boiling solvent
time of addition: 8 h
additional reaction time: 15 h
yield: 0.3 g (4%)

3.4.2 Synthesis of Amides [32c, 47a, 48–57]

The synthesis of mono- and bicyclic amides from acid chlorides and amines under dilution conditions leads to a series of host molecules significant in supramolecular chemistry.

Two methods can be offered when using bifunctionalized educts: either the direct one-step synthesis starting with a diacid chloride and a diamine, or the stepwise reaction to an open-chained intermediate which is closed subsequently.

Therefore the direct cyclization of 45 and 46 under high dilution conditions yields the tetraamide 48 in 35%, whereas in the two-step synthesis via 47 the cycle 48 is obtained in 49% yield [49].

The amide protons of the cyclic products can form hydrogen bonds, e.g., in the cyclophane 49 which is able to recognize nucleotide bases (50) [50a].

The bicyclic cavity of the pyrocatechol type 54 which is shown in Fig. 2 represents a recent example for a modular two-step synthesis.

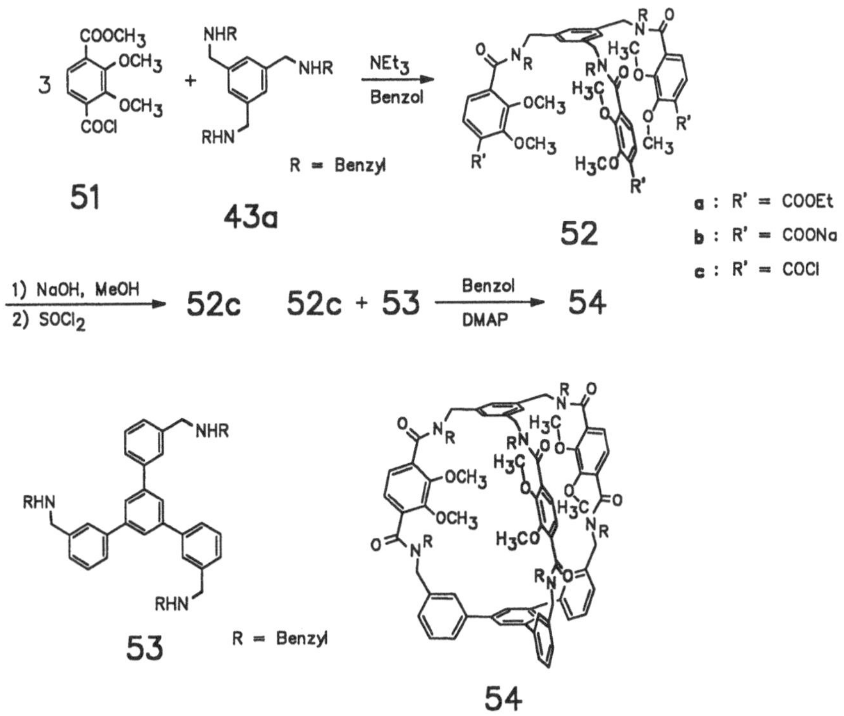

Fig. 2. Modular two-step synthesis of the bicyclic cavity **54** [51]

The podand **52a** as well as the macrocycle **54** show the ability to complex sodium ions with their methoxy functions. The macrocycle exceeds the open-chained compound with regard to the complexation constant by a factor of six. While **52a** complexes only $Na^{\oplus}$, **54** shows increasing complex constants for $K^{\oplus}$ and $Cs^{\oplus}$ with a distinct selectivity for the larger cesium ion. The order of magnitude of the constants of the alkali-metal complexes corresponds to those of the crown ethers [51].

Experimental procedure for **54** [51]:

starting components: a) 2.70 g (4.07 mmol) triamine **53** in 250 ml benzene; b) 2.27 g (2.04 mmol) triacid chloride **52c** in 250 ml benzene
reaction medium: 50 mg (0.41 mmol) 4-(dimethylamino)pyridine in 2 l benzene
reaction temperature: boiling solvent, subsequently room temperature
time of addition: 8 h
additional reaction time: 15 h
yield: 0.58 g (17%)

As early as 1984 Vögtle et al. [52] succeeded in synthesizing a bicyclic cavity with pyrocatechol units (**55**) which is able to complex $Fe^{3\oplus}$ ions with its hydroxy functions. The stability constant of the iron complex exceeds the one of the hitherto best complexing agent of iron, enterobactin, by the factor 10^7 (for 55 lg k = 59).

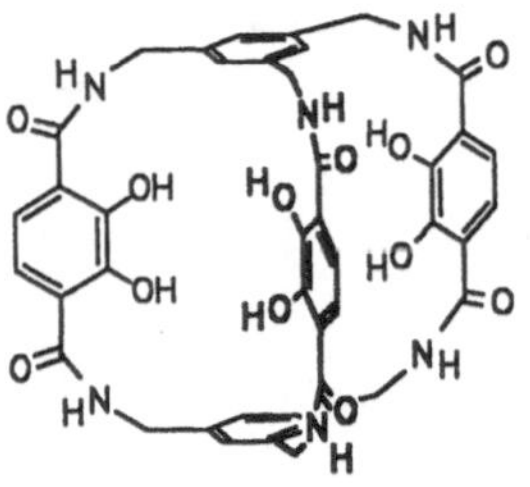

55

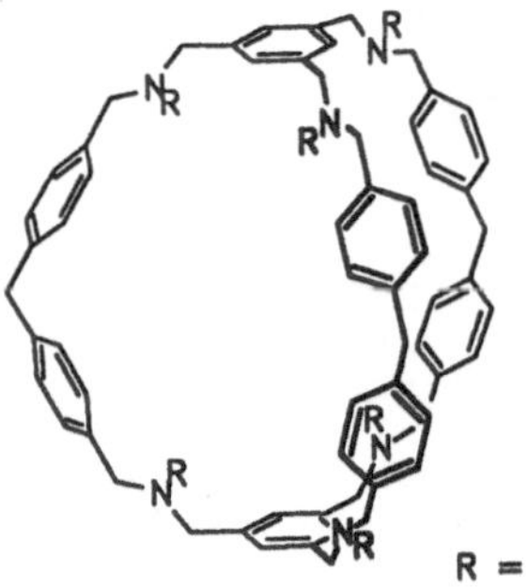

56

R = CH$_3$

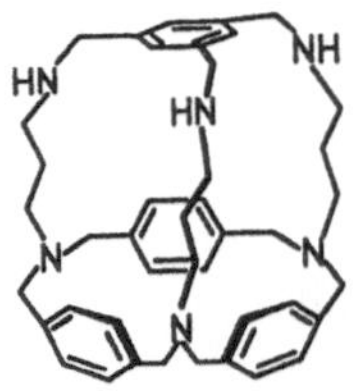

57

The macrobicyclic hexamine 56 which could be synthesized by the same group [53] is able to include plain, disc-like, aromatic guests in its hydrophobic cavity and thus to transport them into an acidic aqueous phase. 56 complexes e.g., triphenylene, perylene, and acenaphthylene, whereas the linear anthracene is not complexed.

Anions, e.g., nitrate, can be complexed by the protonated form of the cyclophane 57. The protonated ligand (57 · 6 H$^{\oplus}$) seems to form a 1:1 complex with the nitrate ion; the stoichiometry suggests that the anion is enclosed inside the cavity of the host [54].

Of course, the peptides [55] and peptide alkaloids [48, 56] belong to the field of macrocyclic amides, as well. The cytotoxic ulicyclamide 58, for example, can be prepared in very diluted solution from the open-chained penta-peptide by reaction with diphenylphosphorazidate in DMF/triethylamine in high yield.

Schmidt et al. [56d] report on a cyclization under reductive conditions: By treatment of the Z-protected amine 59 with palladium and hydrogen in diluted solution the cyclopeptide 60 could be obtained in 80% yield.

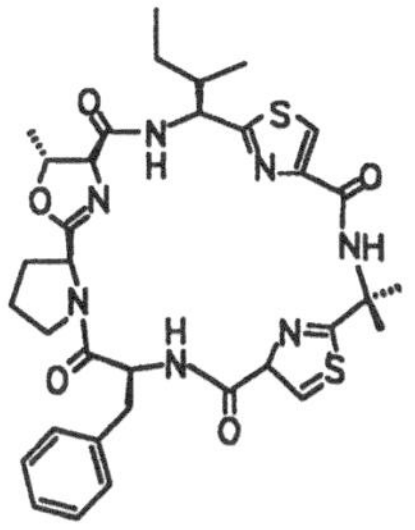

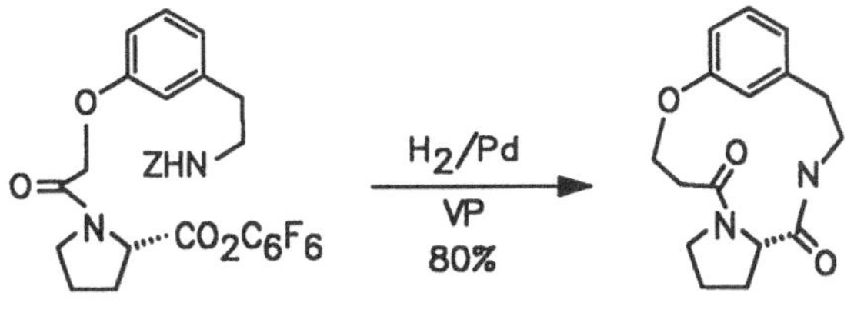

58   **59**   **60**

3.5 Synthesis of Macrolides [58–67]

A simple possibility for the synthesis of esters, the reaction of an acid chloride with an alcohol, was used by Schrage and Vögtle [58] for a two-step synthesis of the macrocycle **63** from the alcohol **61** and the acid chloride **62**. Compound **63**, an example from the field of host/guest chemistry, forms a cavity, as studied with CPK-models, which could include planar, aromatic guests. Crystals obtained from benzene/n-heptane point to a 1:2 stoichiometry of **63** and benzene according to NMR-spectroscopic data. However, whether this is a molecular inclusion complex or just a clathrate is not yet known.

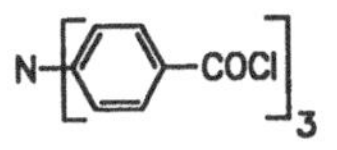

62

61

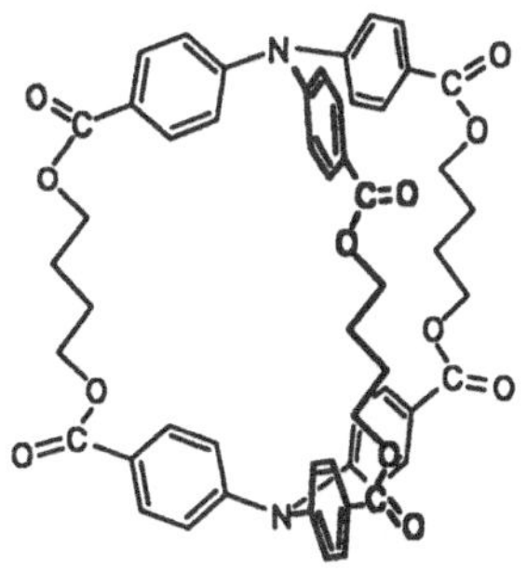

63

Another synthetic strategy known from peptide chemistry consists in the transformation of an acid into an activated ester and its subsequent reaction with an alcohol. An example of this strategy is the preparation of the 32-membered macrolide tetranactine **65** from the acid **64** which is activated by 3-cyano-4,6-dimethylpyridine-2-thiol **66** [59].

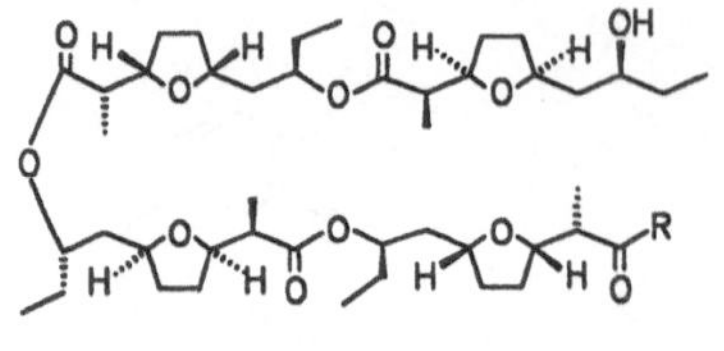

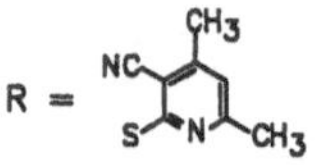

64

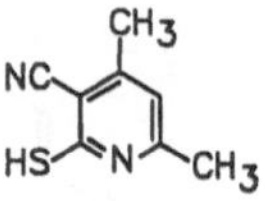

66

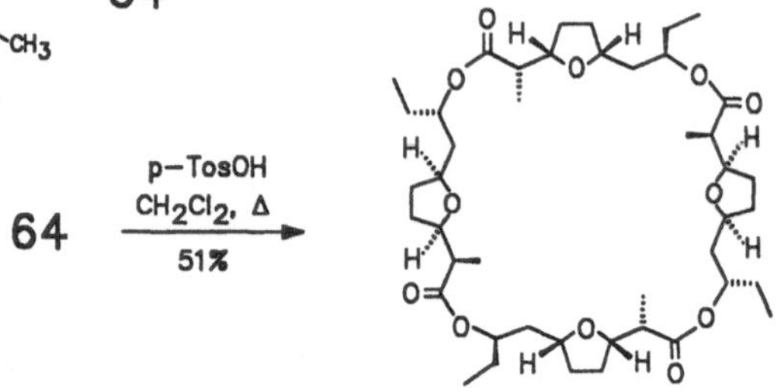

R =

$$64 \xrightarrow[\substack{CH_2Cl_2, \ \Delta \\ 51\%}]{p-TosOH} $$

65

Another example of the activation of a hydroxy acid was described by Rastetter and Phillion [60]: First the *O*-protected hydroxyacid **68** reacts with a thiol group containing crown ether **67**. Then the resulting thioester **69** reacts with potassium *tert*-butoxide to give the alkoxide. At the same time a complexation of the potassium ion by the [18]crown-6 part of the molecule occurs. Thus, the alkoxide ion comes close to the carbonyl group of the molecule, so that nucleophilic attack leading to ring formation is facilitated (cooperation of dilution principle, template effect, and ion pair interaction).

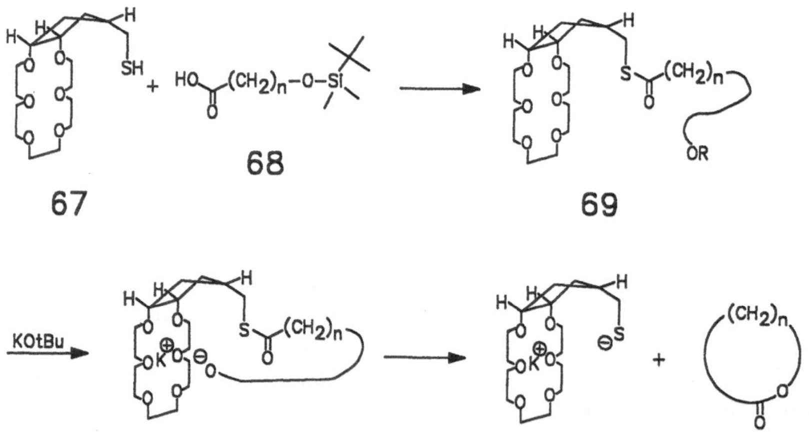

68

67

69

Regarding the temperature dependence of the formation of oligomers in the macrolide synthesis under Yamaguchis conditions for macrocycles, Seebach et al. [61] obtained an interesting result for the lactonization of (R)- and (S)-3-hydroxybutyric acid **70**: At room temperature the cyclic penta-, hexa-, and heptamers **74, 73**, and **72** were formed in a ratio of 1:1:1 in 50% total yield as the only isolable cyclic products (Fig. 3), while at 110 °C the ratio of **74** to **73** to **72** was shifted to 59:30:9 in favor of the pentamer **74**.

Experimental procedure for **72–74** [61]:

starting components: a) 0.5 g (4.80 mmol) (R)-3-hydroxybutyric acid **70** in 10 ml THF; b) 0.87 ml (6.20 mmol) triethylamine and 0.67 ml (4.80 mmol) 2,4,6-trichlorobenzoyl chloride (**71**) in 10 ml toluene
reaction medium: DMAP in 200 ml toluene
reaction temperature: room temperature
time of addition: 4 h
additional reaction time: 30 min
yield: 50% **74, 73, 72** (1:1:1)

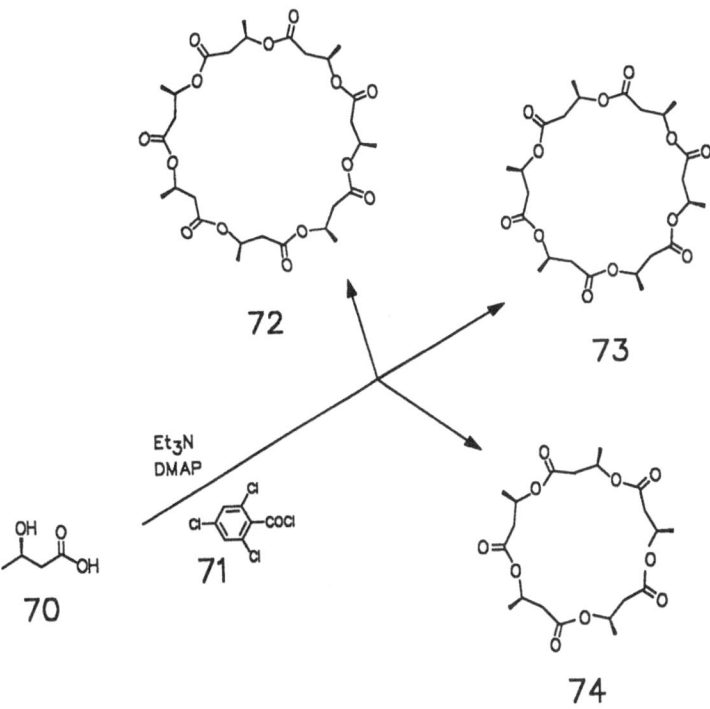

Fig. 3. Product distribution for the cyclization of (R)-3-hydroxybutyric acid [61]

An alternative ester synthesis, the reaction of a carboxylate anion with an alkyl halide, was used by Maciejewski [62] for the preparation of (*E,E*)-1,9-dioxacy-clohexadeca-3,11-dien-2,10-dione (**76**) under dilution conditions. This 16-membe-red dilactone represents a precursor for the synthetic norpyrenophorin **77a**, the physiological activity of which corresponds to the one of the natural products pyrenophorin **77b** and vermiculin **77c**. The lactone **76** can be obtained in 77% yield by dimerization of (*E*)-7-bromo-2-heptenic acid (*75*) in DMF in the presence of potassium carbonate.

75

K$_2$CO$_3$, DMF
16 h, 40°C

76

77

a) R = H
b) R = CH$_3$
c) R = CH$_2$COCH$_3$

Several macrocyclic dilactones such as **80** were prepared by reaction of dicarboxylic acid difluorides (e.g. **78**) with triphenyltin derivatives of diols (e.g. **79**) in excellent yields [63]. The required acid fluorides can be synthesized by reaction of the acids with 2-fluoro-*N*-methylpyridiniumtosylate (**81**) and addition of triethylamine. The triphenyltin derivatives of the diols were prepared in situ from Ph$_3$Sn$-$O$-$SnPh$_3$ and the diols.

FOC$-$(CH$_2$)$_3$$-$COF

78

Ph$_3$SnO   OSnPh$_3$

79

80

81

The macrocyclization step proceeds at 80 °C at moderate dilution conditions because tin exerts a template effect in the course of the reaction, so that good yields are already obtained in 10^{-2} M solutions. If, on the contrary, the reaction temperature is increased to 140 °C, the yield of monomer **80** is drastically reduced in favor of higher molecular, linear oligomers, an effect which can be attributed to the cancellation of the template effect.

Lactone syntheses under dilution conditions were carried out photochemically, too [64], e.g., the synthesis of (+)-aspicilin (**84**), an 18-membered macrolide isolated from a lichen source. Quinkert et al. [64a] transformed the *ortho*-chinol acetate **82** ($c = 2 \times 10^{-3}$ M) by irradiation in toluene with addition of DABCO to the mixture of the diastereomeric lactones **83 a, b** in 62% yield which is the key step in the synthesis of (+)-aspicilin (**84**).

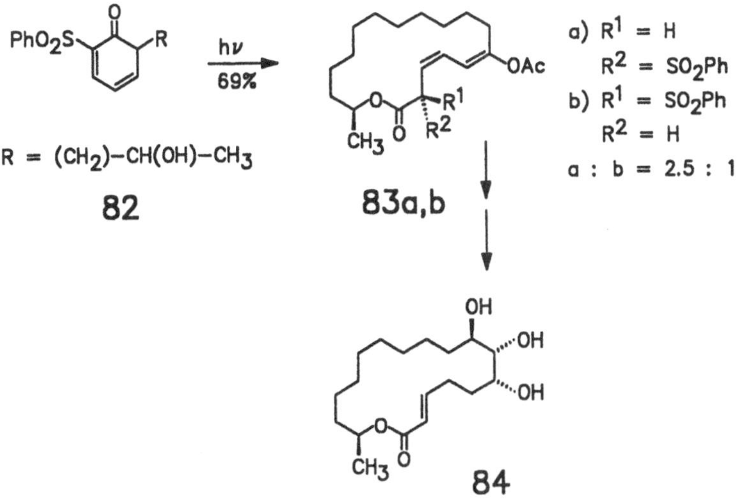

First of all, a photochemical ring opening of the chinolacetate to the ketene occurs; in the subsequent thermic cyclization the tertiary amine serves as a nucleophilic catalyst to form the lactone.

3.6 Formation of C−C Single Bonds [68–74]

A large number of reactions for the formation of C−C bonds exist which can often be used under dilution conditions to form cyclic systems. Of all these reactions some are discussed here in detail:

The application of the Wurtz coupling for the synthesis of [2.2](2,6)pyridinophane (**85**) from 2,6-bis(bromomethyl)pyridine (**86**) was already attempted in the fifties by Baker et al. [68]. Yet, the desired cyclophane could not be found. However, the intramolecular cyclization of **87** with phenyl lithium or butyl lithium in ether led to **85** in 3 and 28% yield, respectively [68]. The *inter*molecular cyclization of **86** with phenyl lithium in dioxane gives a 9% yield of **85**, too [69].

Compounds **88** [70d] and **89a** [70a] could be synthesized by alkylation of malonic esters [70] in 26 and 0.6% yield, respectively.

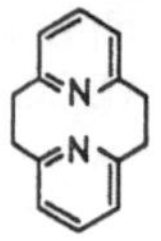

85 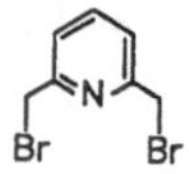 **86** 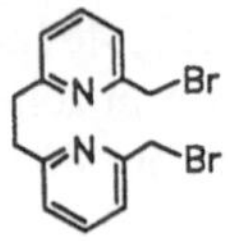 **87**

E = CO₂CH₃

a) R = CO₂Et
b) R = CO₂H

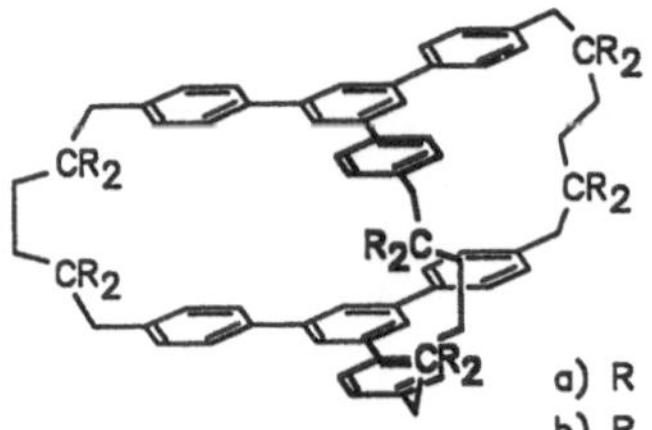

88 **89**

Macrocycle **89b** is capable of including benzene, toluene, and mesitylene into its cavity in alkaline aqueous solution.

Intramolecular acyloin condensation [71] of the diester **90** under dilution conditions with sodium in toluene and with addition of trimethylsilyl chloride led to the bis(trimethylsilyl)en-diol ether **91** (29% yield) which was transformed subsequently in several steps to [1.1.1.1]paracyclophane **92** [71a].

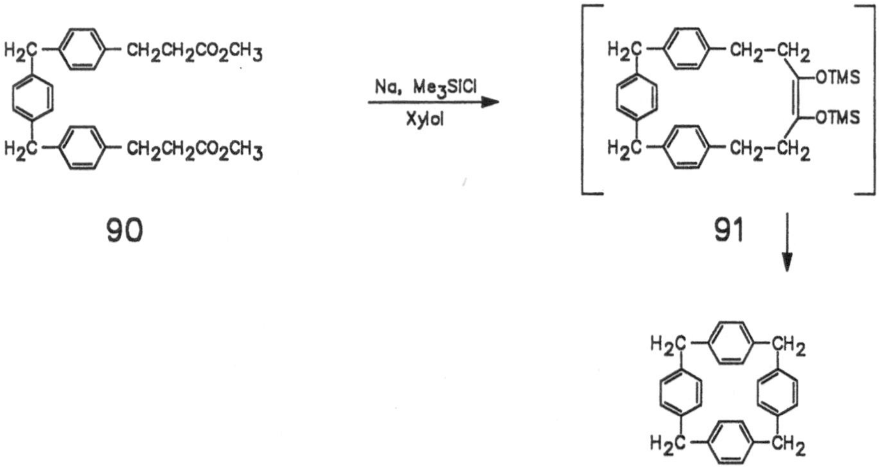

90

Na, Me₃SiCl
Xylol

91

92

Bis(benzyl) halides react with *p*-toluenesulfonyl methylisocyanide (TosMIC, **94**) in the two-phase system sodium hydroxide/methylene chloride and in presence of tetra-*n*-butylammonium bromide as phase transfer catalyst to [3.3]cyclophanes which contain a carbonyl function in each bridge [72]. The furanophane **95** is accessible from **93** and **94** in 39% yield [72b].

Another standard method for the formation of $C-C$ bonds is the coupling of acetylenes which was applied in many cases under dilution conditions to synthesize macrocycles [73]. Examples for this method are **96** and **97** [73 a, e].

3.7 Formation of $C=C$ Double Bonds [75–78]

The Wittig [75] and the McMurry reaction [76] are standard methods to form $C=C$ double bonds. For example, the bridged biphenylene **98** was prepared in diluted solution by Wittig reaction [75a]. Vogel synthesized the bridged [14]annulene **100** with its phenanthrene perimeter from dialdehyde **99** by McMurry reaction [76c]. "π-Spherand" **101** could be prepared likewise. Because of its π-orbitals arranged inwards, **143** becomes an eight-electron donor, so that it is capable of enclosing an $Ag^{\oplus}$ ion into its cavity and thus complexing it [76g].

Flexibilene (**103**), a 15-membered diterpene, which was not isolated until 1976 from the coral *Sinularia flexibilis* [77], could be obtained in 1987 by addition of a 10^{-2} M solution of the dicarbonyl compound **102** to $TiCl_4$/zinc within 32 h in 53% yield [76f].

102

103

The reaction of keto phosphonates with sodium hydride as base is applicable to the synthesis of α,β-unsaturated carbonyl compounds, even under the conditions of the dilution principle [78]. Thus the keto phosphonate **104** could be transformed by this cyclization reaction into the mixture of the epimers **105** in 28–36% yield.

104   **105**

3.8 Intramolecular Cycloadditions Under Dilution Conditions
[79–81]

Pericyclic reactions were used successfully, too, to build up macrocyclic ring systems as found, for example, in natural peptides and peptide alkaloids. These reactions are carried out in general at moderate dilution conditions (10^{-2}–10^{-3} M). An example of a Diels-Alder reaction is the cyclization of **106** to **107**, where the basic framework of cytochalasane B (**108**), a substance found in fungi, is built up in yet 27% yield [79].

The bridged paracyclophanes **111** and **112** could be obtained from the aziridines **109** by 1,3-dipolar cycloaddition with remarkable diastereoselectivity [80]: First

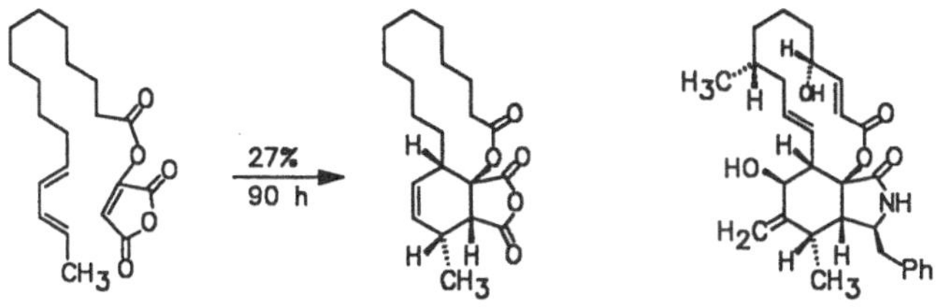

106   107   108

the unstable azomethinylide **110** is prepared by thermolysis of **109** which then reacts as an 1,3-dipole to give **111** and **112**. Starting with the *trans*-aziridines, only two of the four possible diastereomers could be obtained in a diastereoselective manner, whereby the yields increased with increasing chain length (Fig. 4).

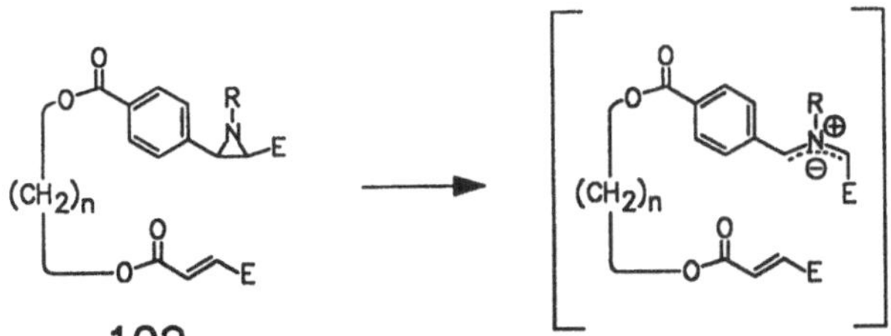

109

110

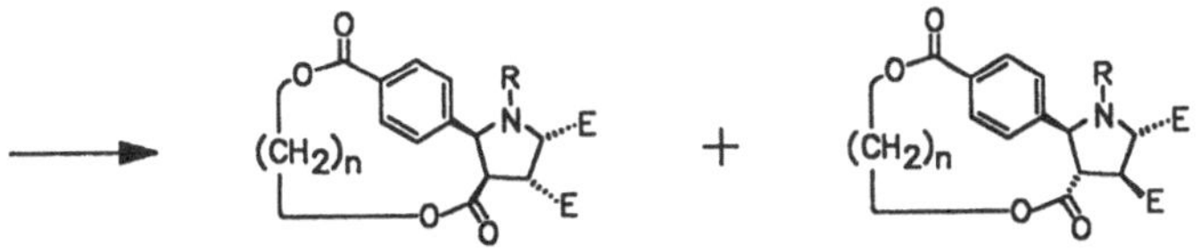

111   +   112

Fig. 4. Macrocyclization by 1,3-dipolar cycloaddition [80]

3.9 Ring Expansion Reactions [67, 82–84]

An interesting way of achieving macrocyclization consists in building-up of bicyclic systems with "normal" ring sizes at first and subsequent cleavage of the bridge between both rings. This method was applied, for example, by Mahajan and Resck [67] in the synthesis of the 12-membered lactone **115** from **113** via the bicyclic intermediate **114**.

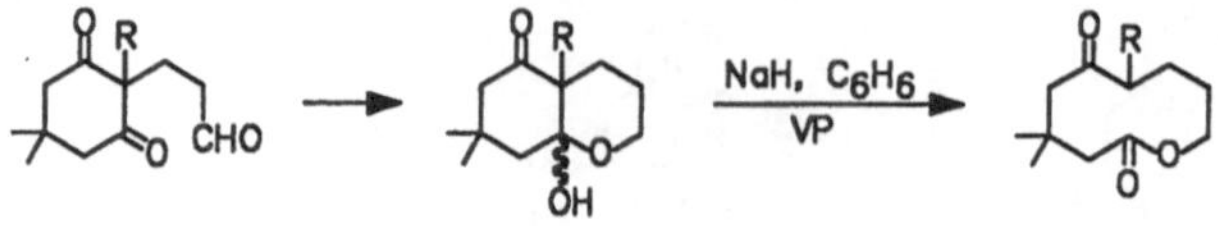

113   114   115

Reactions of this kind are not always carried out under dilution conditions [82]; and as a consequence, this results in appreciable amounts of dimeric products, too.

The radical ring extension of α-alkyl-β-tin substituted cyclohexanones in the presence of azobis(isobutyronitril) (AIBN) and tributyltin hydride in diluted solution results in good to excellent yields of the ring-extended products [83] (Fig. 5).

Using this method the 12-membered ketone **117** can be isolated as the only product in 72% yield from **116**.

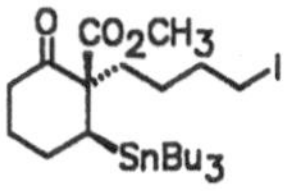

116 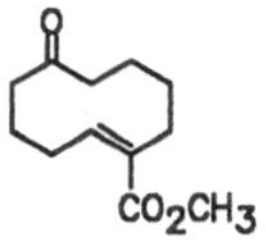   117

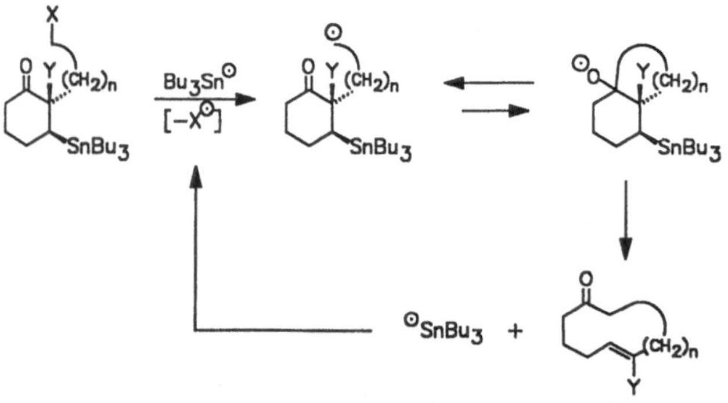

Fig. 5. Radical ring expansion [83]

Experimental procedure for **117** [83]:

reaction of 0.4 mmol **116** in 80 ml dry benzene (c = 5×10^{-3} M) with 0.2 equiv.
AIBN and 0.1 equiv. tributyltin hydride
reaction temperature: boiling solvent
reaction time: 50 h
yield: 72%

3.10 Formation of Cyclic Disulfides [85, 86]

Up to now the easy oxidizability of thiols to disulfides has rather rarely been used
for the synthesis of macrocycles. **118** could be prepared from 2,3,5,6-tetrafluoro-
1,4-benzene dithiol by oxidation with DMSO in 95% yield [85]; the bicyclic
cyclophane **119** was synthesized by Whitesides through oxidation of 1,3,5-
tris(mercaptomethyl)benzene with iodine [86]. The high yield of 68% is not only
attributed to the high dilution but also to the low solubility of the macrocycle
being formed which precipitates during the reaction.

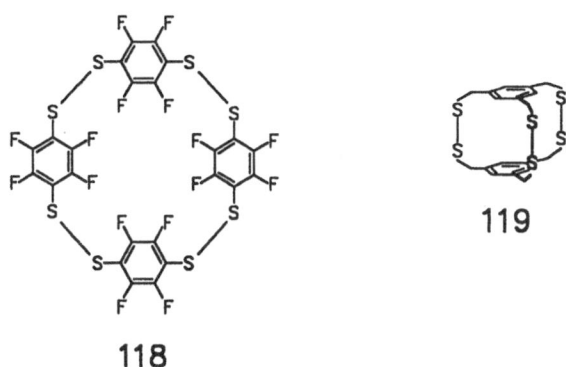

118

119

Experimental procedure for **119** [86]:

starting components: a) 1 g (4.7 mmol) 1,3,5-tris(mercaptomethyl)benzene in
400 ml ethanol; b) 1.76 g (6.93 mmol) iodine in 400 ml ethanol
reaction medium: 600 ml ethanol
reaction temperature: room temperature
time of addition: 3 h
additional reaction time: –
yield: 0.68 g (68%)

4 Synthesis of Macrocycles Without Use of the Dilution Principle: "Disguised Dilutions" [9f, 18c, 87–90]

A certain disadvantage of the dilution principle consists in the sometimes long reaction times and the required large amounts of solvent. Repeatedly this disadvantage led to efforts to carry out cyclizations with only small amounts of solvent. Using these modifications it is still important to keep the concentration of the reacting species low. This can be done in different ways:

Heterogeneous dilutions [87]: Here the dilution effect can be achieved by using a multiple-phase system (solid-liquid or liquid-liquid). If a component with low solubility is applied, it dissolves successively, its stationary concentration is extremely low [18c]. Similar effects can be obtained by using phase transfer catalysts in liquid-liquid two-phase systems.

Dilution reactions in concentrated solution [88]: A method applied successfully by Schneider [88] in the synthesis of tosylaza macrocycles consists in the reaction of tosylamides with bromides in DMF using potassium carbonate as base. Here K_2CO_3 compared with Cs_2CO_3 has the advantage of lower solubility (ca. 10^{-5} M). Thus more concentrated solutions of the reactants can be added to a suspension of the base in DMF. The low solubility of K_2CO_3 causes only a small amount of the reactive tosylamide anion to be present at all times, securing the dilution effect.

High-pressure syntheses [89]: By application of high pressure the viscosity of the reaction medium is raised, with the result that the components of the reaction are "frozen" in the solution: molecules which are not so far from one another in the reaction mixture stay close together during the reaction and thus react preferentially with one another. An example for this is the reaction of **120** with **121** which results in the 12-membered cyclic dication **122** in almost quantitative yield [89a]. 18-Membered rings could be obtained in almost the same yield, too.

The application of high pressure (9–10 kbar) enabled Stoddart [89 b] to synthesize the cycles **123** and **124** by repetitive "structure directed" Diels-Alder reactions; the synthesis of **124** under "normal" dilution conditions resulted in a substantially lower yield of the desired product.

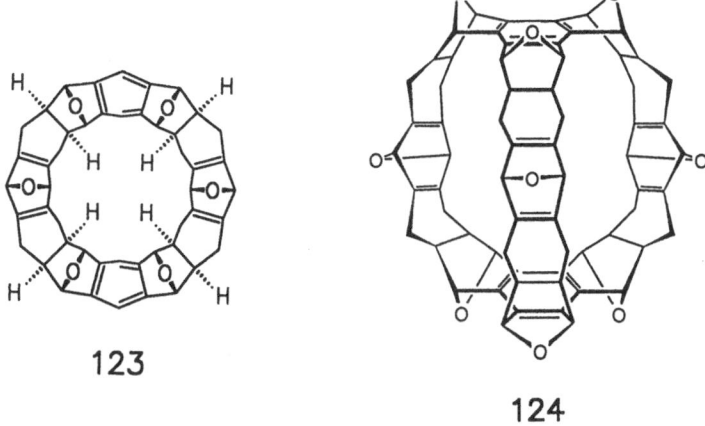

123

124

A similar result can be obtained by using donor-acceptor-templates [87 i, k, 90, 91]: The final cyclization step in the synthesis of the [2] catenane **128** succeeded in 70% yield starting from **125, 126,** and **127** [90a]:

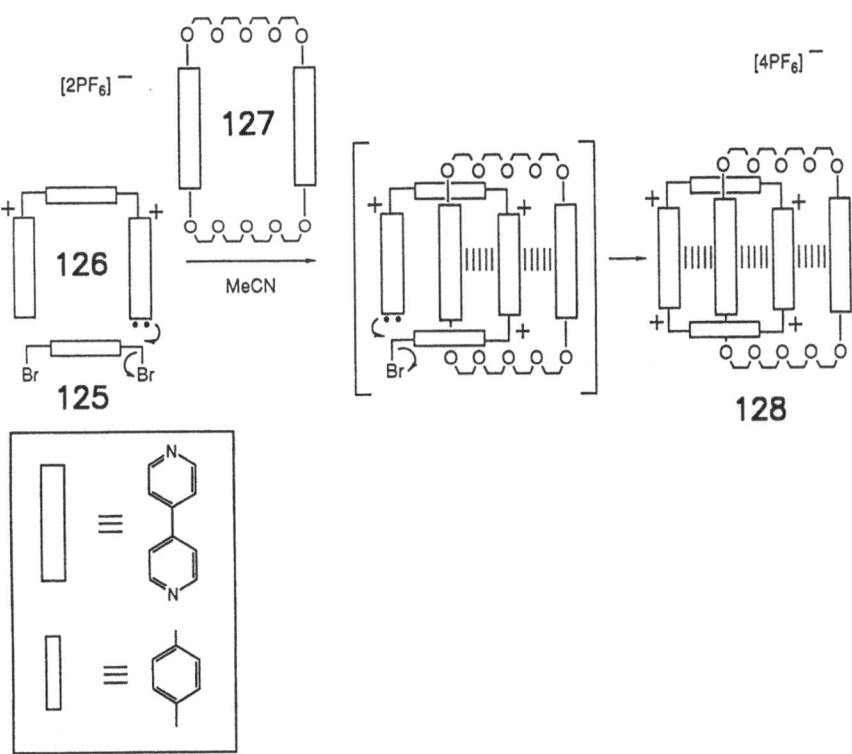

5 Concluding Remarks

As the examples mentioned above show, a variety of experimental procedures for cyclizations under dilution conditions exist. Up to this day no generaly applicable procedure is known; also, the determination of cyclization parameters to a given target molecule is based on the experiences made by synthesizing similar compounds. Each new structure requires precise conditions for its synthesis; often a small "window" has to be hit with regard to the reaction conditions.

The synthesis of macrocycles with the aid of the dilution principle still remains a dominating cyclization method although there have been many attempts to avoid the standard procedure, i.e., the simultaneous addition of reactants to the reaction medium (see Sect. 4). The dilution principle can be avoided in some cases [90, 92] by application of "template-directed reactions" and/or by use of the "rigid group principle".

6 Acknowledgement

The authors kindly thank Dipl.-Chem. Joachim Dohm, Jochen Schulz, and Christian Seel for their assistance in preparing the final version of this progress report.

7 References

1. Vögtle F (1989) Supramolekulare Chemie − Eine Einführung, Teubner, Stuttgart
2. Weber E, Vögtle F (1987) Nachr Chem Tech Lab 35: 1149
3. a. Rossa L, Vögtle F (1983) Top Curr Chem 113: 1;
   b. Tabushi I, Yamamura K (1983) Top Curr Chem 113: 145;
   c. Kiggen W, Vögtle F (1987) In: Izatt RM, Christensen JJ (eds), Synthesis of macrocycles: The design of selective complexing agents, Wiley, New York, p 309;
   d. Vögtle F (1972) Chem-Ztg 96: 396;
   e. Karbach S, Löhr W, Vögtle F (1981) J Chem Res (S) 314; (1981) J Chem Res (M) 3579
4. Christen HR, Vögtle F (1990), Organische Chemie − Von den Grundlagen zur Forschung, Bd II, Salle-Sauerländer, Frankfurt/Aarau
5. Ruggli P (1912) Liebigs Ann Chem 392: 92
6. a. Ziegler K (1955) In: Houben-Weyl 4/2, p 729;
   b. Diederich F, Staab HA (1978) Angew Chem 90: 383; (1978) Angew Chem, Int Ed Engl 17: 372
7. a. Meurer K, Vögtle F, Mannschreck A, Stühler G, Puff H, Roloff A (1984) J Org Chem 49: 3484;
   b. Vögtle F, Przybilla KJ, Mannschreck A, Pustet N, Büllesbach P, Reuter H, Puff H (1988) Chem Ber 121: 823;
   c. Vögtle F, Struck J, Puff H, Woller P, Reuter H (1986) J Chem Soc, Chem Commun 1248;
   d. Vögtle F, Mittelbach K, Struck J, Nieger M (1989) ibid 65;
   e. Przybilla KJ, Vögtle F (1988) Chem Ber 122: 347
8. a. Klieser B, Rossa L, Vögtle F (1984) Kontakte (Darmstadt) 1: 3;

b. Koepp E, Vögtle F (1987) Synthesis 177;

c. Vögtle F, Klieser B (1982) Synthesis 294;

d. Navarro P, Rodriguez-Franco MI, Samat A (1987) Synth Commun 17: 105;

e. Kruizinga WH, Kellogg RM (1981) J Am Chem Soc 103: 5183

9. a. Review: Laider DA, Stoddart JF (1980) In: Patai S (ed), The chemistry of functional groups, Suppl E, part 1, Wiley, New York, p 1–15;

   b. Bowsher BR, Rest AJ (1981) J Chem Soc, Dalton Trans 1157;

   c. Kulstad S, Malmsten LA (1980) Tetrahedron 36: 521;

   d. Shanzer A, Mayer-Shochet N (1980) J Chem Soc, Chem Commun 176;

   e. Poon CK, Che CM (1979) J Chem Soc, Chem Commun 861;

   f. Dietrich-Buchecker CO, Guilhelm J, Khemiss AK, Kintzinger JP, Pascard C, Sauvage JP (1987) Angew Chem 99: 711; (1987) Angew Chem, Int Ed Engl 26: 661;

   g. Dietrich-Buchecker CO, Khemiss A, Sauvage JP (1986) J Chem Soc, Chem Commun 1376;

   h. Shanzer A, Libman J (1983) J Chem Soc, Chem Commun 846;

   i. Saigo K, Lin RJ, Kubo M, Youda A, Hasegawa M (1986) J Am Chem Soc 108: 1996;

   k. Healy MDS, Rest AJ (1978) Adv Inorg Chem Radiochem 21: 1

10. Galli C, Mandolini L (1982) J Chem Soc, Chem Commun 251

11. a. Hammerschmidt E, Schlüter H, Vögtle F (1980) J Chem Res (S) 86; ibid (M) 1083;

    b. Mattice WL, Newkome GR (1982) J Am Chem Soc 104: 5942

12. Fastrez J (1987) Tetrahedron Lett 28: 419

13. Chojnowski J, Rubinsztajn S, Wilczek L (1983) J Chem Soc, Chem Commun 69

14. Hammond PJ, Beer PD, Hall CD (1983) J Chem Soc, Chem Commun 1161

15. a. Review: Illuminati G, Mandolini L (1981) Acc Chem Res 14: 95;

    b. Illuminati G, Mandolini L, Masci B (1981) J Am Chem Soc 103: 4142;

    c. Winnik MA (1981) Chem Rev 81: 491;

    d. Rodgers SJ, Ng CY, Raymond KN (1985) J Am Chem Soc 107: 4094

16. Vögtle F, Nieger M, Przybilla KJ, Franken S (1988) Angew Chem 100: 987; (1988) Angew Chem, Int Ed Engl 27: 976

17. a. Sato T, Torizuka K, Komaki R, Atobe H (1980) J Chem Soc, Perkin II 561;

    b. Rheingold ID, Schmidt W, Boekelheide V (1979) J Am Chem Soc 101: 2121;

    c. Kißener W, Vögtle F (1985) Angew Chem 97: 782; (1985) Angew Chem, Int Ed Engl 24: 794;

    d. Mitchell RH, Chaudhary M, Kamada T, Slowey PD, Williams RV (1986) Tetrahedron 42: 1741;

    e. Schriver JW, Thomas TA (1987) J Am Chem Soc 109: 4121;

    f. Struck J (1986) PD thesis, University of Bonn;

    g. Bodwell G, Ernst L, Hopf H (1989) Chem Ber 122: 1013

18. a. Bolm C, Sharpless KB (1988) Tetrahedron Lett 29: 5101;

    b. Bodwell G, Ernst L, Haenel MW, Hopf H (1989) Angew Chem 101: 509; (1989) Angew Chem, Int Ed Engl 28: 455;

    c. Sendhoff N, Weißbarth KH, Vögtle F (1987) Angew Chem 99: 794; (1987) Angew Chem, Int Ed Engl. 26: 777

    d. Sendhoff N, Kißener W, Vögtle F, Franken S, Puff H (1988) Chem Ber 121: 2179;

    e. Tashiro M, Yamato T, Kobayashi K, Arimura T (1987) J Org Chem 52: 3196;

    f. Hisatome M, Yoshihashi M, Yamakawa K, Iitaka Y (1987) Bull Chem Soc Jpn 60: 2953;

    g. Pascal Jr RA, Winans CG, Van Engen D (1989) J Am Chem Soc 111: 3007;

    h. Cordes AW, Lin ST, Lin LH (1990) Acta Crystallogr C46: 170

19. a. Kawashima T, Kurioka S, Tohda Y, Ariga M, Mori Y (1985) Chem Lett 1289;

    b. Funhoff DJH, Staab HA (1986) Angew Chem 98: 757; (1986) Angew Chem, Int Ed Engl 25: 742

20. a. Rau H, Lüddecke E (1982) J Am Chem Soc 104: 1616;

    b. Butcher Jr JA, Dutta AK (1986) Tetrahedron Lett 27: 3341;

    c. Vögtle F, Saitmacher K, Peyerimhoff S, Hippe D, Puff H, Büllesbach P (1987) Angew Chem 99: 459; (1987) Angew Chem, Int Ed Engl 26: 470;

    d. Billen S, Vögtle F (1989) Chem Ber 122: 1113

21. Dittrich U, Grützmacher HF (1985) Chem Ber 118: 4404

22. Meurer K, Luppertz F, Vögtle F (1985) Chem Ber 118: 4433

23. Ostrowicki A, Vögtle F (1988) Synthesis 1003

24. Sergeev VA, Ovchinnikov YE, Nedel'Kin VI, Astankov AV, Andrianova OB, Shklover VE, Zamaev IA, Struchkov YT (1988) Izv Akad Nauk SSSR, Ser Khim 1625

25. a. Mashraqui SH, Keehn PM (1983) J Org Chem 48: 1341;

    b. Otsubo T, Kohda T, Misumi S (1980) Bull Chem Soc Jpn 53: 512;

    c. Machida H, Tatemitsu H, Otsubo T, Sakata Y, Misumi S (1980) Bull Chem Soc Jpn 53: 2943;

    d. Nakazaki M, Yamamoto K, Toya T (1981) J Org Chem 46: 1611;

    e. Kißener W (1985) PD thesis, University of Bonn;

    f. Staab HA, Diederich F (1983) Chem Ber 116: 3487;

    g. Grützmacher HF, Husemann W (1987) Tetrahedron 43: 3205

26. a. Brudermüller M, Musso H, Wagner A (1988) Chem Ber 121: 2239;

    b. Staab HA, Wahl P, Kay KY (1987) Chem Ber 120: 541

27. Hojjatie M, Mulidharan S, Freiser H (1989) Tetrahedron 45: 1611

28. Higuchi H, Tani K, Otsubo T, Sakata Y, Misumi S (1987) Bull Chem Soc Jpn 60: 4027

29. a. Vögtle F, Ley F (1978) Chem Ber 111: 2445;

    b. Papkalla T (1989) PD thesis, University of Bonn

30. Meurer KP (1984) PD thesis, University of Bonn

31. a. Walba DM, Armstrong JD, Perry AE, Richards RM, Homan TC, Haltiwanger RC (1986) Tetrahedron 42: 1883;

    b. Walba DM, Richards RM, Hermsmeier M, Haltiwanger RC (1987) J Am Chem Soc 109: 7081

32. a. Dijkstra PJ, van Steen BJ, Hams BHM, den Hertog Jr HJ, Reinhoudt DN (1986) Tetrahedron Lett 27: 3183;

    b. Cram DJ (1986) Angew Chem 98: 1041; (1986) Angew Chem, Int Ed Engl 25: 1039;

    c. Alston DR, Slawin AMZ, Stoddart JF, Williams DJ, Zarzycki R (1987) Angew Chem 99: 697; (1987) Angew Chem, Int Ed Engl 26: 692

33. Saward E, Diederich F (1987) Tetrahedron Lett 28: 5111

34. Chambron JC, Sauvage JP (1987) Tetrahedron 43: 895

35. Marshall JA, Lebreton J, DeHoff BS, Jenson TM (1987) Tetrahedron Lett 28: 723

36. Dietrich-Buchecker CO, Sauvage JP (1989) Angew Chem 101: 192; (1989) Angew Chem, Int Ed Engl 28: 189

37. a. Heinz W, Räder HJ, Müllen K (1989) Tetrahedron Lett 30: 159;

    b. Richards TI, Layden K, Warminski EE, Milburn PJ, Haslam E (1987) J Chem Soc, Perkin Trans I 2765;

    c. Chen CS, Chao HE, Wang SJ (1987) Synth Commun 17: 1431;

    d. van Staveren CJ, Aarts VMLJ, Grootenhuis PDJ, Droppers WJH, van Eerden J, Harkema S, Reinhoudt DN (1988) J Am Chem Soc 110: 8134;

    e. Diederich F, Hester MR, Uyeki MA (1988) Angew Chem 100: 1775; (1988) Angew Chem, Int Ed Engl 27: 1705

38. a. Bencini A, Bianchi A, Garcia-Espana E, Giusti M, Mangani S, Micheloni M, Orioli P, Paoletti P (1987) Inorg Chem 26: 1243;

    b. Bencini A, Bianchi A, Garcia-Espana E, Giusti M, Michelou M, Paoletti P (1987) Inorg Chem 26: 681

39. Marecek JF, Burrows CJ (1986) Tetrahedron Lett 27: 5943

40. Zask A, Gonnella N, Nakanishi K, Turner CJ, Imajo S, Nozoe T (1986) Inorg Chem 25: 3400

41. Odashima K, Itai A, Iitaka Y, Koga K (1980) J Am Chem Soc 102: 2504

42. Mori K, Odashima K, Itai A, Iitaka Y, Koga K (1984) Heterocycles 21: 388
43. Hosseini MW, Lehn JM (1987) J Am Chem Soc 109: 7047
44. Heyer D, Lehn JM (1986) Tetrahedron Lett 27: 5869
45. Losensky HW, Spelthann H, Ehlen A, Vögtle F, Bargon J (1988) Angew Chem 100: 1225; (1988) Angew Chem, Int Ed Engl 27: 1189
46. Ebmayer F, Vögtle F (1989) Chem Ber 122: 1725
47. a. Vriesema BK, Buter J, Kellogg RM (1984) J Org Chem 49: 110;
    b. Hosseini MW, Lehn JM (1986) Helv Chim Acta 69: 587;
    c. Dietrich B, Hosseini MW, Lehn JM, Sessions RB (1985) Helv Chim Acta 68: 289;
    d. Breslow R, Czarnik AW, Lauer M, Leppkes R, Winkler J, Zimmerman S (1986) J Am Chem Soc 108: 1969;
    e. Saigo K, Lin RJ, Kubo M, Youda A, Hasegawa M (1986) Chem Lett 519;
    f. Wilcox CS, Cowart MD (1986) Tetrahedron Lett 27: 5563;
    g. Jazwinski J, Lehn JM, Lilienbaum D, Ziessel R, Guilhem J, Pascard C (1987) J Chem Soc, Chem Commun 1691;
    h. Chang CA, Ochaya VO (1988) J Org Chem 53: 5
48. Schmidt U, Lieberknecht A, Haslinger E (1985) In: Brossi E (ed) The alcaloids. Academic Press, New York, vol 26, p 299
49. Larkins HL, Hamilton AD (1986) Tetrahedron Lett 27: 2721
50. a. Hamilton AD, Van Engen D (1987) J Am Chem Soc 109: 5035;
    b. Hirst SC, Hamilton AD (1990) Tetrahedron Lett 31: 2401;
    c. Hamilton AD, Little D (1990) J Chem Soc, Chem Commun 297
51. Peter-Katalinić J, Ebmeyer F, Seel C, Vögtle F (1989) Chem Ber 122: 2391
52. a. Kiggen W, Vögtle F (1984) Angew Chem 96: 712; (1984) Angew Chem, Int Ed Engl 23: 714;
    b. Stutte P, Kiggen W, Vögtle F (1987) Tetrahedron 43: 2065
53. Vögtle F, Müller WM, Werner U, Losensky HW (1987) Angew Chem 99: 930; (1987) Angew Chem, Int Ed Engl 26: 901
54. Fujita T, Lehn JM (1988) Tetrahedron Lett 29: 1709
55. a. Schmidt U (1986) Pure Appl Chem 58: 295;
    b. Pastuszak J, Gardner JH, Singh J, Rich DH (1982) J Org Chem 47: 2982;
    c. Rothe M, Mästle W (1982) Angew Chem 94: 223; (1982) Angew Chem, Int Ed Engl 21: 220;
    d. Schmidt U, Utz R, Lieberknecht A, Griesser H, Potzolli B, Bahr J, Wagner K, Fischer P (1987) Synthesis 233;
    e. Baker R, Castro JL (1989) J Chem Soc, Chem Commun 378;
    f. Kato S, Hamada Y, Shioiri T (1986) Tetrahedron Lett 27: 2653;
    g. Sugiura T, Hamada Y, Shioiri T (1987) Tetrahedron Lett 28: 2251;
    h. Evans DA, Ellman JA (1989) J Am Chem Soc 111: 1063;
    i. Schmidt U, Kroner M, Griesser H (1988) Tetrahedron Lett 29: 3057
56. a. Schmidt U, Weller D (1986) Tetrahedron Lett 27: 3495;
    b. Schmidt U, Griesser H, Lieberknecht A, Talbiersky J (1981) Angew Chem 93: 271; (1981) Angew Chem, Int Ed Engl 20: 280;
    c. Schmidt U, Lieberknecht A, Griesser H, Talbiersky J (1982) J Org Chem 47: 3261;
    d. Schmidt U, Lieberknecht A, Bökens H, Griesser H (1983) J Org Chem 48: 2680;
    e. Schmidt U, Schanbacher U (1984) Liebigs Ann Chem 1205;
    f. Schmidt U, Bökens H, Lieberknecht A, Griesser H (1983) Liebigs Ann Chem 1459;
    g. Schmidt U, Lieberknecht A, Griesser H, Hänsler J (1982) Liebigs Ann Chem 2153
57. a. Biernat JF, Bochenska M, Bradshaw JS, Koyama H, Lindk G, Lamb JD, Christensen JJ, Izatt RM (1987) J Incl Phen 5: 729;
    b. Nakatsuka S, Masuda T, Sakai K, Goto T (1985) Tetrahedron Lett 26: 5735;
    c. Rodgers SJ, Ng CY, Raymond KN (1985) J Am Chem Soc 107: 4094;
    d. Werner U, Müller WM, Losensky HW, Merz T, Vögtle F (1986) J Incl Phen 4: 379;

Fritz Vögtle et al.

    e. Neugebauer FA, Fischer H (1986) Tetrahedron Lett 27: 5367;

    f. Murakami Y, Aoyama Y, Kikuchi J, Nishida K (1982) J Am Chem Soc 104: 5189;

    g. McMurry TJ, Rodgers SJ, Raymond KN (1987) J Am Chem Soc 109: 3451;

    h. Hider RC (1984) Struct Bonding 58: 25;

    i. Grammenudi S, Vögtle F (1986) Angew Chem 98: 1119; (1986) Angew Chem, Int Ed Engl 25: 1122;

    k. Schurmann G, Diederich F (1986) Tetrahedron Lett 27: 4249;

    l. Wallon A, Peter-Katalinić J, Werner U, Müller WM, Vögtle F (1990) Chem Ber 123: 375;

    m. Dubowchik GM, Hamilton AD (1986) J Chem Soc, Chem Commun 665;

    n. Dubowchik GM, Hamilton AD (1987) J Chem Soc, Chem Commun 293;

    o. Collman JP, Chong AO, Jameson GB, Oakley RT, Rose E, Schmittou ER, Ibers JA (1981) J Am Chem Soc 103: 516;

    p. Collman JP, Brauman JI, Collins TJ, Iverson BL, Lang G, Pettman RB, Sessler JL, Walters MA (1983) J Am Chem Soc 105: 3038;

    q. Wambach L, Vögtle F (1985) Tetrahedron Lett 26: 1483;

    r. Ireland C, Scheuer PJ (1980) J Am Chem Soc 102: 5688;

    s. Lagarias JC, Houghten RA, Rapoport H (1978) J Am Chem Soc 100: 8202;

    t. McMurry TJ, Hosseini MW, Garrett TM, Hahn FE, Reyes ZE, Raymond KN (1987) J Am Chem Soc 109: 7196;

    u. Uemori Y, Kyuno E (1987) Inorg Chim Acta 138: 9

58. Schrage H, Vögtle F, Steckhan E (1988) J Incl Phen 6: 157

59. Schmidt U, Werner J (1986) Synthesis 986

60. Rastetter WH, Phillion DP (1981) J Org Chem 46: 3209

61. Seebach D, Brändli U, Schnurrenberger P, Przybylski M (1988) Helv Chim Acta 71:155

62. Maciejewski L, Martin M, Ricart G, Brocard J (1988) Synth Commun 18: 1757

63. Picard D, Cazaux L, Tisnes P (1986) Tetrahedron 42: 3503

64. a. Quinkert G, Heim N, Glenneberg J, Döller U, Eichhorn M, Billhardt UM, Schwarz C, Zimmermann G, Bats JW, Dürner G (1988) Helv Chim Acta 71: 1719;

    b. Quinkert G, Heim N, Glenneberg J, Billhardt UM, Autze V, Bats JW, Dürner G (1987) Angew Chem 99: 363; (1987) Angew Chem, Int Ed Engl 26: 362;

    c. Wasserman HH, Gambale RJ, Pulwer MJ (1981) Tetrahedron 37: 4059;

    d. Quinkert G, Billhardt UM, Jakob H, Fischer G, Glenneberg J, Nagler P, Autze V, Heim N, Wacker M, Schwalbe T, Kurth Y, Nats JW, Dürner G, Zimmermann G, Kessler H (1987) Helv Chim Acta 70: 771

65. a. Jolley ST, Bradshaw JS (1980) J Org Chem 45: 3554;

    b. Feldman KS, Lee YB (1987) J Am Chem Soc 109: 5850;

    c. Miljkovic D, Kuhajda K (1987) Tetrahedron Lett 28: 5737;

    d. Schmidt U, Heermann D (1979) Angew Chem 91: 330; (1979) Angew Chem, Int Ed Engl 18: 308;

    e. Schmidt U, Werner J (1986) J Chem Soc, Chem Commun 996;

    f. Schmidt U, Dietsche M (1981) Angew Chem 93: 786; (1981) Angew Chem, Int Ed Engl 20: 771;

    g. Gerlach H, Thalmann A (1974) Helv Chim Acta 57: 2661;

    h. Kurihara T, Nakajima Y, Mitsunobu O (1976) Tetrahedron Lett 2455

66. a. Bestmann HJ, Schobert R (1985) Angew Chem 97: 784; (1985) Angew Chem, Int Ed Engl 24: 791;

    b. Bestmann HJ, Schobert R (1985) Angew Chem 97: 783; (1985) Angew Chem, Int Ed Engl 24: 790;

    c. Scott PW, Harrison IT (1981) J Org Chem 46: 1914;

    d. Koert U, Fernholz E (1988) Git Fachz Lab 32: 627;

    e. Yoshida M, Harada N, Nakamura H, Kanematsu K (1988) Tetrahedron Lett 29: 6129;

    f. Cazaux L, Duriez MC, Picard C, Tisnes P (1989) Tetrahedron Lett 30: 1369;

34

g. Leeper FJ, Smith DHC (1988) Tetrahedron Lett 29: 1325;

h. Leygue N, Picard C, Tisnes P, Cazaux L (1988) Tetrahedron 44: 5845

67. Mahajan JR, Resck IS (1980) Synth Commun 998
68. Baker W, Buggle KM, McOmie JFW, Watkins DAM (1958) J Chem Soc 3594
69. Vögtle F, Körsgen UU, Puff H, Reuter H (1989) Chem Ber 122: 343
70. a. Merz T, Wirtz H, Vögtle F (1986) Angew Chem 98: 549; (1986) Angew Chem, Int Ed Engl 25: 567;

    b. Casadei MA, Galli C, Mandolini L (1981) J Org Chem 46: 3127;

    c. Shinmyozu T, Sakai T, Uno E, Inazu T (1985) J Org Chem 50: 1959;

    d. Deslongchamps P, Lamothe S, Lin HS (1987) Can J Chem 65: 1298

71. a. Miyahara Y, Inazu T, Yoshino T (1983) Tetrahedron Lett 24: 5277;

    b. Park CH, Simmons HE (1972) J Am Chem Soc 94: 7184

72. a. Kurosawa K, Suenaga M, Inazu T, Yoshino T (1982) Tetrahedron Lett 23: 5335;

    b. Shinmyozu T, Hirai Y, Inazu T (1986) J Org Chem 51: 1551

73. a. Wirtz H, Vögtle F, Puff H, Woller P (1986) J Incl Phen 4: 135;

    b. Miller SP, Whitlock Jr HW (1984) J Am Chem Soc 106: 1492;

    c. Sheridan RE, Whitlock Jr HW (1986) J Am Chem Soc 108: 7120;

    d. Ojima J, Fujita S, Masumoto M, Ejiri E, Kato T, Kuroda S, Nozawa Y, Tatemitsu H (1987) J Chem Soc, Chem Commun 534;

    e. Ojima J, Ejiri E, Kato T, Kuroda S, Hirooka S, Shibutani M (1986) Tetrahedron Lett 27: 2467;

    f. Jarvi ET, Whitlock HW (1982) J Am Chem Soc 104: 7196

74. a. Marshall JA, Gung WY (1989) Tetrahedron Lett 30: 309;

    b. Brocchini SJ, Eberle M, Lawton RG (1988) J Am Chem Soc 110: 5211;

    c. Catoni G, Galli C, Mandolini L (1980) J Org Chem 45: 1906;

    d. Morgan B, Dolphin D (1985) Angew Chem 97: 1000; (1985) Angew Chem, Int Ed Engl 24: 1003;

    e. Staab HA, Matzke G, Krieger C (1987) Chem Ber 120: 89

75. a. Wilcox Jr CF, Weber KA (1986) J Org Chem 51: 1088

    b. Yvergnaux F, Le Floc'h Y, Grée R (1989) Tetrahedron Lett 30: 7397

76. a. McMurry JE (1983) Acc Chem Res 16: 405;

    b. Ojima J, Yamamoto K, Kato T, Wada K, Yonegawa Y, Ejiri E (1986) Bull Chem Soc Jpn 59: 2209;

    c. Vogel E, Püttmann W, Duchatsch W, Schieb T, Schmickler H, Lex J (1986) Angew Chem 98: 727; (1986) Angew Chem, Int Ed Engl 25: 720;

    d. McMurry JE, Haley GJ, Matz JR, Clardy JC, Duyne GV, Gleiter R, Schäfer W, White DH (1986) J Am Chem Soc 108: 2932;

    e. Yamamoto K, Shibutani M, Kuroda S, Ejiri E, Ojima J (1986) Tetrahedron Lett 27: 975;

    f. McMurry JE, Matz JR, Kees KL (1987) Tetrahedron 43: 5489;

    g. McMurry JE, Haley GJ, Matz JR, Clardy JC, Mitchell J (1986) J Am Chem Soc 108: 515

77. Herin M, Colin M, Tursch B (1976) Bull Soc Chim Belges 85: 801
78. a. Stork G, Nakamura E (1979) J Org Chem 44: 4010;

    b. Raddatz P, Winterfeldt E (1981) Angew Chem 93: 281; (1981) Angew Chem, Int Ed Engl 20: 286;

    c. Nicolaou KC, Chakraborty TK, Daines RA, Simpkins NS (1986) J Chem Soc, Chem Commun 413;

    d. Nicolaou KC, Daines RA, Chakraborty TK (1987) J Am Chem Soc 109: 2208

79. Bailey SJ, Thomas EJ, Vather SM, Wallis J (1983) J Chem Soc, Perkin I 851
80. Eberbach W, Heinze I, Knoll K, Fritz H, Borle F (1988) Helv Chim Acta 71: 404
81. Juriew J, Skorochodowa T, Merkuschew J, Winter W, Meier H (1981) Angew Chem 93: 285; (1981) Angew Chem, Int Ed Engl 20: 269
82. Scott PW, Harrison IT (1981) J Org Chem 46: 1914
83. Baldwin JE, Adlington RM, Robertson J (1989) Tetrahedron 45: 909

84. Vagt U, Haase M, Konusch J, Tochtermann W (1987) Chem Ber 120: 769
85. Raasch MS (1979) J Org Chem 44: 2629
86. Houk J, Whitesides GM (1989) Tetrahedron 45: 91
87. a. Regen SL, Kimura Y (1982) J Am Chem Soc 104: 2064;
    b. Trost BM, Warner RW (1982) J Am Chem Soc 104: 6112;
    c. Kimura Y, Regen SL (1983) J Org Chem 48: 1533;
    d. Gonzalez A, Holt SL (1981) J Org Chem 46: 2594;
    e. Singh P, Jain A (1988) Inidian J Chem, Sect B 27: 937;
    f. Singh H, Kumar M, Singh P, Kumar S (1988) J Chem Res (S) 132;
    g. Singh H, Kumar M, Singh P (1989) J Chem Res (S) 94; (1989) ibid (M) 0675;
    h. Geuder W, Hünig S, Suchy A (1983) Angew Chem 95: 501; (1983) Angew Chem, Int Ed Engl 22: 489;
    i. Geuder W, Hünig S, Suchy A (1986) Tetrahedron 42: 1665
88. Schneider HJ, Busch R (1986) Chem Ber 119: 747
89. a. Jurczak J, Ostazewski R, Salanski P (1989) J Chem Soc, Chem Commun 184;
    b. Kohnke FH, Slawin AMZ, Stoddart JF, Williams DJ (1987) Angew Chem 99: 941; (1987) Angew Chem, Int Ed Engl 26: 892;
    c. Kohnke FH, Mathias JP, Stoddart JF (1989) Angew Chem Adv Mater 101: 1129; (1989) Angew Chem, Int Ed Engl 28: 1103;
    d. Ashton PR, Isaacs NS, Kohnke FH, Stagno d'Alcontres G, Stoddart JF (1989) Angew Chem 101: 1269; (1989) Angew Chem, Int Ed Engl 28: 1261
90. a. Ashton PR, Goodnow TT, Kaifer AE, Reddington MV, Slawin AMZ, Spencer N, Stoddart JF, Vicent C, Williams DJ (1989) Angew Chem 101: 1404; (1989) Angew Chem, Int Ed Engl 28: 1396;
    b. Bühner M, Geuder W, Gries WK, Hünig S, Koch M, Poll T (1988) Angew Chem 100: 1611; (1989) Angew Chem, Int Ed Engl 27: 1553
91. The authors kindly thank Prof. Stoddart for his permission to use drawings of the molecules **123 – 128**
92. Rodriguez-Ubis JC, Alpha B, Plancherel D, Lehn JM (1984) Helv Chim Acta 67: 2264

The "Cesium Effect":
Syntheses of Medio- and Macrocyclic Compounds

Andreas Ostrowicki, Erich Koepp and Fritz Vögtle

Institut für Organische Chemie und Biochemie der Universität Bonn, Gerhard-Domagk-Straße 1, D-5300 Bonn 1, FRG

Table of Contents

Topics in Current Chemistry, Vol. 161
© Springer-Verlag Berlin Heidelberg 1991

Fritz Vögtle et al.

This overview highlights the application of the "cesium effect" in synthetic reactions. Special attention is drawn to macrocyclizations. More recent examples published between 1983 and 1990 are selected critically. In a short introduction cesium salts which are usually applied in organic chemistry are introduced, followed by many concrete examples of the use of cesium salts in ring closure reactions leading to macro- and oligomacrocyclic compounds. It is shown that by using cesium salts instead of the corresponding sodium and potassium compounds in ring closure reactions yields can often be dramatically increased. The review is concluded with attempts to explain the special effects of cesium salts in organic synthesis.

1 Introduction

The synthesis of large and medium-sized rings has been of constant interest to organic chemists. Whereas "normal" ring compounds bearing five to seven ring members usually have a high tendency of formation and therefore are obtained in good yields, the preparation of larger ring compounds, in particular medium-membered ones bearing 8–12 ring atoms, often faces considerable difficulties.

Besides methods using high dilution conditions [1] or a pre-orientation of the reactants [2] ("preorganization" [3]) or template effects [4, 5], ring closure reactions under assistance of cesium ions have been applied during the past years. The advantages of cesium assisted reactions depend on the good accessibility of organic cesium salts and the high yields of cyclization reactions, which often proceed without the method of high dilution [1].

The following contribution is intended to continue our earlier review [6] dealing with the then known cesium assisted ractions; only some very important earlier cesium assisted reactions are described here for the sake of completeness: After a short description of the properties of some cesium compounds used in this respect in organic synthesis, those reactions are discussed that have been published more recently and which proceed under the intermediate formation of organic cesium salts at oxygen, nitrogen or sulphur functions. Only those cyclization reactions are considered which allow a direct comparison of the yields obtained with cesium compounds or cesium metal. The description starts with reactions leading to the formation of C—C bonds and proceeds to the syntheses of ethers, lactones, amines and sulfides. The discussion ends with a summary of the attempts to explain the "cesium effect".

2 Cesium Salts Used in Organic Chemistry and for the "Cesium Effect"

In organic chemistry cesium fluoride and to a lesser extent the carbonate and the hydroxide have gained great importance. These three cesium compounds are applied mainly as basic reagents. Further applications of cesium fluoride are e.g. the preparation of fluoroalkanes by S_N-reactions using fluoride anions [7a] and desilylation reactions [7b].

2.1 Cesium Fluoride

The basicity of fluoride anions under aprotic conditions is based on the stability of the H—F bond (approx. 569 kJ/mol; cf. H—Cl approx. 432 kJ/mol) [8]. In general, the basicity of ionic fluorides depends strongly on the solvent used, on the water content and on the counterion. Whereas ionic fluorides in protic solvents usually exhibit only weak basicity, under aprotic reaction conditions they may allow deprotonation of weak CH-acidic compounds like DMSO or acetonitrile [9]. **Cesium** fluoride is the alkali metal fluoride exhibiting the highest reactivity [10, 11], but often the potassium analog is preferred for its lower price and lower hygroscopicity. The low solubility of all alkali metal fluorides in aprotic solvents is disadvantageous, and as a consequence, reactions proceed mainly in a hetero-geneous way on the surface of the undissolved fluoride [11]. Addition of crown compounds [12] or other phase transfer catalysts [13] therefore may lead to significant rate accelerations or even different reaction paths [13].

2.2 Cesium Carbonate and Cesium Hydroxide

In the series of the alkali metal carbonates and -hydroxides the **cesium** compounds are the strongest bases [14]. For reasons of simpler handling the less hygroscopic carbonate is often preferred to the hydroxide. In dipolar aprotic solvents, carboxylic acids [15], phenols [16], thiols [17, 18] and sulfonamides [19] are easily deprotonated by **cesium** carbonate, whereas with carbamates such as e.g. benzyloxycarbonyl- ("Z"-)protected amino acids no reaction occurs [20].

3 C—C Bond Formation

The use of **cesium** compounds in C—C bond formation reactions has so far been limited to a few cases. Vögtle and Kißener [21] observed that the use of **cesium** metal in the Müller-Röscheisen procedure [22] of the Wurtz coupling reac-tion of *p*-xylylene dibromide, intended to yield $[2_n]$paracyclophanes **1**, leads to the $[2_3]$paracyclophane **1b** with the highest yield compared to other alkali metals. The yields of the other cyclic compounds **1c–e** thereby are much lower compared to the application of sodium metal (Fig. 1) [23].

Vögtle and Mayenfels studied the influence of the bases **cesium** hydroxide and sodium hydroxide on twofold aldol condensations leading to macrocyclic dichal-cones **4** [24]:

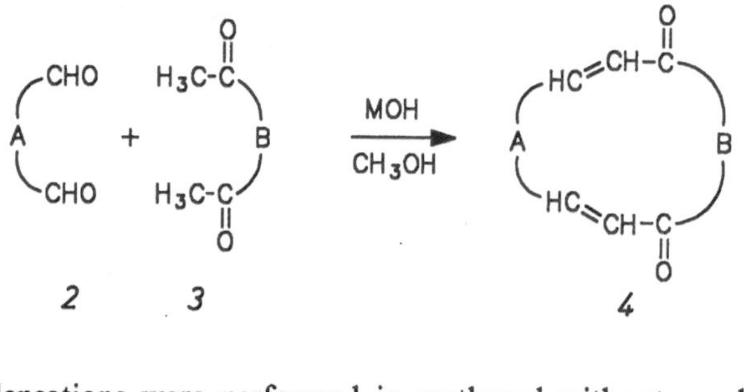

The condensations were performed in methanol without conditions of high dilution [1]. Differing arene units **2a–c** and **3a–c** were introduced. In the reactions

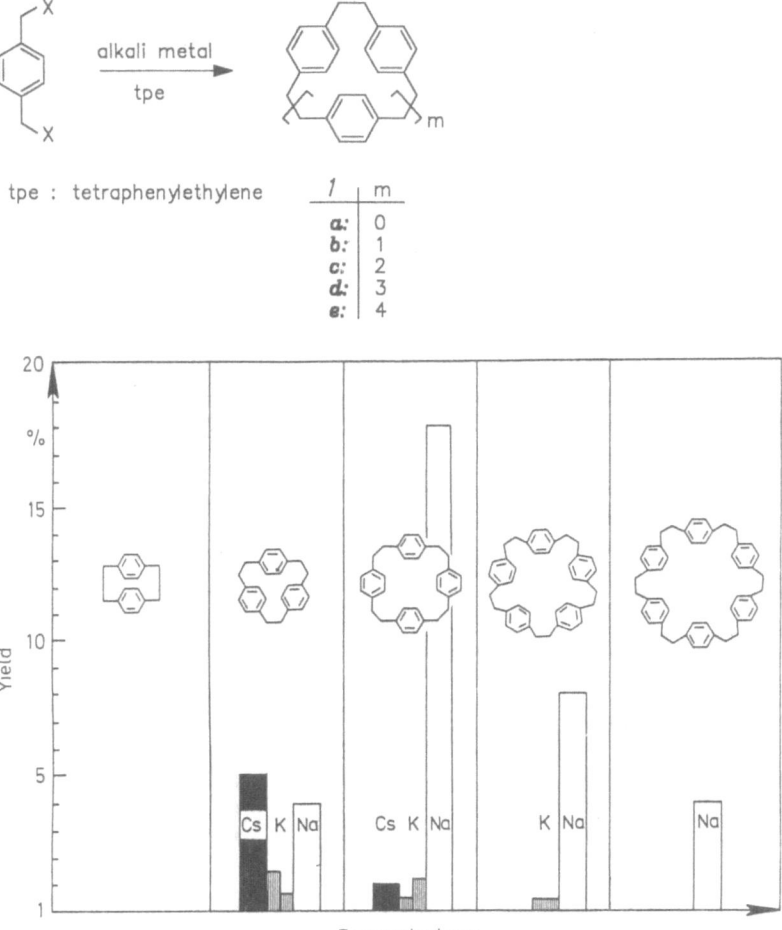

tpe : tetraphenylethylene

1	m
a:	0
b:	1
c:	2
d:	3
e:	4

Fig. 1. Oligomer formation in the [2$_n$]paracyclophane series: dependency on the alkali metal applied [21]

Table 1. Yields in the aldol condensations leading to macrocyclic dichalcones **4** with the bases NaOH and **CsOH** [24]

No.	2	3	Yield [%]		Arene Unit		
	arene unit		NaOH	**CsOH**	a	b	c
4a	a	a	41	48–60			
4b	c	a	23	29			
4c	b	b	12	5			

of conformationally rigid arene units like **2a/3a** or **2c/3a**, only minor yield improvements from 41 % (NaOH) to 48–60 % (**CsOH, 4a**) and from 23 % (NaOH) to 29 % (**CsOH, 4b**) were found. In contrast to this behaviour conformationally flexible arene units such as **2b/3b** did not show any yield-improvements (Table 1).

4 C—N Bond Formation

4.1 Synthesis of Aza Macrocycles

Kellogg et al. [19] for the first time used **cesium** salts of aliphatic tosylamides **5** for the preparation of N-tosyldiazacycloalkanes **7**. The well known method of Stetter [25] and Richman [26] which has often been used for the synthesis of aza crown compounds using sodium or potassium salts of tosylamides in the cyclization with bromoalkanes, usually leads to medium-sized azacycloalkanes in comparably low yields. The **cesium** salts of the tosylamides **5** obtained from the reaction of **5** with Cs_2CO_3 were reacted with the dibromides **6** in an analogous way to the procedure of Stetter [25] using DMF as the solvent, but without application of dilution techniques.

$$TosNH(CH_2)_nNHTos \quad \xrightarrow[\text{2) } Br(CH_2)_mBr]{\text{1) } Cs_2CO_3/DMF}$$

5　　　　　　　**6**

7

As demonstrated in Table 2, the application of **cesium** carbonate particularly in the case of large rings (**7c–e**) leads to high yields. Compared to the method of Richman [26] using sodium methylate as the base, much higher yields of the

Table 2. Yields of macrocycles **7** and **8** in the **cesium** assisted cyclization compared to the classical method of Stetter and Richman [19]

No.	**5**	**6**	Ring Members	Yields [%]	
	n	m		Cs_2CO_3[a]	$NaOCH_3$[b]
7a	5	5	12	30	—
7b	5	10	17	65	40–50
7c	10	6	18	95	—
7d	10	10	22	76	—
7e	10	16	28	60	30–40
8			18	66	—

[a] Analytically pure product. — [b] Yields of raw product

macrocycles were obtained: The 28-membered ring **7e**, e.g., obtained in 60% yield according to Kellogg using Cs_2CO_3, could only be isolated in 30–40% yield according to the method using $NaOCH_3$ as base.

In general, yields decrease in **cesium** assisted cyclizations changing to smaller ring sizes (**7a**) or by introduction of hetero atoms into the ring skeleton. The "**cesium** method" is also well suited for the preparation of cyclophane compounds like the 23-membered macrocycle **9** or the tetraoxa cycle **10**, which were obtained in 60 and 46% yields respectively, starting with the ditosylates [19].

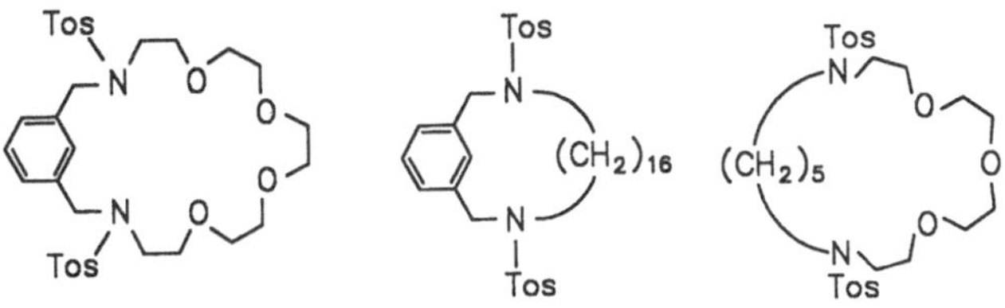

    8   **9**   **10**

A comparison of the yields of the cycles **7b** and **11** obtained by application of different alkali metal carbonates demonstrates the superiority of the **cesium** assisted cyclization:

Table 3. Comparison of the yields of **7b** and **11** in cyclizations with different alkali metal carbonates M_2CO_3 [19, 27]

No.	Yields [%] $M^{\oplus}$				
	$Li^{\oplus}$	$Na^{\oplus}$	$K^{\oplus}$	$Rb^{\oplus}$	$Cs^{\oplus}$
7b [19]	0	10	10	70[a]	quant.[b]
11 [27]	0	21	75	—	75

[a] Yield of raw product. — [b] Nmr pure product

Under identical reaction conditions by using Cs_2CO_3 it was possible to isolate **7b** in quantitative yield as an NMR pure product, whereas by application of Rb_2CO_3 only an impure product in 70% yield was obtained (Table 3) [19]. By addition of other alkali metal carbonates, even after ten days of reaction time, only unsatisfactory yields below 10% could be obtained, whereas in the synthesis of the tris(amide) **11** applying K_2CO_3 as base the yields of the **Cs** procedure were reached [27]. This was attributed to the insufficient basicities of these carbonates, which did not allow complete deprotonation of these tosylamides [19].

M = Li, Na, K, **Cs**

Polyaza macrocycles like **15** are of interest due to their capability of forming complexes with transition metals [28] and anions, e.g. carboxylates [29], phosphates [30] and even ATP [31].

The chiral tosylaza ring compound **14** was obtained in 80% yield in a **cesium** assisted cyclization without application of dilution techniques [1] starting with the building blocks **12** and **13** [31, 32].

The 32-membered ditopic hexaaza monocycle **16** was obtained in 47% yield by Lehn and Hosseini [33] in an analogous direct cyclization reaction using Cs_2CO_3 in DMF as solvent.

4.2 Synthesis of Macrobicyclic Aza Compounds

A significant "cesium effect" was observed in reactions forming C—N bonds leading to macrobicyclic compounds. Lehn et al. [34] synthesized the nona-tosylaza macrocycle **19** in a one step reaction by a threefold bond formation combining the two triply functionalized building blocks **17** and **18**. The cryptand **19** was obtained in 27% yield when Cs_2CO_3 was applied; replacement of the **cesium** salt by potassium carbonate led to a decreased yield of 19% [34].

Vögtle et al. [35] prepared the tris(azo) macrobicyclic compound **22**, which was of interest with respect to host compounds with photochemically modifiable cavity size and shape (photo-switching inclusion compounds). The reaction proceeded in one step starting with 1,3,5-tris(bromomethyl)benzene (**20**) and 3,3'-bis(tosylamino)azobenzene (**21**) under application of the dilution principle and assistance of Cs_2CO_3 in 12% yield. Considering the sixfold bond formation in one step, this yield is remarkable. For explanation, the rigid group principle can also be regarded to be effective here. With potassium carbonate as the base in this cyclization reaction, **22** was obtained in a lower yield of only 4% [36].

5 Formation of C—O Ether Bonds

5.1 Synthesis of Crown Compounds with Cs_2CO_3 as Base

Cesium phenolates were introduced by Kellogg [16] for the synthesis of crown compounds, after crown ether diesters had been obtained in good yields from the **cesium** salts of aromatic carboxylic acids and oligoethylene glycol dihalides [37]. The preparation of the crown ether **25** was achieved by reaction of the cesium phenolate **23** with the dibromo compound **24** in DMF without application of high dilution conditions [1]. The monoesters **28** were obtained in an analogous way be reaction of the **cesium** salicylate **26** with the bromides **27** (Fig. 2).

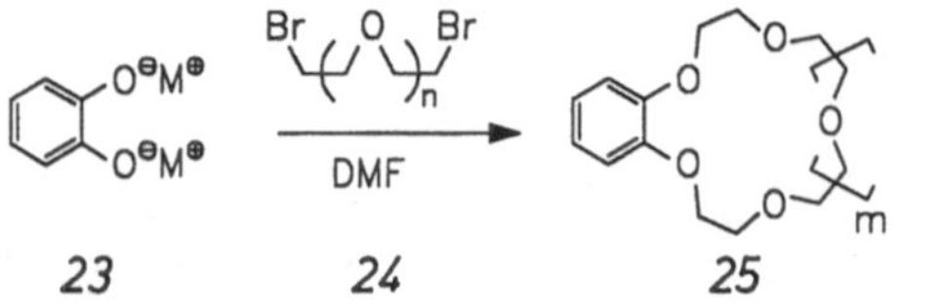

25	m	n	Yields [%], $M^{\oplus}$		38)
			$Cs^{\oplus}$	$Na^{\oplus}$	
a	1	3	50	62	
b	2	4	74	60	
c	3	5	78	-	

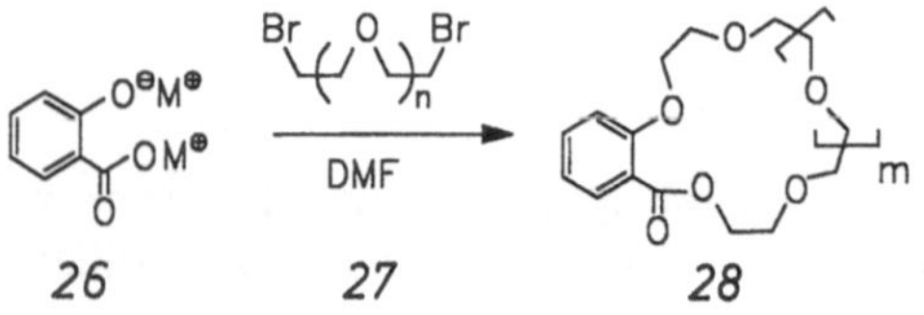

28	m	n	Yields [%], $M^{\oplus}$	
			$Cs^{\oplus}$	$K^{\oplus}$
a	1	3	54	12
b	2	4	68	54

Fig. 2. Yields of the crown compounds **25** and **28** obtained by cyclizations with different alkali metal cations [16]

A comparison of the yields of the crown compounds **28** demonstrates that the application of **cesium** salts leads to significant improvements in yields compared to the use of potassium salts. The yield of **28a** is increased from 12 to 54%, whereas the increase of the yields of the crown ethers **25** by use of **cesium** compounds is not as significant [16] (Fig. 2). Crown compounds of type **25** had been obtained formerly by Pedersen applying the sodium salts [38].

A dramatic **cesium** effect was found by Weber in the case of the synthesis of the tetrabenzo crown ether **31**, which is of interest for applications in ionselective electrodes [39]. The cyclization reactions starting from the **cesium** phenolate obtained from **29** and the tosylate **30** under high dilution conditions in DMF led to **31** in 37% yield. This is a dramatic improvement of the formerly obtained yield of only 6% in the system KOH/n-butanol/ethanol/DMF [40].

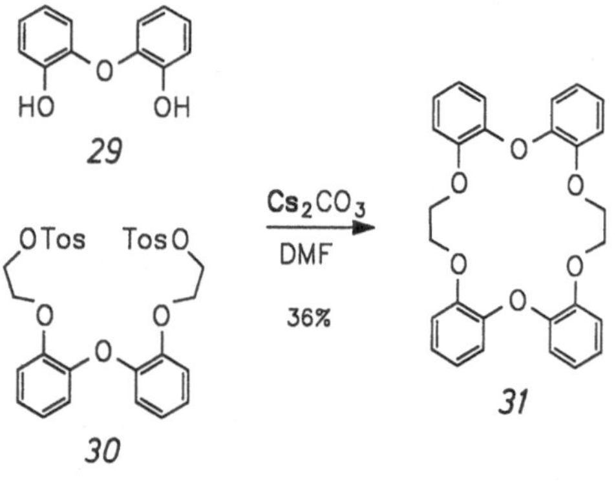

Remarkable yield improvements by **cesium** salts were also observed in the synthesis of crown compounds containing pyridine units in the ring skeleton as shown in formula **32**. This type of pyridino crown compound aroused interest due to its selective formation of inclusion compounds with unbranched aliphatic alcohols such as ethanol or 1-propanol [40]. The yields of these cyclizations, which were carried out as two-component, high-dilution reactions [1] in DMF, were dramatically increased by the application of Cs_2CO_3 for the deprotonation of the phenols. A yield increase of the pyridino crown **32a** from 26% (KOH as base) [40] to 70% (Cs_2CO_3 as base) is reported by Weber [41].

32 a-c

The pyridino crown compounds **32b, c** were obtained by Weber and Vögtle by a systematic variation of the spacer groups A. Again a dramatic increase of the yields could be obtained by the application of the **cesium** method [42, 43]: Whilst **32b** in the system KOH/n-butanol was obtained in only 9.4% and **32c** even only in traces [40], the cyclization with **cesium** carbonate in DMF under high dilution proceeded in the much better yields of 65%, **(32b)** and 33% **(32c)** [42].

5.2 Synthesis of Crown Compounds with CsF as Base

The application of **cesium** fluoride as base in the synthesis of crown compounds from phenols and the ditosylates of polyethylene glycols was first described by Reinhoudt [44]. This method uses the high basicity of weakly solvated ("naked") fluoride anions under aprotic conditions and is based on the formation of very stable H—F bonds (approx. 569 kJ/mol, H—Cl approx. 432 kJ/mol for comparison) [8].

The applicability of **cesium** fluoride for the synthesis of smaller crown ether rings such as benzo[12]crown-4 **(35)** was investigated by Bartsch et al. [45]. The result of this study was that the 12-membered crown ether **35** and its naphthalene analogue **36** can be obtained from the aromatic diols **33** and the tosylate **34** with the base **cesium** fluoride in yields of 29 and 25%, respectively. This means a significant increase in the 4% yield of benzo[12]crown-4 **(35)** reported by Pedersen, who started with sodium phenolate [38].

The preparation of benzo[18]crown-6 (**39**) from catechol **37** (R = H) and the tosylate **38** with different alkali metal fluorides was studied by Reinhoudt et al. [44]. Significant differences in the reactivities of the alkali metal fluorides were observed, whereby the **cesium** salt exhibited the highest reactivity. The benzo[18]-crown-6 (**39**) was obtained in similar high yields of 67 and 60% using RbF and CsF, but the reaction time was shortened from 65 (RbF) to 17 h by application of the **cesium** salt (Table 4) [44].

This method also allows the introduction of substituents, which are sensitive towards nucleophilic bases, into the crown ether ring. In such a way, **39** with an aldehyde function (R=CHO) was obtained in remarkable 57% yield [44].

The yields of crown ethers containing condensed arene units were increased significantly by cyclizations with the application of **cesium** fluoride. Bartsch et al. [45] isolated e.g. the naphthaleno crowns **42** and **43** from the reaction of 1,8-naphthalene diol (**40**) and the corresponding ditosylates of the ethylene glycols **41** using **cesium** fluoride as the base in yields of 63% (with NaOH as base: 28% [46]) and 53% (with K-*tert*-butanolate as base 7% [47]).

Table 4. Yields of the benzo crown 39 in cyclization reactions with different alkali metal fluorides

Base (in CH_3CN)	R	Isolated Yield [%]	Reaction Time[a] [h]
LiF, NaF	H	no reaction	140
KF	H	52	69
RbF	H	67	65
CsF	H	60	17
CsF	CHO	57	23
$(n\text{-}C_4H_9)_4NF$	H	5	19
Cs_2CO_3 (solvent: DMF)	H	74 [16]	96

[a] Until complete reaction of the tosylate (controlled by nmr)

6 Formation of Ester Bonds

Cesium salts of *N*-protected amino acids were introduced by Gisin [15] for the synthesis of ester bonds under mild reaction conditions in the solid phase synthesis according to Merrifield [48]. **Cesium** salts of short chained carboxylic acids like **cesium** propionate found broad application for the selective inversion of the stereochemistry of secondary alcohols which could be performed with **cesium** salts under careful reaction conditions [49, 50].

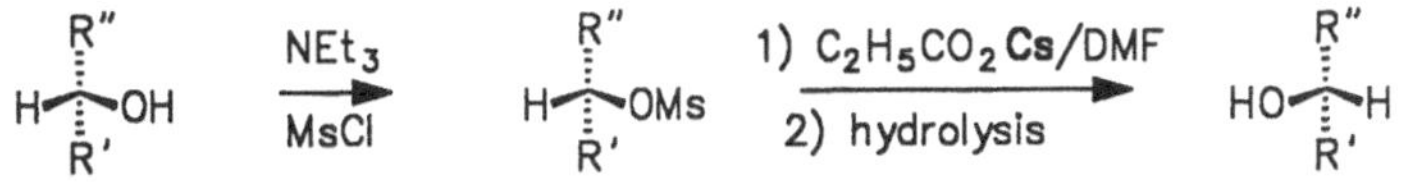

6.1 Synthesis of Heteroaryl Lactones

Cesium salts of substituted pyridine-3,5-dicarboxylic acids were used first by Kruizinga and Kellogg for the synthesis of macrocyclic lactones [51]. Kellogg obtained the bis-lactone **46** in a one-pot reaction of the **cesium** carboxylate **44** and the dibromo compound **45** in 85% yield without application of high dilution conditions [1]. By comparison with other alkali metal carbonates he proved the yield-increasing effect of the **cesium** ions:

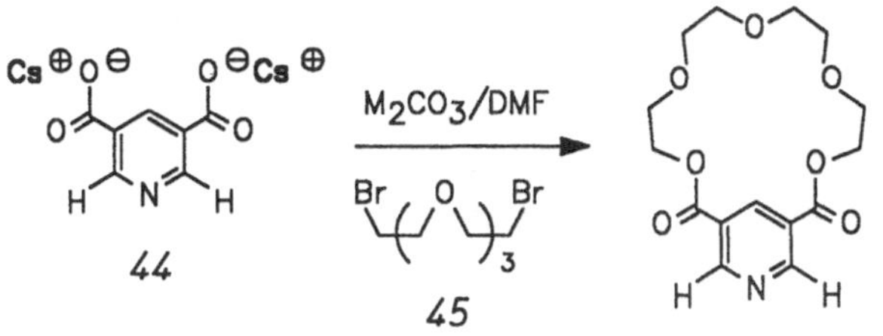

M	yield [%]
Na	5
K	15
Rb	45
Cs	85

In more recent syntheses of chiral macrocycles containing pyridine units like **48**, which are of interest in their reduced dihydropyridine-form as model compounds for the redox system NADH/NAD$^{\oplus}$, the yield increasing effect of added **cesium** salts was also used [52]. The yield of **48** was doubled in the cyclization of bis-L-valine amide **47** with 1,5-dibromopentane in DMF from 41% (Na$_2$CO$_3$) to 80% by application of **Cs$_2$CO$_3$**. The results of cyclizations with several other alkali metal carbonates [52] are shown below.

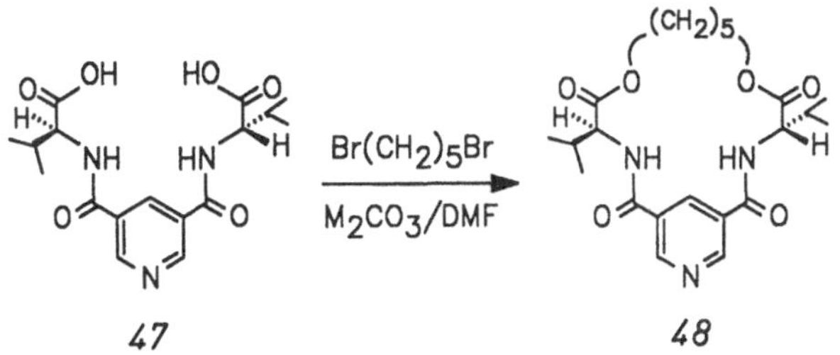

M	yield [%]
Li	21
Na	41
K	64
Rb	70
Cs	80

Lactones with "oligoheteryl" units in the macrocyclic ring skeleton can also be prepared in good yields from the reaction of **cesium** carboxylates with dihalides. Potts [53] obtained the "triheteryl" lactones **51** with a pyridine (X=CH) and a pyrimidine unit (X=N) in 60% (X=CH) and 55% yield (X=N) by cyclizing the **cesium** salt **49** with the dibromo compound **50** using DMF as solvent.

6.2 Synthesis of Arene Lactones

Crown ether lactones with an azulene unit were synthesized by Vögtle and Löhr [54]. They served for studies of the influence of the cation complexation on the absorption spectra of the coloured compounds ("chromoionophores"). The cyclization of the crown ether ring started from the **cesium** salt of the 1,3-azulene dicarboxylic acid **52** and the ditosylate **53a** and **53b** and led, under dilution conditions [1] with DMF as solvent, to the lactones **54a** and **54b** with yields of 66 and 43% respectively.

6.3 Synthesis of Aliphatic Lactones

The high yields in **cesium** assisted crown ether syntheses may be attributed partly to a template effect of the **cesium** ions. In order to estimate the importance of such a template effect in cesium assisted cyclization reactions, Kellogg and Kruizinga [51a] synthesized a series of macrocyclic lactones containing an oligomethylene chain, for which a coordination of **cesium** ions can be excluded.

$$X(CH_2)_n\ CO_2^{\ominus}Cs^{\oplus} \xrightarrow[-CsX]{DMF}$$

55

56   **57**

Table 5 shows the yields of the lactones **56** obtained by an intramolecular cyclization of the ω-halogeno carboxylic acids **55** using **cesium** ions. The yields are compared to those obtained by Corey and Nicolaou [2] using the pyridinethiol procedure.

As a summary of the investigation, the yields of the pyridinethiol procedure can be reached using **cesium** salts with the small as well as the higher ring member numbers (**56g–k**). Thereby the significant simplification of the cyclization procedure due to the **cesium** method, which can be performed as a one pot reaction, should be noted. Only in the preparation of the medium-sized ring compounds **56b–f** the **cesium** method yields higher amounts of the dilides **57**. In these cases, the pyridinethiol method leads to better yields of the desired lactones **56** (Table 5).

A comparison of the yields of the 16-membered ring compound **56i** in cyclization reactions using different metal carbonates is shown in Table 6: In the series of the alkali metal carbonates the best yields are obtained with Cs_2CO_3. $Tl^{\oplus}$ ions, which possess the highest polarizability in the periodic system of the elements [51a], led to yields which are comparable to those obtained with Rb_2CO_3 and K_2CO_3, whereas by using the alkaline earth metal carbonates no reaction could be observed [51].

The stereochemical course of **cesium** assisted cyclizations was investigated in the case of the intramolecular reaction of the enantiomerically pure (R)-mesylate

Table 5. Yields of the lactones **56** and of the dilides **57** according to the **cesium** method [51] and according to the method of Corey and Nicolaou [2]

56	55		Cs_2CO_3		Yield [%] Pyridinethiol Method	
	n	X	56	57	56	57
a	4	I	70	4	—	—
b	5	Br	0	88	87	7
c	8	Br	0	95	—	—
d	9	I	23	55	—	—
e	10	I	33	54	64	30
f	11	I	62	30	76	7
g	12	I	77	18	79	6
h	13	I	72	13	—	—
i	14	I ·	83	17	—	—
k	15	I	85	15	88	5

Table 6. Yields of the makrolides **56i** and of the dilides **57i** in cyclization reactions with different metal carbonates [51]

$M^{\oplus}$	$Li^{\oplus}$	$Na^{\oplus}$	$K^{\oplus}$	$Rb^{\oplus}$	$Cs^{\oplus}$	$Tl^{\oplus}$	$Mg^{2\oplus}$	$Sr^{2\oplus}$	$Ba^{2\oplus}$
56i	0	54	67	68	80	64	0	0	0
57i	0	10	9	12	12	8	0	0	0

58 under formation of the lactone **59** [51]. The cyclization with **cesium** carbonate in DMF led in 80% chemical yield exclusively to the (*S*)-enantiomer of the lactone, which points to a S_N2 mechanism of the reaction [51]. The analogous procedure, using potassium or rubidium carbonate, led to the lactone **59** only in 28% and 54% yield, respectively, a determination of the enantiomeric purity was not possible due to impurities [51].

7 C—S Bond Formations

The preparation of thiamacrocycles of the cyclophane type by substitution of thiolates on suitable substrates usually leads to the desired products in good yields [55], favoured by the rigid group principle [56] and by application of the dilution principle [1]. On the other hand yields of conformative flexible thiacycloalkanes under similar conditions as a rule are not satisfactory and often low [57].

7.1 Synthesis of Dithiacycloalkanes

The yields of macrocyclic dithiacycloalkanes **61** in cyclization reactions were significantly increased by the introduction of **cesium** thiolates by Buter and Kellogg [17, 58]. The synthesis of the cyclic sulfides was performed by the reaction of dibromoalkanes with the **cesium** thiolates **60** under dilution conditions [1] in DMF as solvent.

Whereas by using the sodium thiolates the 9- to 12-membered thioethers **61a** and **61b** formerly were obtained only in 5.8 [59] and 0.8% [57], by application of the **cesium** effect the yields were increased to 45% and 63%, respectively (Table 7) [17, 58].

$$Br(CH_2)_m Br$$
$$+$$
$$Cs^{\oplus}S^{\ominus}(CH_2)_n S^{\ominus}Cs^{\oplus}$$

$$\xrightarrow{\text{DMF}}$$

60　　　　　　　　　　　　　**61**

The macrocyclic bis-sulfide **61 d** containing 27 ring members and the 36-membered bis-sulfide **61 e** were obtained analogously in high yields of 80–90 and 90% for the first time (Table 7).

Table 7. Yields of many-membered dithiacycloalkanes **61** according to the **cesium** method [17]

No.	m	n	Yields [%]	
			Cs_2CO_3	Na-thiolates
61a	3	4	45	5.8 [59]
61b	5	5	63	0.8 [57]
61c	10	10	85[a]	69 [60]
61d	10	10	80–90[a]	—
61e	16	18	90[b]	—

[a] Determination by ^{1}H-NMR — [b] Raw yield.

7.2 Synthesis of Thia-Crown Compounds

The investigation of thia crown ethers, e.g. trithiacyclononane (**64**, "9S3"), whose complexation properties towards transition metal cations were of interest [61, 62], was complicated for a long time by the low yields obtained from the corresponding cyclization reactions [63]. The combined application of the **cesium** effect and of the high dilution principle [1] in the reaction of 1,2-dichloroethane and the dithiol **63** led to an increase of the yield of **64** from 4.1% (tetraalkylammonium methanolate as the base) [64] to 50% [65, 66].

63 HS　S　SH　　　Cs_2CO_3
　　　+　　　　　$\xrightarrow{\text{DMF}}$
　　　Cl　Cl

62　　　　　　　　　　**64**

Though "9S3" (**64**) in the meantime has also been obtained in 60% yield by cyclization with a molybdenum template [67], the **cesium** method is nevertheless advantageous on account of the simplicity of the one-pot-cyclization procedure [65].

The 11-membered chiral tris-sulfide **67** was obtained in a high yield of 80% by Kellogg et al. [16] from the reaction of the D-tartaric acid derivative **66** with 3-thiapentanedithiol under dilution conditions in DMF.

In a similar way the yields of the 12- to 14-membered tetrathia crown compounds **69a–c** were significantly increased by Kellogg and Buter [17], when the dithiols **68** were reacted with 1,2-dibromoethane and 1,3-dibromopropane under dilution conditions [1] by the addition of equimolar amounts of Cs_2CO_3 (Table 8).

Yield increases in a similar order of magnitude as in the synthesis of the lower homologues were achieved also in the preparation of hexathia[18]crown-6 ("18S6", **72**) by the **cesium** effect". Cooper et al. [70] obtained **72** from the dithiol **70** and the dichloro compound **71** in a one-component dilution principle reaction [1] in a yield of 76% using a suspension of **cesium** carbonate in DMF. When the reaction was carried out under analogous conditions with potassium carbonate, **72** was isolated in a much lower yield of only 38% [70].

7.3 Synthesis of Strained Thiacyclophanes

With the introduction of ring contraction reactions for the preparation of [2.2] — phanes such as the sulfone pyrolysis [71] or the photoextrusion of sulfur in thiophilic solvents [72], the synthesis of thia[3$_n$]cyclophanes gained in significance.

Table 8. Yields of the tetrathia crowns **69a–c**

No.	n	m	Yields [%], base	
			Cs_2CO_3	NaOR
69a	2	2	88	4 [68], 6.3 [57]
69b	3	3	72	16 [68]
69c	3	3	76	8 [69]

In particular, the yields of strongly ring-strained or sterically hindered dithia[3.3]-cyclophanes were increased by the application of **cesium** salts in the cyclization reactions.

Whereas Vögtle and Ley in the cyclization of 1,3-bis(bromomethyl)benzene **(73)** with thioacetamide [73] under application of different alkali metal carbonates found a shift in the ratio of the "dimeric" [3.3]cyclophane **74** and "trimeric" [3.3.3]phane **75** towards higher amounts of the "trimeric" phane **75** in switching over from smaller alkali metal ions to· **cesium** ions [74], the ring-strained and sterically hindered cyclophanes **76–78** could be obtained only after use of **cesium** carbonate in the solvent DMF [75]. Synthetic attempts with other bases had met without success [76].

$M^{\oplus}$	Yield [%]	
	74	75
$Li^{\oplus}$	25	10
$Na^{\oplus}$	26-33	14-17
$K^{\oplus}$	26-33	14-17
$Rb^{\oplus}$	18-23	22-27
$Cs^{\oplus}$	15-19	25-32

55

Catalytic amounts of **cesium** carbonate were used in the synthesis of the dithia-[3.3]- and oxathia[3.2]pyridinophanes **79** and **80** by Przybilla and Vögtle [76, 77]. The pyridinophanes **79** and **80** were obtained in yields of 51 and 33% when the cyclization was carried out in ethanol under dilution conditions [1].

The oxathia[3.2]naphthalenophane **82** was obtained by Billen and Vögtle [78] in a yield as high as 78% by cyclization of the bis(bromomethyl) compound **81** with the reagent combination Na_2S/Cs_2CO_3 in acetonitrile under dilution conditions. The choice of the solvent was of high importance: If the cyclization was carried out not in acetonitrile [79] but in an ethanol/benzene mixture (1:1), the [3.2]phane **82** could only be obtained in a 2.5% yield.

The highly strained oxathia[3.1]naphthalenocyclophane **84** was first synthesized by Duchêne and Vögtle [80], who used the **cesium** effect in the cyclization of the bromide **83** with the reagent combination Na_2S/Cs_2CO_3 in ethanol, yielding **84** in 5.4% yield. The sulfide **84** was desulfurized to the highly strained target molecule oxa[2.1]naphthalenophane **85**.

The dependence of the yield of cyclization leading to dithia[2.2]cyclophanes as e.g. **88** on the combination of base and solvent was investigated thoroughly by Vögtle and Meurer [81]. By using a combination of alkali metal hydroxides and ethanol/benzene (12:1) as solvent under application of dilution conditions, the highest yields of the dithia[2.2]metacyclophane **88** were achieved. The use of alkali metal carbonates in DMF led to somewhat lower yields of **88** (Table 9). In both solvents the application of **cesium** salts led to the highest yields.

Cyclizations under addition of the alkali hydroxides in DMF or *tert*-butanol/ benzene (12:1) led to a dramatic decrease of the yields. Similarly, the use of catalytic amounts of **Cs₂CO₃** in DMF lowered the yields of **88** from 48% to 10% [81].

A significant "**cesium** effect" was also observed by Meurer and Vögtle [82] in the synthesis of the naphthalenophane **90** formed by reaction of the bis(bromo-methyl) compound **89** and 1,3-benzenedithiole. The cyclization carried out under dilution conditions [1] with CsOH in ethanol/benzene (10:1) produced the naphtha-lenophane **90** in 22% yield, whereas a change to the base NaOH decreased the yield to 7% [82].

Table 9. Yields of the dithiaphane **88** obtained with different solvent/base combinations [81]

$M^{\oplus}$	M_2CO_3/DMF	MOH/DMF	MOH/EtOH/ benzene (12:1)	MOH/*tert*-butanol/ benzene (12:1)
$Li^{\oplus}$	39	—	34	—
$Na^{\oplus}$	36	—	44	5
$K^{\oplus}$	35	—	47	10
$Rb^{\oplus}$	40	15	48	3
$Cs^{\oplus}$	48	18	58	6

The application of reaction conditions as optimized in the preparation of the dithia[2.2]phane **88** allowed Meurer and Vögtle [83] to synthesize the first helically chiral dihetera[2.2]metacyclophanes **91** and **92** in 1.9 and 9% yield. It was impossible to isolate these cyclophanes in all the previous attempts where **cesium** salts had not been applied [84, 85].

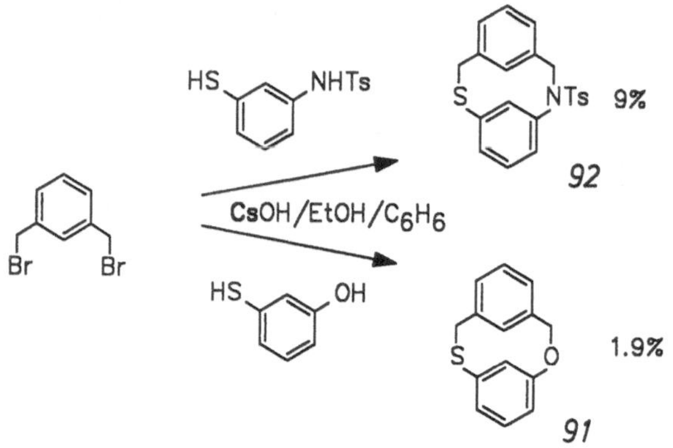

7.4 Synthesis of Manifold Bridged Thiacyclophanes

Vögtle and Klieser [86] succeeded in the first four-(and fivefold) bridging of the benzene ring by cyclization of tetrakis(bromomethyl)benzene (**93**) with the corresponding tetrathiol **94**. Whilst the combination Cs_2CO_3/DMF led to the isomeric phanes **97** in a total yield of 10%, cyclizations with other solvent/base combinations such as e.g. K_2CO_3/DMF only produced the dithia compound **98** [86].

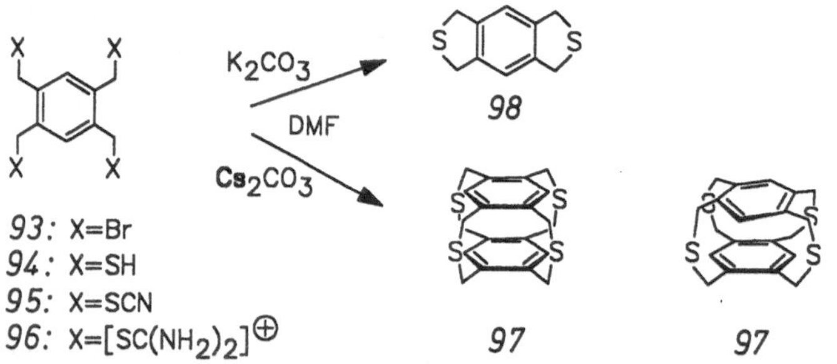

93: X=Br
94: X=SH
95: X=SCN
96: X=[SC(NH$_2$)$_2$]$^{\oplus}$

The preparation of **97** was optimized later by Misumi et al. by coupling of the thiolate **94** formed in situ, using the thiocyanate **95** or the isothiuronium salt **96** with the tetrabromide **93** [87]. The formation of reactive thiolate groups in the course of the reaction maintains a lower concentration of free thiolate anions and thereby restrains side reactions [87].

Using the base CsOH in ethanol/benzene (12:1) as the solvent and the isothiuronium salt **96** increased the yield of the isomeric cyclic compounds **97** to 75%. The base KOH in ethanol/DMSO (74:1) with a total yield of 56% turned out to be less effective.

The synthesis of the triply bridged trithiacyclophane **100** was achieved by Vögtle and Sendhoff in a one-step synthesis starting from the *m*-substituted building blocks **98** and **99** with Cs_2CO_3 as the base [88, 90]. Due to the slow solubility of the starting compounds the reaction was able to be carried out without making use of dilution conditions as a heterogeneous cyclization [91] ("heterogeneous dilution principle") in ethanol/benzene (1:1) and yielded the macrobicyclic tris-sulfide **100** in 37% yield. The use of the base KOH in the cyclization reaction led to a dramatic decrease in yield to 5% [90].

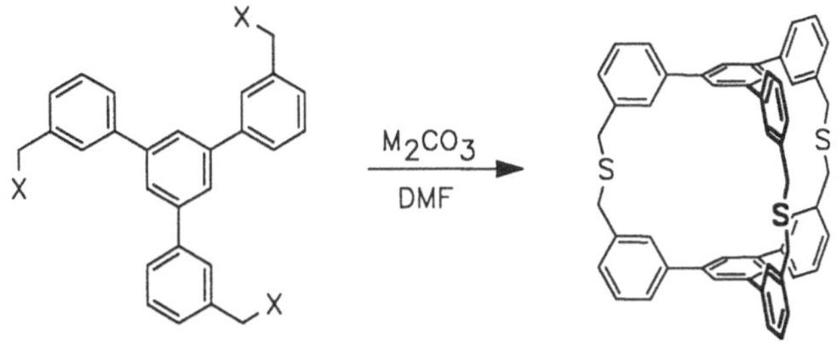

98 : X = Br
99 : X = SH

100

The first sixfold bridge formation in one single step was achieved finally by Kißener and Vögtle [92] after many unsuccessful attempts by cyclization of the hexaphenylbenzene building blocks **101** and **102** under dilution conditions with Cs_2CO_3 as a base. The hexathia macropolycyclic compound **103** was obtained in this way in 0.1% yield. This yield was increased somewhat later by the above mentioned heterogeneous reaction procedure ("heterogeneous dilution principle") up to 0.5% [90].

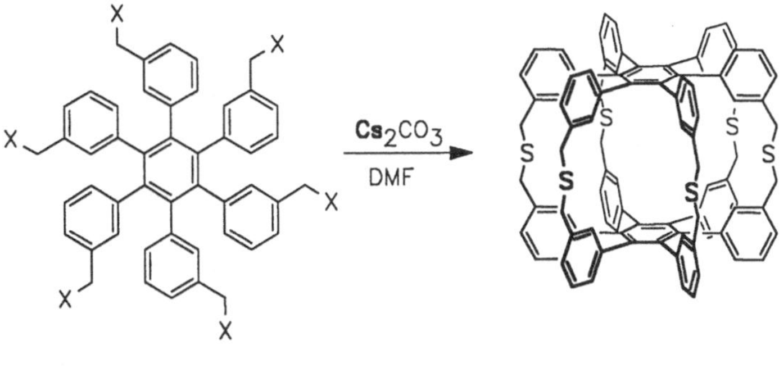

101 : X = Br
102 : X = SH

103

8 Attempts to Interpret the "Cesium Effect"

The particular influence of **cesium** ions on the course of cyclization reactions which is proven by many comparison experiments with different metal cations has been called the "**cesium** effect". As a rule, equimolar amounts of **cesium** compounds are used, which means that one cannot normally speak of a **cesium** catalysis. Nevertheless, a catalytic effect with respect to the **cesium** cation often cannot be excluded.

Many attempts to interpret the "**cesium** effect" proceed from the special position of **cesium** in the periodic system of the elements and in the first main group: **Cesium** forms the cation with the largest ionic radius and follows the $Tl^\oplus$ ion with respect to the largest polarizability (Table 10).

Table 10. Some properties of the alkali metal cations [94]

$M^\oplus$	Ionic radius	Charge/Surface	Polarizability
	[A]	[Z/A²]	[A³]
Li	0.78	0.130	0.03
Na	0.98	0.085	0.30
K	1.33	0.045	1.10
Rb	1.49	0.035	1.90
Cs	1.65	0.030	2.90
Tl	1.40 [93]		4.30 [51a]

Buter and Kellogg proceeded on the assumption that **cesium** salts of carboxylic acids due to their big ionic radii and the high polarizability of the **cesium** cation are present in DMF as weakly solvatized tight ion pairs [51]. This assumption was based on measurements, which showed that **cesium** salts with "soft" anions as e.g. delocalized carbanions in THF exist as contact ion pairs [95], whilst **cesium** salts with "hard" anions form solvent-separated ion pairs due to a stronger anion solvation [96]. Anions which exist as contact ion pairs are usually considered to have a strongly lowered reactivity [51] compared to free ("naked") anions. The reactivity increases further with increasing size of the cations, so that anions in **cesium** salts should take a medium position between the low reactivities in contact ion pairs and the more reactive free anions [51]. Besides direct consequences on the reactivity of the anions in **cesium** salts, electrically neutral contact ion pairs favour an intramolecular course of cyclization reactions and complicate the formation of oligomers, which only can proceed intermolecularly by reaction with another electrically neutral ion pair [51]. These arguments and the determination of "triple ions" [97] $X^\ominus$-$Cs^\oplus$-$X^\ominus$ in the anionic polymerization of styrene ("living polymers") hint to a course of the ring closure reaction at the surface of the **cesium** cation, from which a preorientation of the reactants due to a coordination at the **cesium** ion results (Fig. 3) [51].

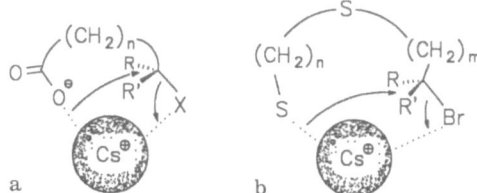

Fig. 3a and b. "Rolling mechanism"-hypothesis at the surface of **cesium** ions in the course of intramolecular cyclizations

Vögtle and Meurer explained the favoured formation of strained ring compounds in the synthesis of cyclophanes by cyclizations with **cesium** thiolates by a preorientation (preorganization) of the reactants at the **cesium** cation [82]. For the **cesium** ion in contrast to the smaller alkali metal cations the capability of forming an 11-membered intermediate is ascribed, which would favour an intramolecular course of the reaction (Fig. 4).

In more recent studies by Kellogg et al. [98] the "**cesium** effect" is explained by solubility and ion pair effects. These authors found by determination of the dependency of the NMR shifts of ^{133}Cs ions on the concentration, that the **cesium** salts investigated in DMSO are completely dissociated, whereas in DMF to some extent ion pairs are present. The determination of the solubilities of the alkali metal propionates proved that the **cesium** salts in DMF as well as in DMSO exhibit the highest solubilities. In reactions of aliphatic mesylates with alkali metal propionates in DMF and DMSO, Kellogg et al. observed a homogeneous reaction course exclusively with the Rb and Cs salts, the highest yields found in every case being with the **cesium** compounds [98]. As a conclusion of these studies, the favoured intramolecular reaction course in many cyclization reactions with **cesium** ions is attributed solely to the presence of solvent separated ion pairs forming reactive anions and to the high solubility of the organic **cesium** salts [98].

More recent attempts to interpret the "**cesium** effects" suggest models of "differential geometry" [99]. So-called periodic zero-potential surfaces ("POPS") and isopotential surfaces ("TFS", "*tangential field surface*") of the **cesium** ions as templates for organic molecules are proposed. According to these model considerations, an orientation of nonpolar molecular substructures at the zero-potential surface ("POPS") and an orientation of polar substructures at the isopotential surface ("TFS"), take place, which should favour an intramolecular

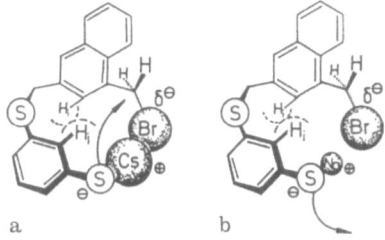

Fig. 4. (a) Formation ("preorganization") of an 11-membered intermediate during cyclization to strained ring compounds by **cesium** cations. **(b)** With the smaller Na$^{\oplus}$ cation the cyclic intermediate cannot be formed (preorganized) because of the steric repulsion of the two H$_i$ atoms [82]

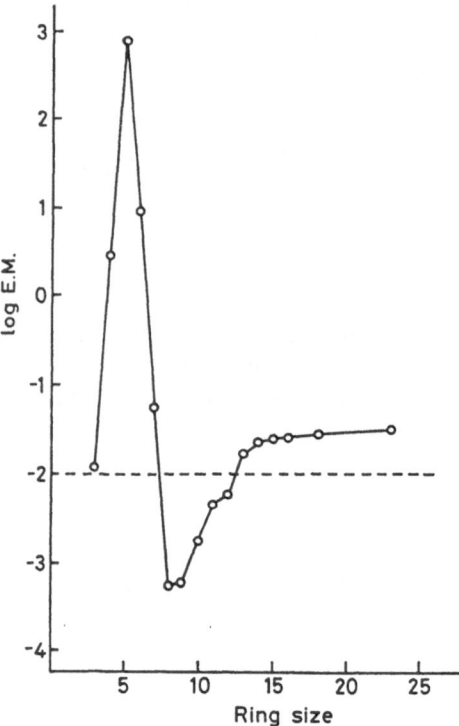

Fig. 5. Profile of the "effective molarity" ("EM") for the lactone formation from bromoalkanoates in DMSO (50 °C). The *broken line* shows the concentration at the start of the reaction (10^{-2} mol/l). Substrates drawn above the broken line react favourably under cyclization at the starting concentration given [102]

reaction course. On account of different potential surfaces of the alkali metal ions due to symmetry reasons, for the **cesium** ion more favourable interactions with organic molecules were predicted compared to the other alkali metal cations [100]. To which degree a transfer of these results of "differential geometry", only exactly valid for solid-state conditions, is possible for dynamic structures in solution, remains uncertain at the present time [100].

Illuminati and Mandolini [101] explained the "**cesium** effect" from a physico-chemical point of view solely in terms of ion pair effects. They introduced the term "effective molarity" ("EM") to characterize the course of the cyclization reaction [101, 102]. The "effective molarity" is defined as the ratio of the velocity constants of the intramolecular reaction (cyclization) and the intermolecular reaction (oligomerization): $EM = k_{intra}/k_{inter}$ [mol/l]. EM therefore is the substrate concentration, at which the cyclization and oligomerization proceed with equal reaction rates. As a consequence, a concentration below the EM in the course of the reaction causes an intramolecular reaction path to be favored [102].

Investigations by Illuminati and Mandolini [103] also hint at a participation of ion pair effects on the particular properties of the **cesium** compounds. These authors studied the inhibition of cyclization reactions by the formation of contact ion pairs. They found that increasing additions of alkali metal bromides inhibit the cyclization of **104** to various extents. Addition of LiBr effects the strongest inhibition of the reaction, whereas **CsBr** leads to the highest reaction rate (Fig. 6).

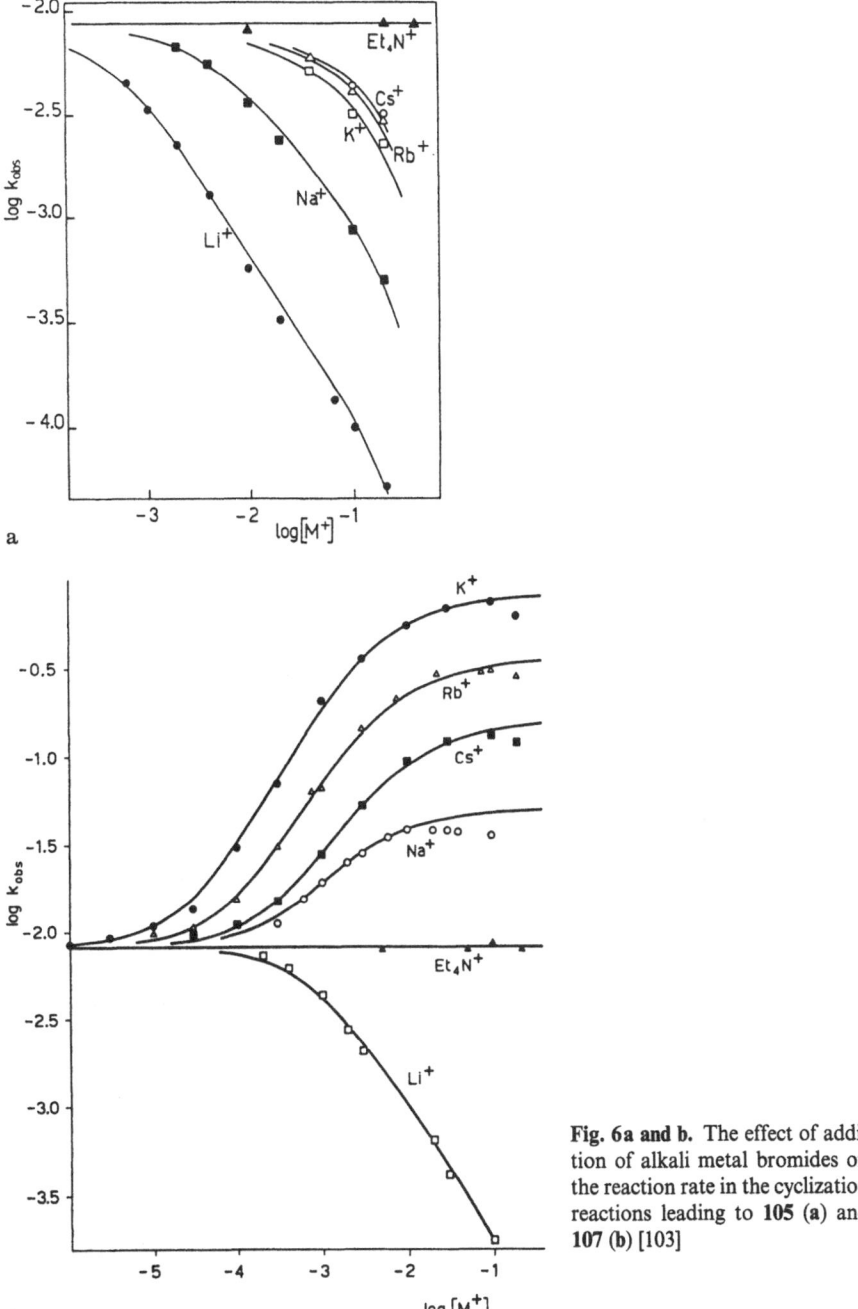

Fig. 6a and b. The effect of addition of alkali metal bromides on the reaction rate in the cyclization reactions leading to **105** (a) and **107** (b) [103]

In the cyclization reaction of benzo[18]crown-6 (107) the reaction rate is also lowered by $Li^{\oplus}$ ions, whereas for the other alkali metal cations an increase in reaction rates was found. With $K^{\oplus}$ ions on account of a template effect the increase of the reaction rate in the crown series was the highest [103].

104   105

106   107

9 Conclusions

Many concrete examples taken from the recent literature demonstrate that the utilization of **cesium** compounds in ring closure reactions with special respect to many membered rings has the advantage of higher yields, or cleaner product formation, or simpler work-up. Especially in research laboratories the **cesium** effect can be very valuable in synthesizing compounds which without the **cesium** effect cannot be formed, e.g. aliphatic compounds [104, 105], crown ethers [106–110], catenanes [111–119], macro(bi)cyclic compounds [120–132] and other branches [133–137]. In some cases the oligomer selectivity can be dramatically influenced. In recent years, the **cesium** effect has stimulated macrocyclization reactions and led to new compounds not available without its application and it is to be expected that the potential of this new synthetic method will bring about more examples of structurally exciting molecules not accessible by standard methods.

The authors kindly thank Dipl.-Chem. Joachim Dohm, Dipl.-Chem. Christian Seel and Dr. S. Pask for their assistance in preparing the final version of this manuscript.

10 References

1. a) Ziegler K (1955) In: Houben-Weyl-Müller (ed), Methoden der organischen Chemie, Thieme, Stuttgart, vol 4 (2), p 729;
   b) Ruggli P (1912) Liebigs Ann. Chem. 392: 92;
   c) Vögtle F (1972) Chem. Ztg. 96: 396;
   d) Rossa L, Vögtle F (1983) Top. Curr. Chem. 113: 1
2. Corey EJ, Nicolaou KC (1974) J. Am. Chem. Soc. 96: 5614
3. Kohnke FH, Mathias JP, Stoddart JF (1989) Angew. Chem. Adv. Mater. 101: 1129; (1989) Angew. Chem. Adv. Mater. Int. Ed. Engl. 28: 1103; Ashton PR, Isaacs NS,

Kohnke FH, Stagno D'Alcontres G, Stoddart JF (1989) Angew. Chem. 101: 1269; (1989) Angew. Chem. Int. Ed. Engl. 28: 1261

4. Ebmeyer F, Vögtle F (1989) Chem. Ber. 122: 1725

5. Ashton PR, Goodnow TT, Kaifer AE, Reddington MV, Slavin AMZ, Spencer N, Stoddart JF, Vicent V, Williams DJ (1989) Angew. Chem. 101: 1404; (1989) Angew. Chem. Int. Ed. Engl. 28: 1396

6. Klieser B, Rossa L, Vögtle F (1984) Kontakte (Darmstadt) 1: 3

7. a) Ichikawa J, Sugimoto K-I, Sonoda T, Kobayashi H (1987) Chem. Lett. 1985;
   b) Padwa A, Gasdaska JR, Haffmanns G, Rebello H (1987) J. Org. Chem. 52: 1027

8. Clark JH (1980) Chem. Rev. 80: 429

9. Rozhkov IV, Knunyant IL (1971) Dokl. Akad. Nauk. SSSR 199: 614; (1972) Chem. Abstr. 76: 7291

10. Burton DJ, Herkes FR (1965) Tetrahedron Lett. 4509

11. Clark JH, Miller JM (1977) J. Am. Chem. Soc. 99: 498

12. a) Liotta CL, Harris HP (1974) J. Am. Chem. Soc. 96: 2250;
    b) Wollenberg RH, Miller SJ (1978) Tetrahedron Lett. 3219

13. a) Sebti S, Foucaud A (1983) Synthesis 546;
    b) Sebti S, Foucaud A (1987) J. Chem. Res. 72

14. Gmelins Handbuch der Anorganischen Chemie, 8th edn, Verlag Chemie, Berlin 1938, vol 25, p 234

15. Gisin BF (1973) Helv. Chim. Acta 56: 1476

16. van Keulen BJ, Kellogg RM, Piepers O (1979) J. Chem. Soc., Chem. Commun. 285

17. a) Buter J, Kellogg RM (1981) J. Org. Chem. 46: 4481;
    b) Buter J, Kellogg RM (1987) Org. Synth. 65: 150

18. Vriesema BK, Lemaire M, Buter J, Kellogg RM (1986) J. Org. Chem. 51: 5169

19. Vriesema BK, Buter J, Kellogg RM (1984) J. Org. Chem. 49: 110

20. Berthet M, Sonveaux E (1983) J. Chem. Soc., Chem. Commun. 10

21. Vögtle F, Kißener W (1984) Chem. Ber. 117: 2538

22. Müller E, Röscheisen G (1957) Chem. Ber. 90: 543

23. Kißener W (1983) Diplomarbeit Univ. Bonn

24. Vögtle F, Mayenfels P, Luppertz F (1984) Synthesis 580

25. Stetter H, Roos EE (1954) Chem. Ber. 87: 566; Stetter H, Roos EE (1955) Chem. Ber. 88: 1390

26. Richman JE, Atkins TJ (1974) J. Am. Chem. Soc. 96: 2268

27. Chavez F, Sherry AD (1989) J. Org. Chem. 54: 2990

28. Melson GA (1979) Coordination Chemistry of Macrocyclic Compounds, Plenum, New York

29. Hosseini MW, Lehn J-M (1982) J. Am. Chem. Soc. 104: 3525

30. Kimura E (1985) Top. Curr. Chem. 128: 113

31. Marecek JF, Burrows CJ (1986) Tetrahedron Lett. 27: 5943

32. Wagler TR, Burrows CJ (1987) J. Chem. Soc., Chem. Commun. 277

33. Hosseini MW, Lehn J-M (1986) Helv. Chim. Acta 69: 587

34. Dietrich B, Hosseini MW, Lehn J-M, Session RB (1985) Helv. Chim. Acta 68: 289

35. Losensky H-W, Spelthann H, Ehlen A, Vögtle F, Bargon J (1988) Angew. Chem. 100: 1225

36. Losensky H-W (1988) Dissertation, University Bonn

37. Kruizinga WH, Kellogg RM (1979) J. Chem. Soc., Chem. Commun. 286

38. Pedersen CJ (1967) J. Am. Chem. Soc. 89: 7017

39. Weber E (1985) Chem. Ber. 118: 4439

40. Weber E, Vögtle F (1980) Angew. Chem. 92: 1067; (1980) Angew. Chem. Int. Ed. Engl. 19: 1030

41. Weber E, Josel H-P, Puff H, Franken S (1985) J. Org. Chem. 50: 3125

42. Weber E, Vögtle F, Josel H-P, Newkome GR, Puckett WE (1983) Chem. Ber. 116: 1906

43. Weber E, Köhler H-J, Reuter H (1989) Chem. Ber. 122: 959

44. Reinhoudt DN, de Jong F, Tomassen HPM (1979) Tetradedron Lett. 2067

45. Czech BP, Czech A, Bartsch RA (1985) J. Heterocycl. Chem. 22: 1297

46. Pedersen CJ (1970) J. Am. Chem. Soc. 92: 391

47. Leppkes R, Vögtle F (1983) Chem. Ber. 116: 215

48. Merrifield RB (1985) Angew. Chem. 97: 801
49. Kruizinga WH, Strijtveen B, Kellogg RM (1981) J. Org. Chem. 46: 4321
50. a) Huffman JW, Desai RC (1983) Synth. Commun. 13: 553
    b) Willis CL (1987) Tetrahedron Lett. 28: 6705;
    c) Torisawa Y, Okabe H, Ikegami S (1984) Chem. Lett. 1555
51. a) Kruizinga WH, Kellogg RM (1981) J. Am. Chem. Soc. 103: 5183;
    b) Piepers O, Kellogg RM (1978) J. Chem. Soc., Chem. Commun. 383
52. Talma AG, Jouin P, de Vries JG, Troostwijk CB, Werumeus Buning GH, Waninge JK, Visscher J, Kellogg RM (1985) J. Am. Chem. Soc. 107: 3981
53. Potts KT, Cipullo MJ (1982) J. Org. Chem. 47: 3038
54. Löhr H-G, Vögtle F, Schuh W, Puff H (1984) Chem. Ber. 117: 2839
55. Baker W, McOmie JFW, Ollis WD (1951) J. Chem. Soc. 200
56. Newkome GR, Sauer JD, Roper JM, Hager DC (1977) Chem. Rev. 77: 513
57. Ochrymowycz LA, Mak CP, Michna JD (1974) J. Org. Chem. 39: 2079
58. Buter J, Kellogg RM (1980) J. Chem. Soc., Chem. Commun. 466
59. Musker WK, Wolford TL, Roush PB (1978) J. Am. Chem. Soc. 100: 6416
60. Stetter H, Wirth W (1960) Liebigs Ann. Chem. 631: 144
61. Cooper SB (1988) Acc. Chem. Res. 21: 141
62. Rawle SC, Yagbasan R, Prout K, Cooper SR (1987) J. Am. Chem. Soc. 109: 6181
63. Ochrymowycz LA, Gerber D, Chongsawangvirod R, Leung AK (1977) J. Org. Chem. 42: 2644
64. Setzer WN, Ogle CA, Wilson GS, Glass RS (1983) Inorg. Chem. 22: 266
65. Blower PJ, Cooper SR (1987) Inorg. Chem. 26: 2009
66. Hartman JR, Cooper SR (108) J. Am. Chem. Soc. 1986: 1202
67. Sellmann D, Zapf L (1984) Angew. Chem. Int. Ed. Engl. 23: 807
68. Rosen W, Busch DM (1970) Inorg. Chem. 9: 262
69. Rosen W, Busch DH (1969) J. Am. Chem. Soc. 91: 4694
70. Wolf RE, Hartman JAR, Ochrymowycz LA, Cooper SR (1989) Inorg. Synth. 25: 122
71. Vögtle F, Rossa L (1979) Angew. Chem. 91: 534
72. Bruhin J, Jenny W (1973) Tetrahedron Lett. 1215
73. Hammerschmidt E, Bieber W, Vögtle F (1978) Chem. Ber. 111: 2445
74. Vögtle F, Ley F (1983) Chem. Ber. 116: 3000
75. Vögtle F, Klieser B (1982) Synthesis 294
76. Przybilla K-J, Vögtle F, Nieger M, Franken S (1988) Angew. Chem. 100: 987
77. Przybilla K-J, Vögtle F (1989) Chem. Ber. 112: 347
78. Billen S, Vögtle F (1989) Chem. Ber. 122: 1113
79. Ostrowicki A, Vögtle F (1988) Synthesis 1003
80. Duchêne K-H, Vögtle F (1985) Angew. Chem. 97: 866
81. Meurer K (1984) Dissertation, University Bonn
82. Meurer K, Luppertz F, Vögtle F (1985) Chem. Ber. 118: 4433
83. Meurer K, Vögtle F, Mannschreck A, Stühler G, Puff H, Roloff A (1984) J. Org. Chem. 49: 3484
84. Vögtle F, Neumann P (1970) Tetrahedron Lett. 115
85. Wieder W (1973) Diplomarbeit, University Würzburg
86. Klieser B, Vögtle F (1982) Angew. Chem. 94: 632; (1982) Angew. Chem. Int. Ed. Engl. 21: 618
87. Asoh T, Tani K, Higuchi H, Kaneda T, Tanaka T, Sawada M, Misumi S (1988) Chem. Lett. 417
88. Vögtle F (1970) Liebigs Ann. Chem. 735: 193
89. Sendhoff N, Weißbarth K-H, Vögtle F (1987) Angew. Chem. 99: 794
90. Sendhoff N, Kißener W, Vögtle F, Franken S, Puff H (1988) Chem. Ber. 121: 2179
91. Schneider H-J, Busch R (1986) Chem. Ber. 119: 747
92. Kißener W, Vögtle F (1985) Angew. Chem. 97: 782; (1985) Angew. Chem. Int. Ed. Engl. 24: 794
93. Holleman-Wiberg (1976) Lehrbuch der Anorganischen Chemie, 81–90th edn, de Gruyter, Berlin, p 736

94. Lehn J-M (1973) Struct. Bonding 16: 1
95. a) Bors DA, Kaufman MJ, Streitwieser A (1985) J. Am. Chem. Soc. 107: 6975
    b) Hogen-Esch TE, Smid J (1966) J. Am. Chem. Soc. 88: 318
96. Exner JH, Steiner EC (1974) J. Am. Chem. Soc. 96: 1782
97. Bhattacharyya DN, Lee CJ, Smid J, Swarc M (1965) J. Phys. Chem. 69: 619
98. Dijkstra G, Kruizinga WH, Kellogg RM (1987) J. Org. Chem. 52: 4230
99. von Schnering HG, Nesper R (1987) Angew. Chem. 99: 1097
100. Blum Z (1989) Acta Chem. Scand. 43: 248
101. a) Illuminati G, Mandolini L, Masci B (1977) J. Am. Chem. Soc. 99: 6308
     b) Illuminati G, Mandolini L (1981) Acc. Chem. Res. 14: 95
     c) Galli C, Mandolini L (1984) J. Chem. Soc., Perkin Trans. 2 1435
102. Galli C, Mandolini L (1982) J. Chem. Soc., Chem. Commun. 251
103. Mandolini L, Masci B (1984) J. Am. Chem. Soc. 106: 168; Illuminati G, Mandolini L, Masci
     B (1983) J. Am. Chem. Soc. 105: 555
104. Barbier M (1982) J. Chem. Soc., Chem. Commun. 668
105. Toda T, Kitagawa Y (1987) Angew. Chem. 99: 366
106. van der Leij M, Oosternik HJ, Hall RH, Reinhoudt DN (1981) Tetrahedron 37: 3661
107. Baker DS, Gold V (1983) J. Chem. Soc., Perkin Trans. 2 1129
108. Richards TP, Hamilton AD (1985) J. Chem. Soc., Chem. Commun. 1998
109. Roesky HW, Schmidt HG (1985) Angew. Chem. 97: 711
110. Sone T, Sato K, Ohba Y (1989) Bull. Chem. Soc. Jpn. 62: 838
111. Dietrich-Buchecker CO, Sauvage J-P (1989) Angew. Chem. 101: 192
112. Dietrich Buchecker CO, Sauvage J-P (1983) Tetrahedron Lett. 5095
113. Dietrich-Buchecker CO, Sauvage J-P, Kern J-M (1984) J. Am. Chem. Soc. 106: 3042
114. Dietrich Buchecker CO, Sauvage J-P, Weiss J (1986) Tetrahedron Lett. 27: 2257
115. Dietrich-Buchecker CO, Khemiss A, Sauvage J-P (1986) J. Chem. Soc., Chem. Commun.
     1376
116. Dietrich-Buchecker CO, Guilhelm J, Khemiss AK, Kintzinger J-P, Pascard C, Sauvage
     J-P (1987) Angew. Chem. 99: 711
117. Mitchell DK, Sauvage J-P (1988) Angew. Chem. 100: 985
118. Sauvage J-P, Weiss J (1985) J. Am. Chem. Soc. 107: 6108
119. Chambron J-C, Sauvage J-P (1986) Tetrahedron Lett. 27: 865
120. Dharanipragada R, Ferguson SB, Diederich F (1988) J. Am. Chem. Soc. 110: 1679
121. Diederich F, Dick K, Griebel D (1985) Chem. Ber. 118: 3588
122. Diederich F, Hester MR, Uyeki MA (1988) Angew. Chem. 100: 1775
123. Jiminez L, Diederich F (1989) Tetrahedron Lett. 30: 2759
124. Rubin Y, Dick K, Diederich F, Georgiadis TM (1986) J. Org. Chem. 51: 3270
125. Schürmann G, Diederich F (1986) Tetrahedron Lett. 27: 4249
126. Wilcox CS, Cowart MD (1986) Tetrahedron Lett. 27: 5562
127. Petti MA, Shepodd TJ, Dougherty DA (1986) Tetrahedron Lett. 27: 807
128. Heyer D, Lehn J-M (1986) Tetrahedron Lett. 27: 5868
129. Rodriguez-Ubis J-C, Alpha B, Plancherel D, Lehn J-M (1984) Helv. Chim. Acta 67: 2264
130. Dietrich-Buchecker CO, Sauvage JP (1983) Tetrahedron. Lett. 5091
131. Navarro P, Rodriguez-Franco MI, Samat A (1987) Synth. Commun. 105
132. Cram DJ, Karbach S, Kim YH, Baczynski L, Kalleymeyn GW (1985) J. Am. Chem. Soc.
     107: 2572; Sherman JC, Cram DJ (1989) J. Am. Chem. Soc. 111: 4527; Cram DJ, Karbach S,
     Kim H-E, Knobler CB, Maverick EF, Ericson JL, Helgeson RC (1988) J. Am. Chem. Soc.
     110: 2229
133. Staab HA, Alt R (1984) Chem. Ber. 117: 850
134. Vögtle F, Saitmacher K, Peyerimhoff S, Hippe D, Puff H, Büllesbach P (1987) Angew. Chem.
     99: 459
135. Koepp E, Vögtle F (1987) Synthesis 177
136. Spichiger-Ulmann M, Augustynski J (1986) Helv. Chim. Acta. 69: 632
137. Lerchen H-G, Kunz H (1985) Tetrahedron Lett. 26: 5257

Synthesis of (Strained) Macrocycles by Sulfone Pyrolysis

Joachim Dohm and Fritz Vögtle

Institut für Organische Chemie und Biochemie der Universität Bonn,
Gerhard-Domagk-Str. 1, W-5300 Bonn 1, FRG

Table of Contents

Among the methods for the synthesis of strained macrocycles, the ring contraction of unstrained cyclic precursors by thermal elimination of sulfur dioxide ("sulfone pyrolysis") is of general importance because it offers access to a large diversity of macrocycles not equalled by other methods. Highly strained macrocycles as well as macrocycles containing labile moieties or functional groups can be synthesized. Several bonds can be created simultaneously and yields are comparably high.

In order to fully acknowledge the scope of this method, recent pyrolyses are listed and arranged according to structural features. Experimental parameters can be adapted to reactivity and stability of the compounds and are listed for purpose of easy comparison.

By reviewing the reactions listed, it becomes obvious that the potential of this synthetic method is not yet fully explored.

1 Introduction

The pyrolytic elimination of sulfur dioxide from cyclic sulfones to give (strained) macrocycles (sulfone pyrolysis) counts among the most important reactions in cyclophane chemistry. There is hardly any research group dealing with cyclophanes which did not make use of the sulfone pyrolysis for the synthesis of such molecules. Since its first application in cyclophane synthesis over twenty years ago [1], the sulfone pyrolysis enabled researchers to synthesize a wealth of strained cyclophanes and developed into a reaction of general importance.

In 1979, Rossa and Vögtle published a review on sulfone pyrolysis which did not only cover examples from cyclophane chemistry but also discussed the pyrolysis of acyclic sulfones, of small-membered cyclic sulfones as well as questions of mechanisms and practical aspects of the reaction [2].

Even though only little new knowledge of theory and practice of the reactions exists, the surprising variety of successful sulfone pyrolyses described since 1979 justifies a new progress report dealing with this synthetic method as a continuation (part II) of the first review mentioned above. The following survey deals with the application of the sulfone pyrolysis on the synthesis of cyclophanes and focusses on the variety of structural types of molecules that can be synthesized via sulfone pyrolysis.

2 General Synthetic Strategies for Strained Compounds of the Cyclophane Type

The synthesis of strained cyclic compounds with carbon-bridges of variable length relies on two different strategies for the introduction of molecular strain.

2.1 Synthesis via Direct Formation of C–C Bonds

Some methods introduce strain directly by generating the new C–C bond in the cyclization step [3]. Wurtz-type cyclizations (modification by Müller-Röscheisen [4]) and cyclizations using phenyllithium [5] or other organometallic reagents [3] belong to this group. Often these methods suffer from low selectivity and large quantities of oligomeric by-products. In addition, they do not work when being applied for the synthesis of extremely strained compounds and do not tolerate functional groups, which might react with the organometallic reagents applied in the cyclization (e.g., carbonyl groups or halide substituents).

Scheme 1. Direct cyclization yielding the strained target molecule

71

2.2 Synthesis via Ring Contraction of a Less-Strained Cyclic Precursor

As an alternative to direct cyclization the molecular strain can be generated stepwise by first synthesizing a less strained cyclic precursor. This precursor is constructed in a way as to contain ring elements, which in a second step can be eliminated, thus bringing about a ring contraction and an increase in strain leading to the strained target compound. Even though this method requires more synthetic steps, the high yields of these steps, the high selectivity and tolerance of functional groups more than compensate for this disadvantage.

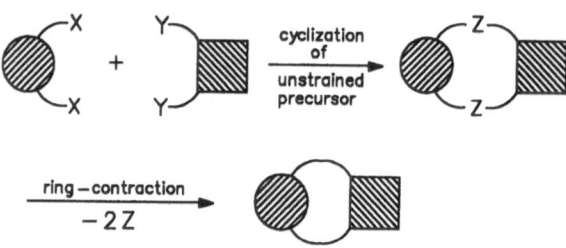

Scheme 2. Cyclization and subsequent ring contraction for the introduction of molecular strain

3 Cyclizations to Sulfides and Methods for Desulfurization

The latter method via cyclization and subsequent ring contraction at one or several centres in most cases makes use of cyclizations to sulfides [3]. These cyclizations follow S_N mechanisms and are easy to carry out giving high yields of the cyclic sulfides, which are easily isolable [6, 7]. For the ring contraction step, in this case a desulfurization step, a variety of methods is at hand:

- Oxidation of the cyclic sulfides to cyclic sulfones and subsequent thermal extrusion of sulfur dioxide in the gas phase ("sulfone pyrolysis").
- Oxidation of cyclic sulfides to cyclic sulfoxides and subsequent pyrolysis of the sulfoxides [8]; less common method only applicable in a few cases.
- Oxidation of the cyclic sulfides to cyclic sulfones and subsequent photochemical extrusion of sulfur dioxide from suspensions of the sulfones in benzene [9]; this method is only applicable for the synthesis of [2.2] phanes without photolabile functional groups.
- Photochemical desulfurization by irradiation of the sulfides in thiophilic solvents, e.g., trialkylphospites [10].
- Direct pyrolysis of the cyclic sulfides without prior chemical modification [11]; special method without general importance in cyclophane chemistry.

- Wittig rearrangement of the cyclic sulfides and methylation of the products [12] or Stevens rearrangement (after methylation at the sulfur atom) [13]; in both cases the rearranged product is subsequently treated with Raney-nickel to remove the sulfur, thus yielding sulfur-free cyclophanes; only certain functional groups tolerate the reaction conditions for rearrangement and desulfurization.
- Thermal desulfurization with $Fe(CO)_5$ in high-boiling solvents, e.g., toluene [14]; this method was only applied in a few cases.
- Cyclization to selenides and subsequent thermal or photochemical extrusion of selenium [15]; this method rarely applied has additional disadvantages in terms of costs, toxicity and yield.

3.1 Sulfone Pyrolysis

3.1.1 Advantages of the Sulfone Pyrolysis

Besides photochemical desulfurization, sulfone pyrolysis allows for the broadest range of applications. The compounds which have been synthesized belong to classes as different as heterophanes, multilayered phanes, nonbenzenoid phanes, molecules containing molecular cavities or "phenylenicenes". The pyrolysis of sulfones in the gas phase in comparison with other methods has many advantages:

- Even "one-sided" or "both-sided" non-benzylic sulfones can be desulfurized, an advantage which has been made use of many times as the examples in this overview will show.
- One or several sulfone groups can be removed at a time thus creating up to five new C–C bonds in one step.
- Unsymmetrical target compounds can be synthesized from the corresponding sulfones as well.
- It gives high yields (up to 95%!).
- It tolerates many functional groups.
- Only small amounts of by-products form; in a few cases partially desulfurized sulfones have been isolated, which can be submitted to pyrolysis once again.
- The intramolecular course of the reaction in most cases selectively leads to the desired product and only in a few cases rearrangements do occur.
- Even target molecules having extreme strain can be synthesized; in this way a considerable distortion of the rigid adamantane skeleton could be achieved for the first time [16].
- There is no need for expensive, toxic, or labile reagents.
- The work-up procedure simply consists of taking up the reaction products in a solvent and subsequent chromatography. No by-products or reagents need to be removed.
- The starting materials are mostly crystalline and easily accessible in usually high yields (e.g., by application of the dilution principle [6] and/or the cesium effect [7]).

- The scale of the reaction can be varied between a few milligrams and several grams.
- The reaction conditions can be varied in a wide range thus allowing for optimization according to the reactivity of the substrates used.

3.1.2 Experimental Procedure for the Sulfone Pyrolysis

Only rarely observations on the influence of reaction conditions on the outcome of the reaction have been reported (cf. [2]). Complete sublimation of the starting material is important for the elimination to take place in the gas phase. Therefore conditions first have to be chosen in a way that complete sublimation takes place and the reaction time in the hot region of the apparatus remains short ("flash pyrolysis"). The temperature for sublimation correspondingly is lower than the temperature for pyrolysis. Since temperatures for sublimation are rarely stated in the original papers they are not listed in this review.

Staab et al. in one case studied the influence of the pyrolysis temperature on the yield of the reaction [17]. They varied the pyrolysis temperature in steps of 20 °C in a range of 500–580 °C. Yields varied between 28% and 86% which is surprising in consideration of the small range of temperature variation. The optimum yield was achieved using a temperature of 560 °C in this case. In case that a sufficient amount of starting material is available, an optimization of the reaction temperature therefore seems advisable. The temperature range of the examples listed below lies between 450 and 750 °C, most often a value between 500 and 600 °C has been chosen. This value might therefore serve as a guideline for new reactions. Considering the variety of substrates and pressures used, a correlation between temperature and yield cannot be deduced from the examples listed.

The reaction pressures applied range between 0.00013 and 13000 Pa (10^{-6}–100 Torr; 1 Torr = 133.3 Pa), the mean value lying somewhere between 0.13 and 13.3 Pa (10^{-3}–0.1 Torr). Again there is no discernible correlation.

In a few cases it has been reported that yields vary according to the scale of the experiment without quantification of these observations [17]. Optimization of the scale of reaction might be recommendable therefore. According to reactivity, volatility, and lability of the substrate, different reaction times seem advisable, in this case again there do not exist any general tendencies.

For substrates with extremely low volatility, Staab et al. used a modification of the standard procedure [18], which has been successfully applied for the synthesis of kekulene [19]. A closed and evacuated valve filled with the substrate is introduced for 3–5 min in a preheated air-bath (ca. 500 °C). Pyrolysis does occur in the solid phase, not in the gas phase. However, the sulfur-free products sublime more easily than the starting material and after a short time they condensate at a cooling device outside the reaction vessel, where the work-up can be carried out as usual. Yields are low for this modification; nevertheless, for substrates with extemely low volatility (and extremely low solubility of the sulfides as well) this often is the only choice available.

For larger quantities of substance to be pyrolyzed (e.g., [2.2]paracycloplane), Vögtle et al. developed an apparatus for continuous pyrolysis of sulfones which allows for continuous introduction of starting material and continuous removal of products without breaking the vacuum [20].

4 Scope and Arrangement of this Progress Report

This overview lists as complete as possible those sulfone pyrolyses which yielded cyclophanes having been described since 1979.

It encompasses work published up to early 1990. 150 different pyrolysis reactions are classified − a detailed discussion of each work is not intended. Instead − if reported in the original publication − the temperature of pyrolysis, pressure, and yield of the reactions are listed in the reaction equations. For some interesting cases by-products that have been characterized are listed as well. By this treatment the scope of the sulfone pyrolysis becomes obvious.

The arrangement of the reactions follows structural types: First pyrolyses of "both-sided non-benzylic" sulfones, then "one-sided benzylic" sulfones are listed, each arranged by increasing number of sulfone groups to be eliminated and increasing complexity of the substrate.

We hope that the creativity of chemists dealing with synthesis of cyclophanes as well as synthesis of other classes of compounds will be stimulated to try new, creative syntheses of sophisticated, strained molecules. Thus, in view of successful fourfold-pyrolyses and the availability of the modification for sulfones of low volatility [18], sixfold-pyrolyses of sulfones promise some potential. This synthetic method will be undispensable for the synthesis of large, complex, and sophisticated molecules.

5 Pyrolyses of "Both-Sided" Benzylic Sulfones

5.1 Pyrolyses of "Both-Sided" Benzylic Monosulfones

These pyrolyses in general give good yields since only one new bond has to be created during pyrolysis. Still this variation has seldom been applied. Synthesis of the mono-sulfone is in many cases more complicated than synthesis of a disulfone

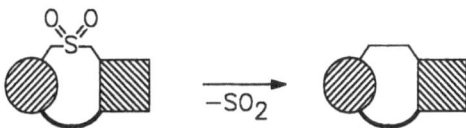

Scheme 3. Pyrolyses of "both-sided" benzylic mono-sulfones

because more of the structural features of the target molecule have to be incorporated into the starting material than in the case of a disulfone. However, this variation is useful for the synthesis of unsymmetrical cyclophanes, i.e., [m.n]phanes.

Using this variation, Tashiro et al. obtained in about equal yield the *tert*-butyl protected [3.2]- and [4.2]phanes 2 and 4, respectively [21].

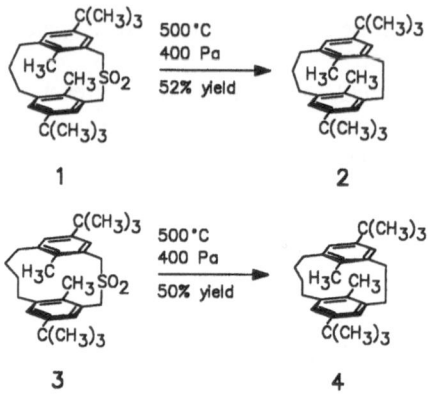

Haenel et al. for their studies of exciplex-interactions used a [3.2](1,4)naphthalenophane, the aromatic planes of which are inclined towards each other about 10° to 20° because of the different lengths of the bridges [22]. Pyrolysis of the mono-sulfone in the gas phase yielded a 1 : 3 mixture of the *syn*- and *anti*-phanes 6 and 7a/b, respectively, that could be separated by chromatography.

Vögtle et al. for the first time synthesized "phenylenicenes", helical compounds with non-annulated benzene rings, by pyrolysis of corresponding mono-sulfones [23]. The parent-compound, the conformationally rigid benzo[2.2]metacyclophane 9 ("triphenylenicene") is helically-chiral and could be separated into enantiomers showing strong optical rotation. In contrast, benzo[2.2]metaparacyclophane 11 turned out to be conformationally mobile not allowing for separation of the enantiomers [24]. Both phenylenicenes were obtained in yields of about 50%.

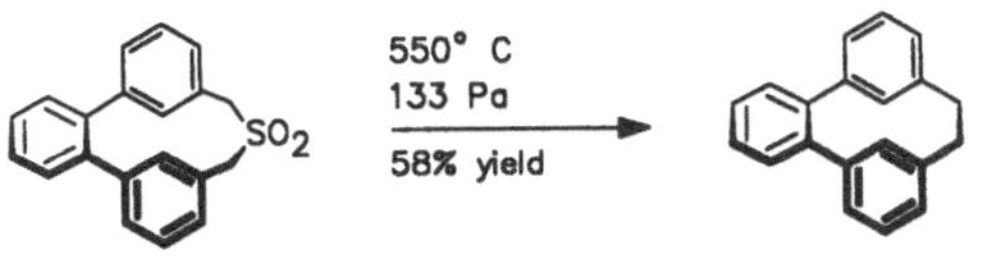

8   **9**

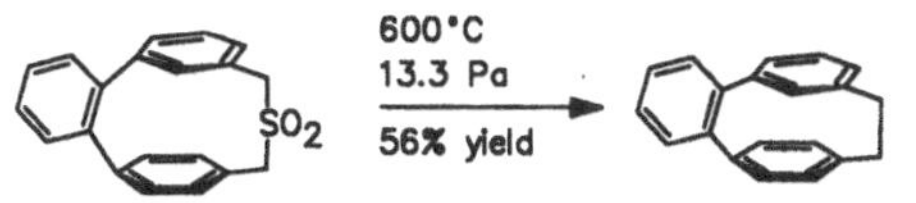

10   **11**

The corresponding thiopheno-annulated cyclophanes were obtained by Vögtle and co-workers as well [25]. The enantiomers of **15** could be separated; their absolute conformation was determined by the method of Bijvoet. The helical pentaphenylenicene **17** was produced similarly [26].

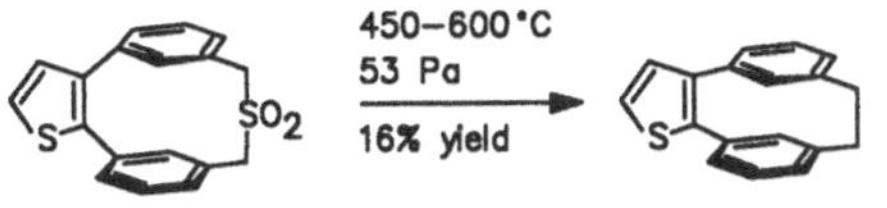

12   **13**

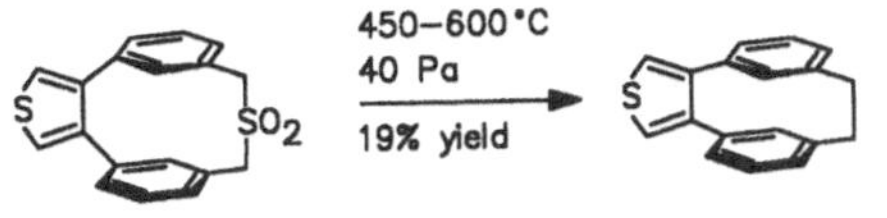

14   **15**

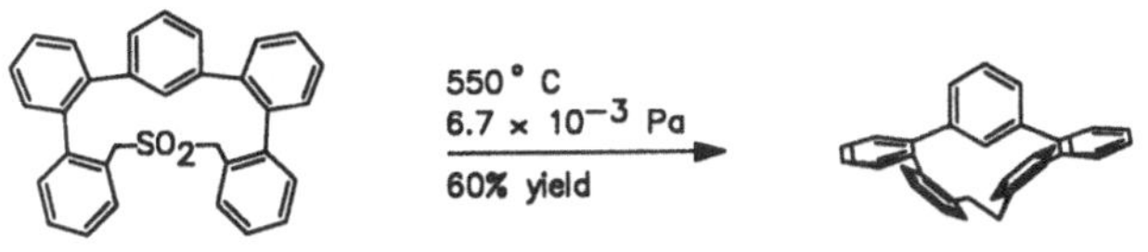

16   **17**

Interestingly, the pyrolyses of "clamped" quaterphenyls **19, 21, 23, 25,** and **27** led to remarkably different yields [27]. **19, 21,** and **27**, which are conformationally rigid, were obtained in significantly higher yields than the conformationally more flexible quaterphenyls **23** and **25**. Possibly the reactive radical-centers, which form after homolytic fission of the bonds next to the sulfone group, stay in closer contact in the more rigid skeletons compared to the more flexible skeletons. Therefore their recombination might be facilitated and the formation of by-products decreased. Whether this is a general tendency remains to be checked.

18 →(600° C, 1.33 Pa, 42% yield)→ **19**

20 →(600° C, 1.33 Pa, 35% yield)→ **21**

22 →(600° C, 1.33 Pa, 5% yield)→ **23**

24 →(600° C, 1.33 Pa, 22% yield)→ **25**

26 →(600° C, 1.33 Pa, 30% yield)→ **27**

In a remarkably high yield of 85% a clamped anthraquinonophane **29** was obtained, which served as a model for investigating the influence of bridging an anthraquinone-unit on UV-spectroscopic properties [28].

28 →(450 – 600° C, 67 Pa, 85% yield)→ **29**

5.2 Pyrolyses of "Both-Sided" Benzylic Disulfones

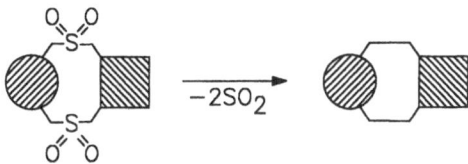

Scheme 4. Pyrolyses of "both-sided" benzylic disulfones ($\bullet$, $\blacksquare$ $\triangleq$ aromatic building blocks, $\ulcorner$ $\triangleq$ CH_2)

Hopf et al. in 1989 synthesized [2.2]orthometacyclophane **31a/b**, that has long been a target of synthetic efforts in cyclophane chemistry [29]. Pyrolysis of the disulfone **30** gave in 16% yield a 1:4 mixture of the *syn/anti* isomers **31a/b** and two methyl-substituted dibenzocycloheptadienes. An earlier attempt to obtain **31** by Vögtle et al. in 1970 via the same method failed because of lack of efficient methods of separation at that time [30]. In continuation of their work Hopf et al. by sulfone pyrolysis also tried the synthesis of [2.2]orthoparacyclophane, the most strained of the [2.2]cyclophane hydrocarbons [31]. However, they did not isolate the desired cyclophane but a spirotriene **33**, that possibly is formed by rearrangement of the intermediate, sulfur-free diradical and subsequent generation of the spirobond.

30	535 °C 1.33 Pa 16 % yield	**31a** : **31b** 4 : 1

32	515 °C 1.33 Pa 40 % yield	**33**

Sato et al. otained a variety of substituted [2.2]metacyclophanes **35a–d** and **37a–e** in high yields [32, 33].

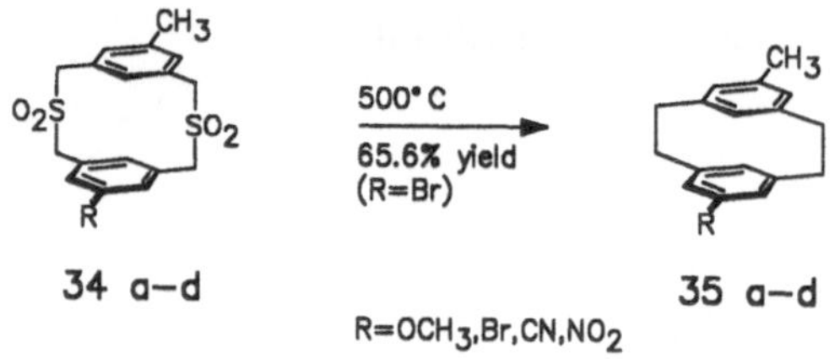

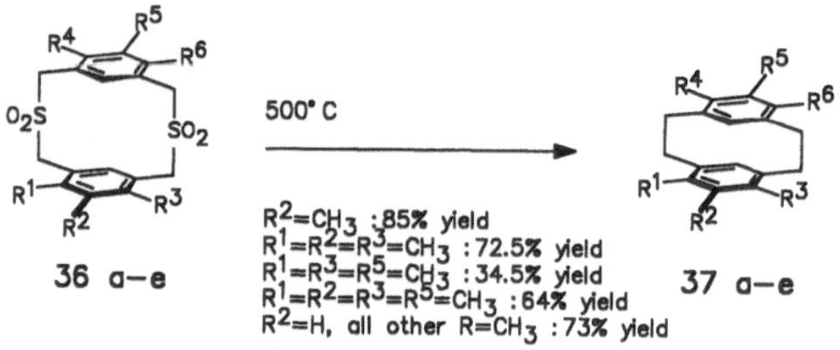

34 a–d

500° C

65.6% yield
(R=Br)

35 a–d

R=OCH$_3$,Br,CN,NO$_2$

36 a–e

500° C

R^2=CH$_3$:85% yield
R^1=R^2=R^3=CH$_3$:72.5% yield
R^1=R^3=R^5=CH$_3$:34.5% yield
R^1=R^2=R^3=R^5=CH$_3$:64% yield
R^2=H, all other R=CH$_3$:73% yield

37 a–e

For investigations of intramolecular charge-transfer interactions, donor- and acceptor-substituted cyclophanes are particularly well suited, since they allow for wide variation of substitutents. Interplanar-distance and orientation of aromatic planes can be tuned by choosing appropriate bridge lengths. *Syn/anti* conformational preferences have an additional influence on charge-transfer interactions. Severals studies of this kind were undertaken by Staab et al. [17, 34–36]. In usually good yields they synthesized methoxy-substituted [2.2]meta- and metaparacyclophanes with bromo-, cyano-, nitro-, or ester-substituents. In the case of methoxy-substituted substrates, often a few percent of quinoid by-products do form, e.g., **40, 41,** and **44,** which have been isolated in some cases.

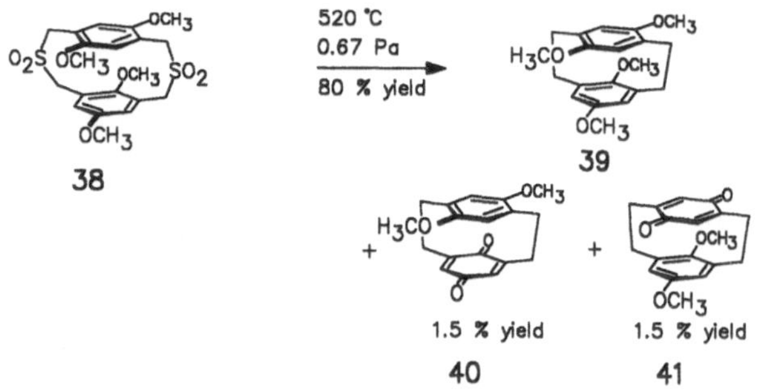

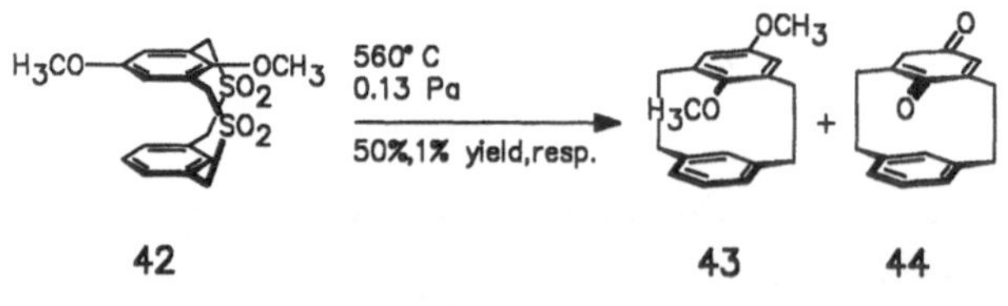

38

520 °C
0.67 Pa

80 % yield

39

+
1.5 % yield
40

+
1.5 % yield
41

42

560° C
0.13 Pa

50%,1% yield,resp.

43   **44**

Only the synthesis of the mono-nitro species **54** but not the one of the corresponding dinitro-compound succeeded. In the latter case, the tetrahydropy-rene-derivative **56** forms by removal of the intraannular groups.

The synthesis of the [2.2]paracyclophanes **59, 60, 62, 64, 66,** and **68** containing the tetracyanobenzene-moiety and different donor-substituents required due to the extremely low volatility of the corresponding sulfones a modification of the experimental standard-procedure (cf. Sect. 3.1.2) [18].

As an intermediate of their synthesis of [2.2]metaparacyclophane-quinones, Tashiro et al. obtained **70**, that could not be completely demethylated with boron tribromide [37]. Using the *tert*-butyl group as a positional protecting group, these authors also tried the synthesis of [2.2]metacyclophanes **72a/b** containing intraannular halide-substituents [38]. The preparation of the difluoro compound

72a was successful, whereas the dichloro-substituted molecule **72b** was only obtained in 5% yield and the dibromo derivative was not accessible at all.

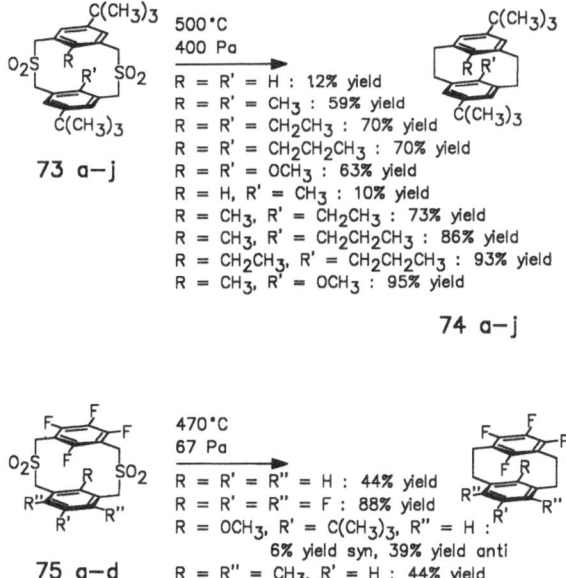

Similar [2.2]metacyclophanes **74a–k** were obtained by pyrolysis as well [39]. The yields of these reactions vary without systematic pattern between 10 and 95%. Several tetrafluoro[2.2]metacyclophanes **76a,c,d** and an octafluoro[2.2]meta-cyclophane with a completely fluorinated aromatic moiety **76b** were also accessible by sulfone pyrolysis [40].

Misumi et al. successfully tried to apply the sulfone pyrolysis to heteroaromatic systems such as **78** [41]. Even though yields were lower than for the carbocyclic analogs, the pyridinophanes **80**, **82**, **84**, **86**, **88**, and **90** have been obtained, which by other methods are difficult to synthesize [42].

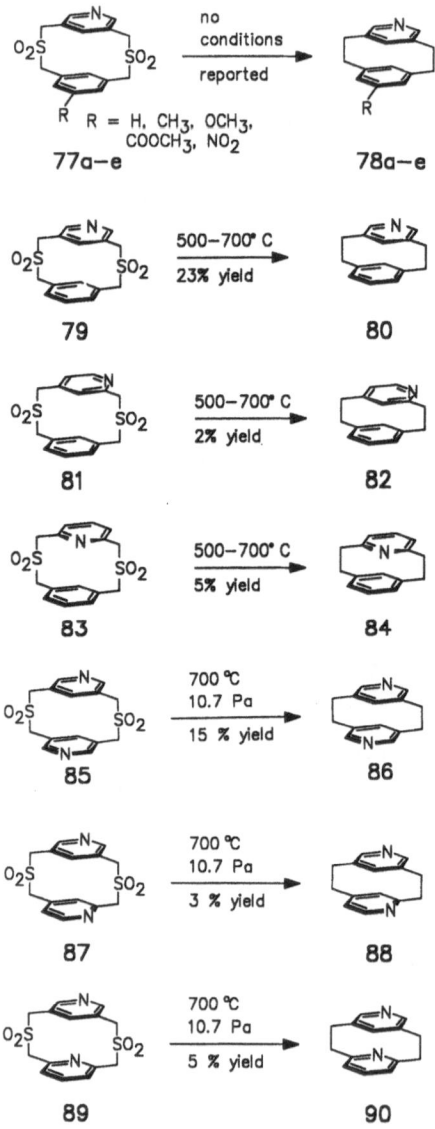

The [2.2]naphthalenophane **92** served in a project by Haenel et al. as a model for investigations of transannular π-interactions in strongly distorted aromatic systems [43]. The naphthalene moiety in **92** is particularly distorted since it is forced together with a *para*-disubstituted benzene moiety. Surprisingly, a by-product **93** containing two naphthalene units was isolated in 13.5% yield. The formation of the bridging C–C bond in the intermediate diradical mono-sulfone is supposedly aggravated in such a way that prior to the bond formation the second sulfone group becomes eliminated. Besides *p*-quinodimethane, a free naphthalene diradical forms that then dimerizes to give the chiral naphthalenophane **93**.

A chiral naphthalenophane **95** with two bromo substituents related to **93** is described in a further publication [44]. At the same time **96** forms, which by halide-metal exchange and subsequent hydrolysis was transformed into the corresponding achiral naphthalenophane. A similar naphthalenophane **98** was synthesized by Boekelheide et al. starting from the corresponding disulfone **97** [45].

Model compounds for amine-arene-exciplexes were constructed by Haenel et al. by positioning pyridine or pyrazine units above naphthalene or anthracene units [46, 47]. On oxidation of the sulfides to give the sulfones, the sterically more easily accessible nitrogen which is pointing "outwards" becomes oxidized as well to give the *N*-oxides **105** and **107**, respectively. On submitting these sulfones to sulfone pyrolysis the *N*-oxides become reduced again thus yielding **106** and **108**, respectively [48].

99 → **100** 500–550 °C, 13.3 Pa, 11 % yield

101 → **102** 500–550 °C, 13.3 Pa, 65 % yield

103 → **104** 500 °C, 40 Pa, 28 % yield

105 → **106** 550 °C, 5.3 Pa, 47 % yield

107 → **108** 550 °C, 5.3 Pa, 36 % yield

Even cyclophanes containing large aromatic units such as pyrene, anthracene, or benzophenanthrene units or partially hydrogenated annulated systems can easily be desulfurized via sulfone pyrolysis, as Staab et al. have shown in a series of works on such compounds [49–53].

109 → **110** 500 °C, 13.3 Pa, 16% yield

111 → **112** + **113** 500 °C, 13.3 Pa, no yield given

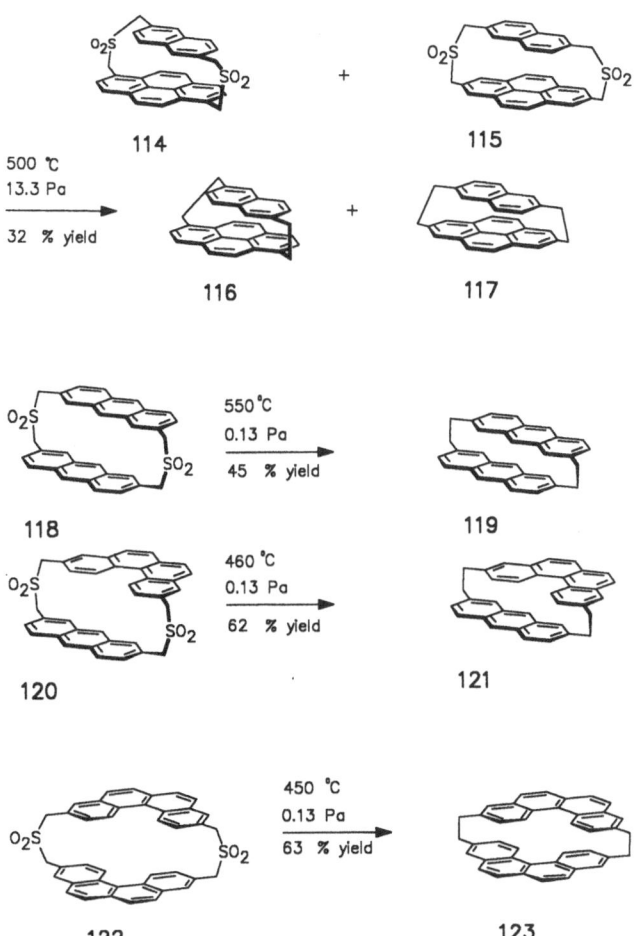

A highlight in this field certainly is the successful pyrolysis of **124** to give **125** in 75% yield, an intermediate in the synthesis of kekulene [19].

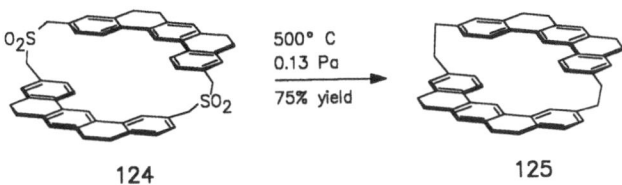

Not only annulated systems but also oligophenyls were successfully prepared by sulfone pyrolysis (cf. **127**, **129**, and **131**) [54, 55].

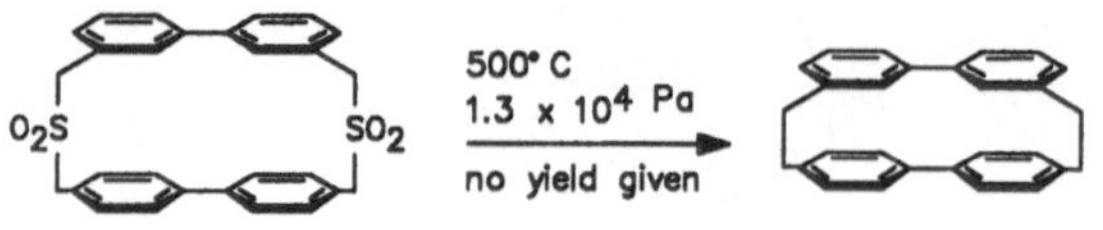
O_2S
SO_2
500° C
1.3 x 10^4 Pa
no yield given

126 127

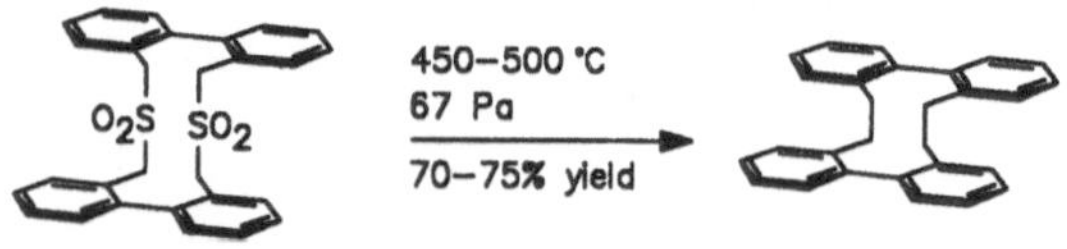
O_2S SO_2
450–500 °C
67 Pa
70–75% yield

128 129

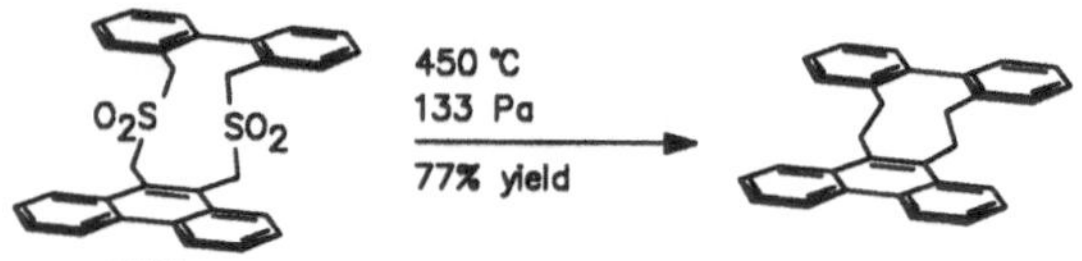
O_2S SO_2
450 °C
133 Pa
77% yield

130 131

A series of stereochemically interesting *p*-terphenylophanes **133**, **135**, **137**, and **139** was described by Vögtle et al. [56]. They served for the study of rotational barriers of the central benzene rings in dependence of the bridging moieties. Especially interesting in this respect is **137** possessing an intraanular phenyl group, obtained by pyrolysis in remarkable 82%.

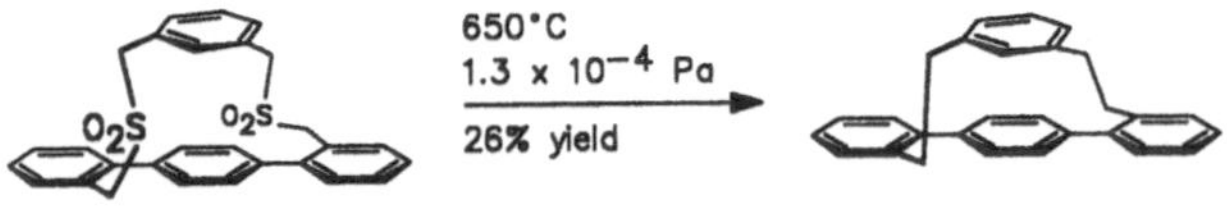
O_2S O_2S
650 °C
1.3 x 10^{-4} Pa
26% yield

132 133

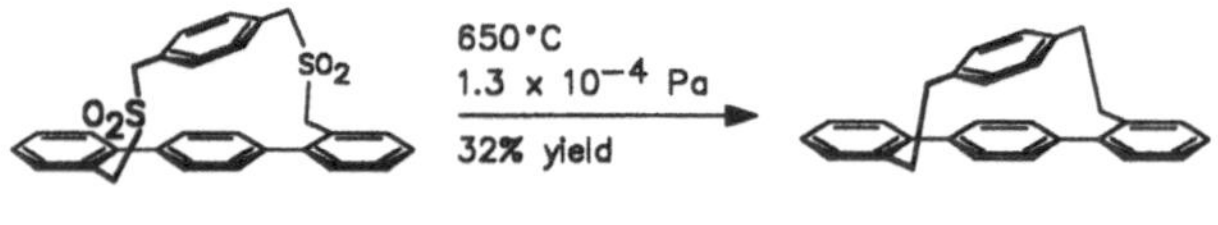
SO_2
O_2S
650 °C
1.3 x 10^{-4} Pa
32% yield

134 135

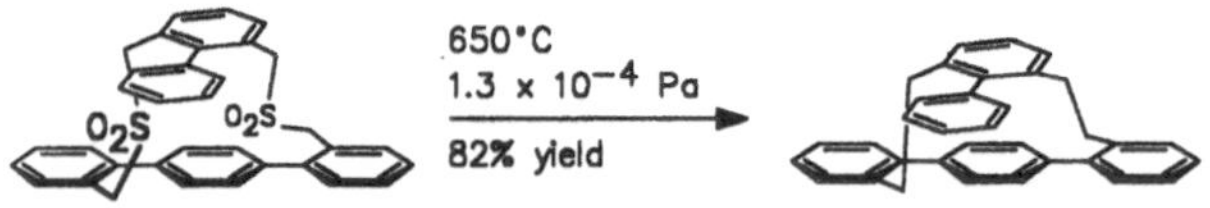
O_2S O_2S
650 °C
1.3 x 10^{-4} Pa
82% yield

136 137

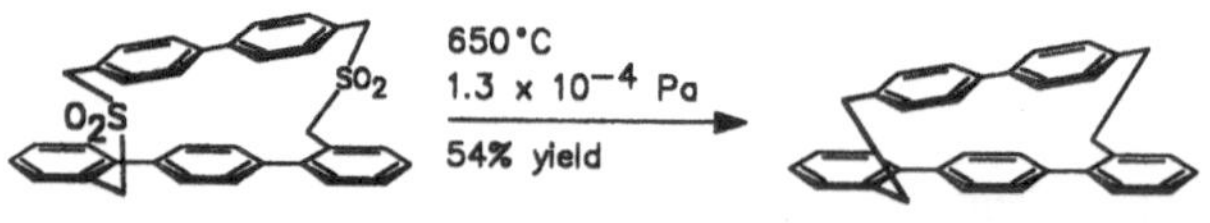
SO_2
O_2S
650 °C
1.3 x 10^{-4} Pa
54% yield

138 139

For related studies other cyclophanes with intraannular phenyl groups were designed by Vögtle et al. [57]. In a surprisingly high yield of 85% the parent compound 8-phenyl[2.2]metacyclophane 141 could be isolated. Stacked benzene rings without direct connections to the aromatic planes lying above and below offers compound 143 with a biphenyl unit. For the purpose of spectral comparison, 146 without an intraannular phenyl group was prepared.

The doubly clamped helix 148, designed by Vögtle et al., which also contains stacked benzene rings was accessible by sulfone pyrolysis from 147 in 37% yield [26].

Further sulfone pyrolyses of both-sided benzylic disulfones by Haenel et al. led to the [2.2]phanes 150 and 151 possessing the fluorene skeleton [58]. By reduction to the 9-fluorenyl anion an aromatic unit can be incorporated into these phanes. Even fluorene units tolerate the conditions for the pyrolysis giving an example for carbonyl-containing phanes that can be synthesized by sulfone pyrolysis.

158 und **160**, anthraquinone derivatives described by Vögtle et al., which also contain carbonyl groups are interesting because of their electrochemical behavior [28].

Labile functional groups form part of the phanes **162**, **164**, and **166** prepared by Tashiro et al. [59]. After complete construction of the molecular skeleton by sulfone pyrolysis, the 1,2,5-thiadiazol unit is transformed by treatment with Grignard-reagents into 1,2-dicarbonyl units forming the bridges.

A non-benzenoid, aromatic unit is incorporated into the tropolonophane **168**, which by Ito et al. has been transformed into a [2.2]tropoquinonophane [60].

5.3 Pyrolysis of "Both-Sided" Benzylic Trisulfones

Surprisingly, no pyrolyses of this type have been carried out during the time covered by this overview; for examples of this type refer to the earlier review from 1979 [2].

5.4 Pyrolyses of "Both-Sided" Benzylic Tetrasulfones

The pyrolysis of tetrasulfones in general leads to lower yields than in the case of disulfones because there are more opportunities for recombination reactions and rearrangements. Nevertheless, there are a few examples for successful pyrolyses of this type that give products which are not or hardly accessible by other methods. Low yields therefore are acceptable in these syntheses.

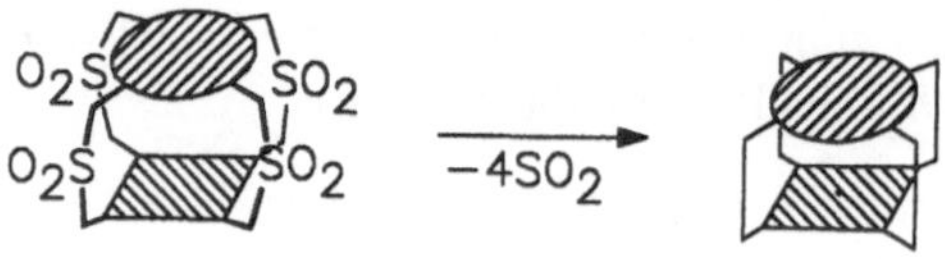

Scheme 5. Pyrolysis of "both-sided" benzylic tetrasulfones

Vögtle et al. submitted a mixture of cyclic tetrasulfones **169** and **170** (the sulfide products of the cyclization were not separated but directly oxidized) to a sulfone pyrolysis at 600 °C [61]. However, the product was shown not to be the desired four-fold-bridged cyclophane but a doubly clamped 1,2-dihydrocyclobutabenzene. Only an increase of the pyrolysis temperature to 750 °C led to successful formation of the desired [2.2.2.2](1,2,4,5)cyclophane **171** in the mixture of products. Manyfold *ortho*-clamped phanes prior to this work were not accessible by sulfone pyrolysis.

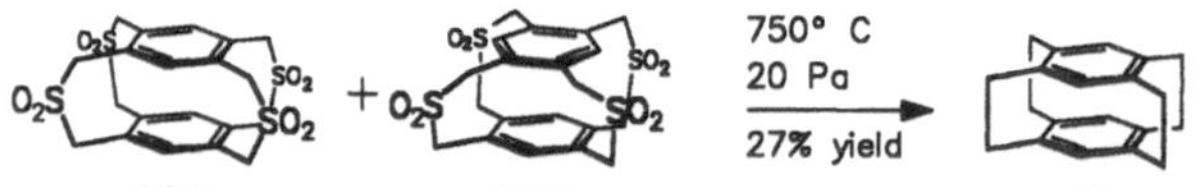

169 170 171

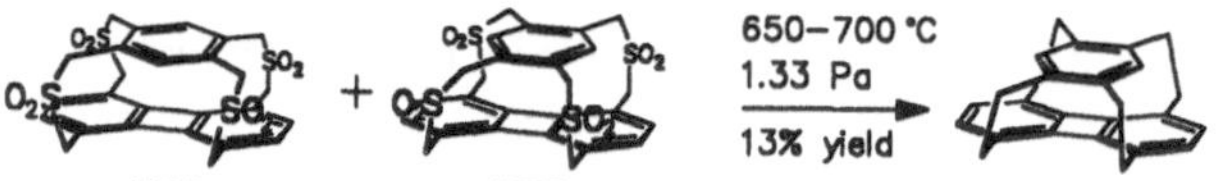

172 173 174

By the same method, a mixture of tetrasulfones was pyrolyzed by Vögtle et al. to yield the extremely strained biphenylenophane **174** [62]. The isomer derived from **172** does not form. A four-fold intraannularly phenyl-substituted tetrasulfone was transformed by Vögtle et al. by sulfone pyrolysis into the sterically extremely crowded phane **176** [57]. The high-temperature NMR spectrum of **176** leads to the conclusion that at temperatures of more than 120 °C the phenyl rings are able to pass the inner region of the macrocycle.

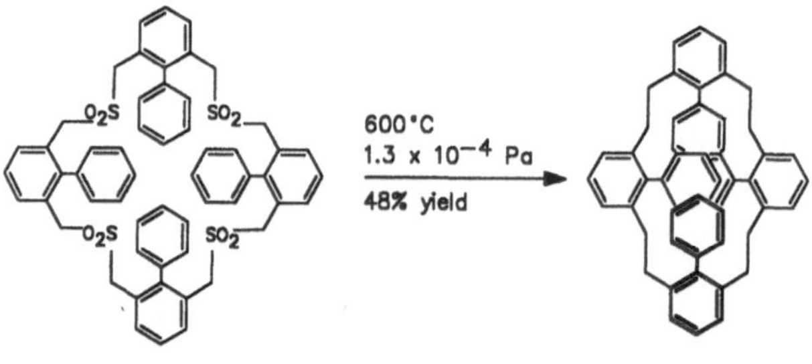

175 176

"Cuppedo-" and "cappedophanes" are names coined by Hart et al. for systems accessible by sulfone pyrolysis, that possess an open ("cuppedo-", in **178**, **180**, and **182**) or closed ("cappedo-", in **184**) molecular cavity [63]. The central phenyl ring in the *m*-terphenyl moiety adopts a position perpendicular to the outer, neighbouring phenyl rings, thus allowing introduction of substituents, which point into the center of the cavity, as exemplified in **182** and **184**.

6 Pyrolyses of "One-Sided Non-benzylic" Sulfones

6.1 Pyrolyses of "One-Sided Non-benzylic" Monosulfones

There are no examples of pyrolyses of this type even though there is no obvious reason against making use of this combination.

6.2 Pyrolyses of "One-Sided Non-benzylic" Disulfones

The pyrolysis of "one-sided non-benzylic" sulfones offers access to phanes having aliphatic bridges to [3.3]phanes or even higher phanes [3].

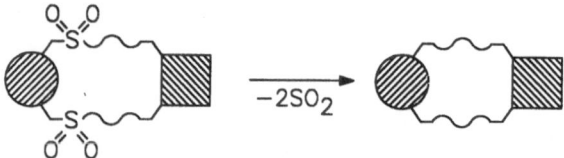

Scheme 6. Pyrolysis of "one-sided non benzylic" disulfones

Misumi et al. by this method bridged a cyano-disubstituted benzene ring with aliphatic chains of 8 to 10 carbon atoms as in **187a/b** [64]. The yield of the less-strained phane having the 10-membered bridge is higher than for the more strained derivative.

$[CH_2]_m$ 630° C
               9.3 Pa $[CH_2]_n$
O_2S CN SO_2 ⟶ CN
       CN   n=8 : 43% yield   CN
             n=10 : 30% yield   CN

m=6,8   n=8,10
186a/b   **187a/b**

Staab et al. investigated clamped benzophenones [65]. The benzene rings in these compounds are arranged nearly "face-to-face" depending on the flexibility of the bridging moiety. For these [m.1]phanes (**189a–e**), yields increase with increasing number of ring-members, whereas yields for the related clamped triphenylmethane systems **191a–d** and **193a/b** behave reverse [66]. Higher number of ring members leads to lower yields in this case. Introduction of a *tert*-butyl group in *p*-position of the exocyclic phenyl ring in **193a/b** lowers yields considerably. Possibly the conditions of pyrolysis lead to cleavage of this substituent during reaction.

O_2S 580 °C
$[CH_2]_n$ 0.013 Pa
O_2S n=6 : 33 % yield $[CH_2]_n$ O
         O   n=7 : 28 % yield
                 n=8 : 35 % yield
                 n=9 : 70 % yield
                 n=10: 76 % yield

188 a–e   **189 a–e**

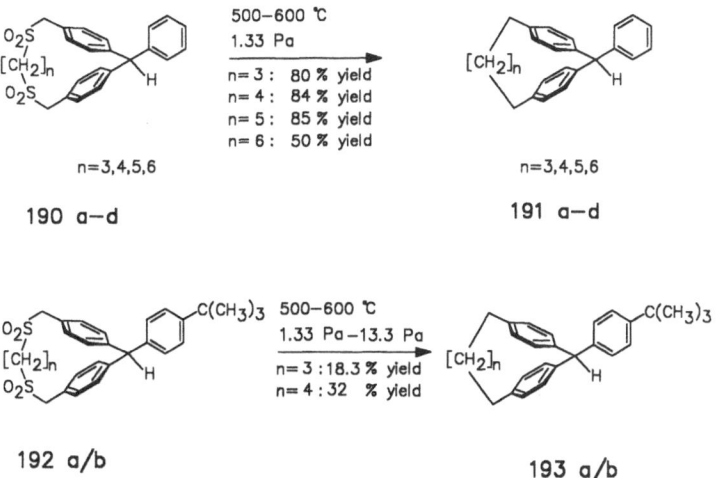

Tropolonophanes **195** and **197** bridged with aliphatic chains are described by Ito et al. [67]. Ether cleavage of the intraannular methoxy substituent can be carried out before or after pyrolysis; however, the hydroxy substituent seems to better tolerate the conditions of pyrolysis leading to higher yields for this order of synthetic steps. The bridging aliphatic chain can be substituted by chains containing benzene rings in different patterns of substitution, thus leading to [3.3]tropolonophanes **199** and **201** [68].

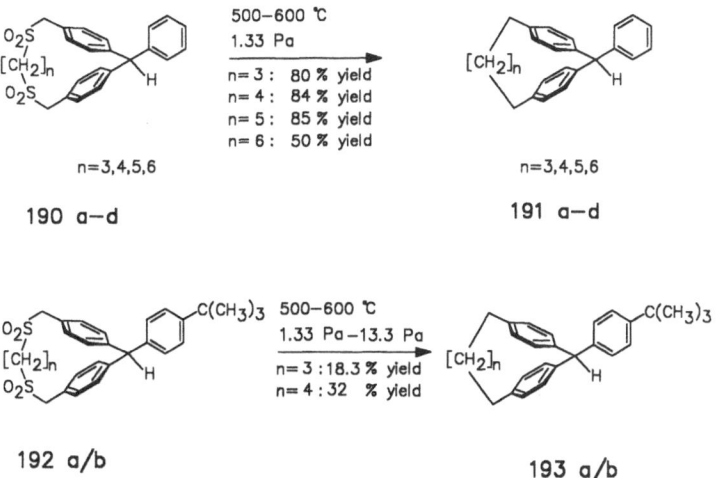

"Araliphanes", phanes containing alicyclic building blocks such as the adamantane building block besides aromatic building blocks, were first described by Vögtle and Dohm [16]. The adamantanophane **203** was accessible by sulfone pyrolysis in a remarkable yield of 50%. The high yield is surprising considering the extreme strain of this compound. This strain not only leads to a stronger distortion of the benzene ring than in the comparable [2.2]metacyclophane but even more remarkably leads to an extreme distortion of the rigid adamantane skeleton as shown by x-ray crystallography. This example again proves the synthetic power of the method since there is no other method by which a comparable distortion of the rigid adamantane could be achieved so far.

Most examples of pyrolyses of "one-sided non-benzylic" disulfones come from the synthesis of [3.3]- and [4.4]phanes. The parent compound [3.3]paracyclophane **205** so far only was obtained by ring-enlarging methods in low yields. Haenel et al. succeeded in synthesizing this compound by sulfone pyrolysis in 75% starting from the readily available corresponding sulfide and sulfone, respectively [69].

The same reaction yielding **205** was carried out by Misumi et al. in 75% yield [70].

They also prepared related phanes by sulfone pyrolysis such as [3.3]metaparacyclophane **207** (40% yield) and [3.3]metacyclophane **209** (52% yield). The synthesis of a cyano-disubstituted [3.3]phane **211** is described in a separate publication by the same authors [71].

For their investigations of intramolecular charge-transfer interactions, Staab et al. designed several donor- and acceptor-substituted [3.3]phanes [18, 34, 72–75]. Compared to the [2.2]phanes they have the advantage of possessing less distorted aromatic moieties, at the same time having interplanar distances allowing for still sufficient π-interaction. Methoxy substituents and ester groups as acceptor units were applied as donor substituents, which later in the synthetic sequence can be transformed into TCNQ units. For the purpose of comparison, compounds having only methoxy or only ester groups were synthesized as well. Depending on the system, quinoid structures such as **218** or **219** are being formed during pyrolysis.

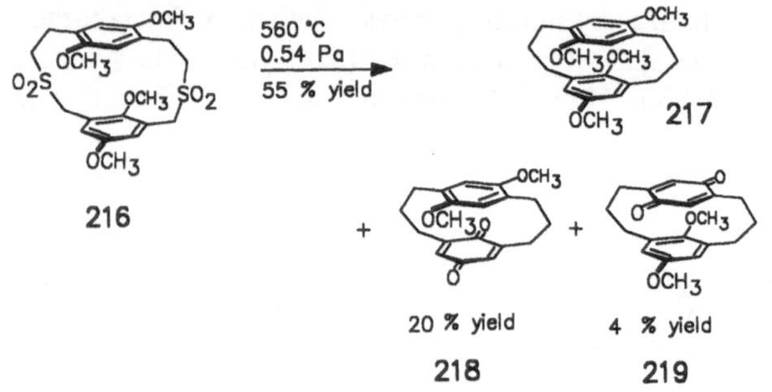

560 °C
0.54 Pa
55 % yield
216
217
20 % yield
218
4 % yield
219

Another feature to be taken into consideration is the ratio of *syn*/*anti* isomers. **222** and **223** as well as **230** and **231** can exist in pseudo-*geminal* or pseudo-*ortho* forms. Assignment was in a few cases only possible by x-ray crystallography.

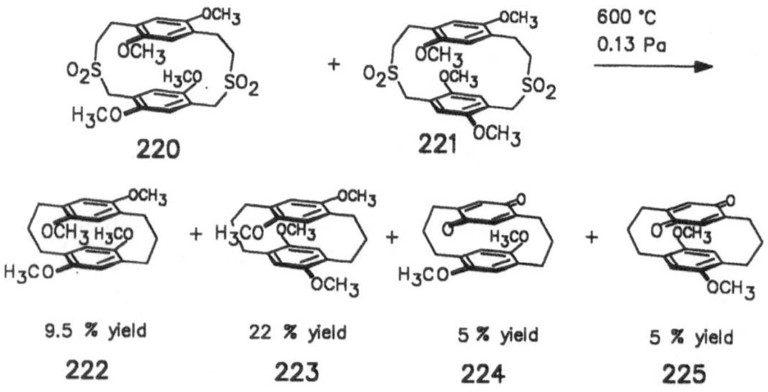

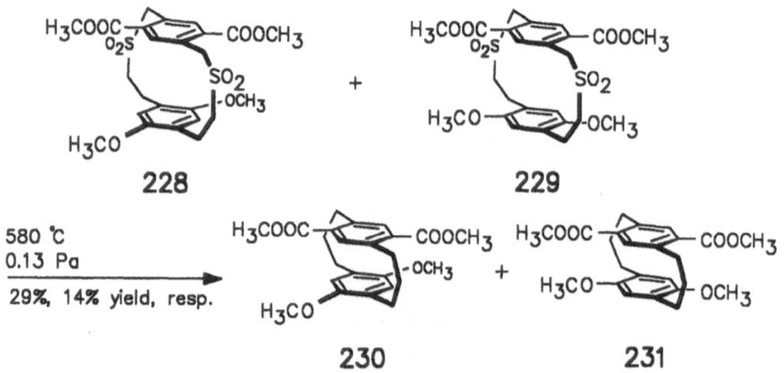

220
221
600 °C
0.13 Pa
9.5 % yield
222
22 % yield
223
5 % yield
224
5 % yield
225

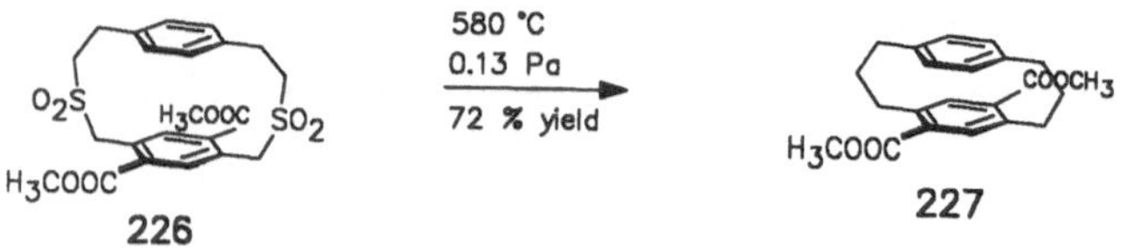

226
580 °C
0.13 Pa
72 % yield
227

228
229
580 °C
0.13 Pa
29%, 14% yield, resp.
230
231

Further interesting systems such as **233**, **235**, and **237** containing a tetra-cyanobenzene unit and different substituents at the opposing benzene ring were prepared by pyrolysis as well.

232 → 233 (580°C, 0.13 Pa, 11% yield)

234 → 235 (580°C, 0.13 Pa, 18% yield)

236 → 237 (580°C, 0.13 Pa, 11% yield)

[4.4]Paracyclophane (**239**), the parent compound of the [4.4]phanes was synthesized by Misumi et al. in 76% yield [70]. Similarly, a dicyano substituted [4.4]phane **241** was obtained in 48% yield [71]. In consideration of the complexity of the system, a 5% yield of the triple layered [4.4][4.4]paracyclophane **243** represents another proof for the potential of the sulfone pyrolysis as a synthetic tool for the construction of sophisticated molecules [70]. The long bridges in **243** minimize distortions in the triple-layered system, thereby offering a perfect model system for the study of transanular π-interactions.

238 → 239 (650°C, 13–67 Pa, 76% yield)

240 → 241 (650°C, 67 Pa, 48% yield)

242   650 °C / 67 Pa / 5 % yield   243

As in the case of the comparable [3.3]phanes, Staab et al. obtained during synthesis of tetramethoxy-substituted [4.4]phanes a mixture of the pseudo-*geminal* isomer **246** and the pseudo-*ortho* isomer **247** and additional quinoid by-products [76].

244 + 245   580–590° C / 0.13 Pa / 45% yield (1:2)   246 + 247

Synthetic efforts of these researchers extended to the synthesis of the [5.5]phanes **249** and **250** and even the [6.6]phane **252** [77]. The high yield of 42% for the [6.6]phane shows that the potential of the sulfone pyrolysis is not yet fully used for similar or higher systems.

248   600°C / 0.13 Pa / 32 % yield   249   1 : 1   250

251   520–540 °C / 1.33 Pa / 42 % yield   252

[3.3]Phanes containing condensed aromatic units are accessible by sulfone pyrolysis as well. Two such phanes, **254** and **256**, are described by Misumi et al. [70].

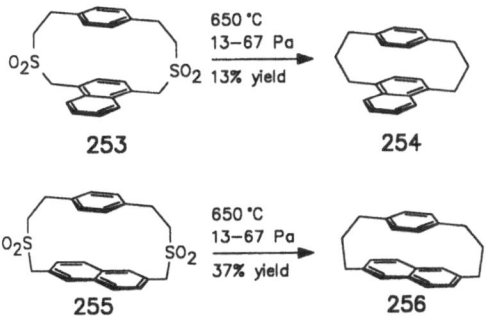

Haenel et al. investigated [3.3]naphthalenophanes as model systems for excimers [78]. Pyrolysis of the sulfone **257** possessing two 2,6-disubstituted naphthalene units led to a 3 : 2 mixture of diastereomers consisting of the achiral phane **259** with mirror-symmetry and of a slightly crossed, chiral phane **258**. A mixture of a slightly crossed and a clearly crossed phane **261** and **262**, respectively, was obtained after pyrolysis of the (2,6)(1,5)sulfone **260**. Pyrolysis of **263** containing two 1,4-disubstituted naphthalene units resulted in a 3 : 2 *syn/anti* mixture of **264** and **265** [79]. The absorption spectra of the *syn* and *anti* isomers show more differences than in the case of the comparable [3.3] or even [2.2]phanes.

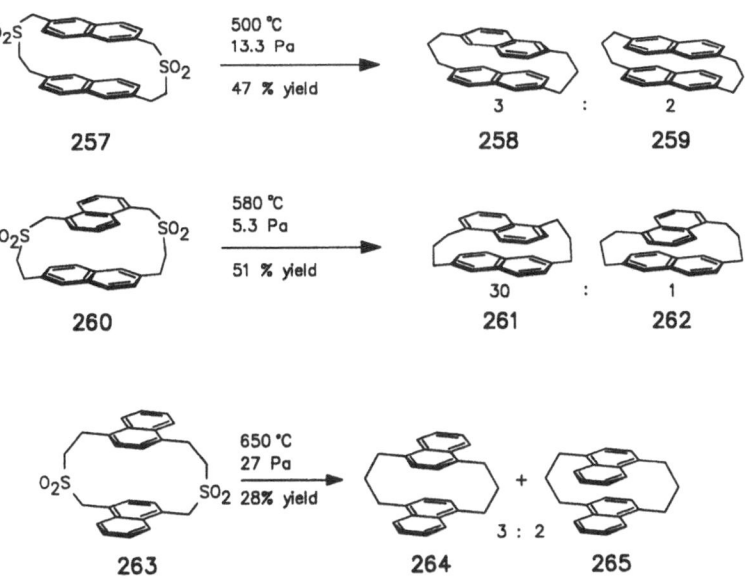

Joachim Dohm and Fritz Vögtle

Dehydrogenation of [3.3]- and [4.4]octahydropyrenophanes **267** and **269** gave fully aromatic pyrenophanes, serving as model systems for excimers in studies by Staab et al. [80].

A C$_4$-bridged anthraquinonophane **271** was described by Vögtle et al. [28].

580–600 °C
2.7 × 10⁻⁴ Pa
25 % yield

266

267

580–600 °C
1.3 × 10⁻⁴ Pa
69 % yield

268

269

450 – 600° C
67 Pa
17% yield

270

271

6.3 Pyrolyses of "One-Sided Non-benzylic" Trisulfones

The parent compound [3.3.3](1,3,5)cyclophane **273** was prepared by Misumi et al. in moderate yield [70]. By formal substitution of a benzene ring in **273** by a CH-group, **275** is obtained, which was synthesized by Pascal et al. [81]. The methine hydrogen atom points inwards and is exactly positioned above the aromatic plane, leading to a strongly high-field shifted ¹H-NMR absorption at −4.03 ppm (!).

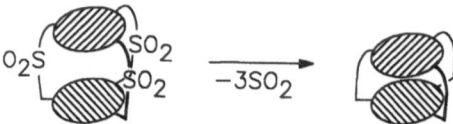

Scheme 7. Pyrolysis of "one-sided non-benzylic" trisulfones

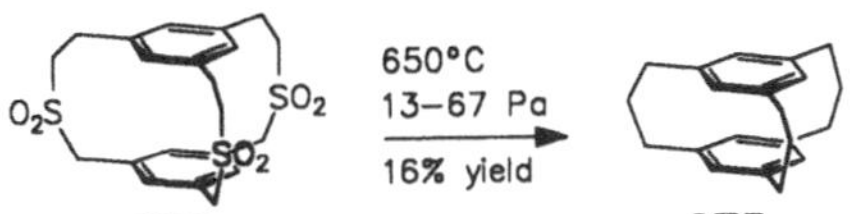

272 273

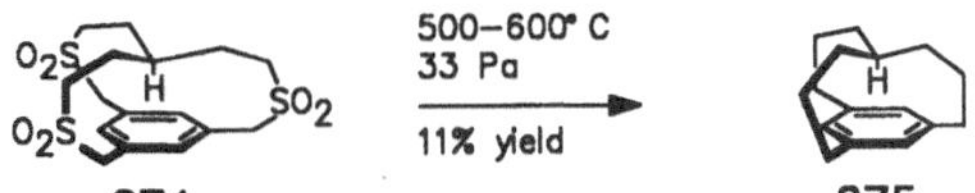

274 275

6.4 Pyrolyses of "One-Sided Non-benzylic" Tetrasulfones

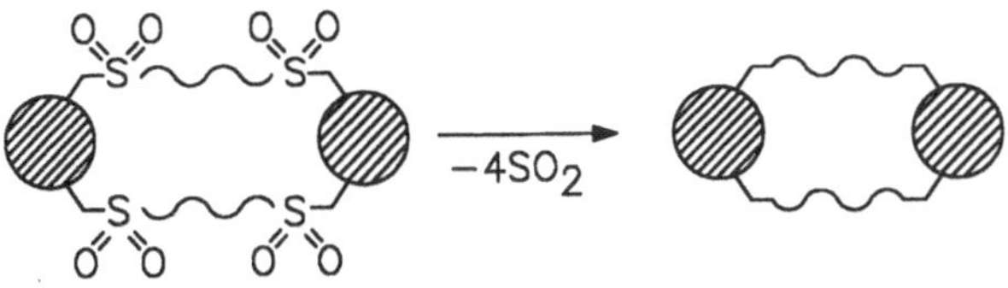

Scheme 8. Pyrolysis of "one-sided non-benzylic" tetrasulfones

As an alternative to the synthesis of the [5.5]phanes, **249** and **250**, respectively, by pyrolysis of the one-sided non-benzylic disulfone **248**, Staab et al. tried the pyrolysis of a one-sided non-benzylic tetrasulfone **276** to give **249** and **250** [77]. The yield of 15% compared to 32% shows that the concept works but suffers from loss of carbon fragments and from increasing rearrangements due to the higher number of bonds to be formed in the pyrolysis of the tetrasulfone.

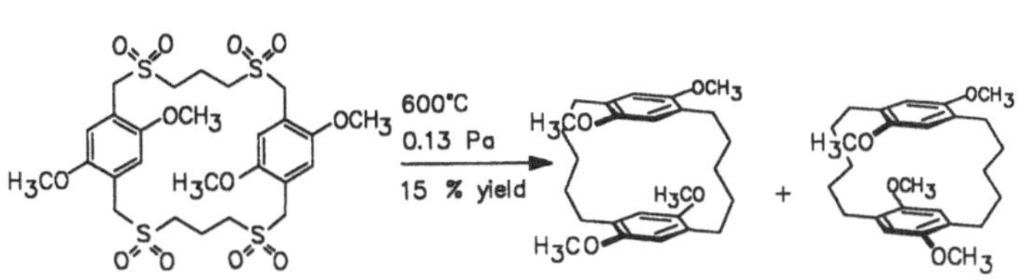

1 1

276 249 250

7 Pyrolyses Under Simultaneous Loss of Fragments

In a few cases pyrolysis and ring contraction occurs under simultaneous loss of carbon-fragments (mostly ethano units because of synthetic reasons). The benzylic side of the disulfone undergoes recombination, whereas the ethano bridge and both the SO_2 groups become cleaved and eliminated.

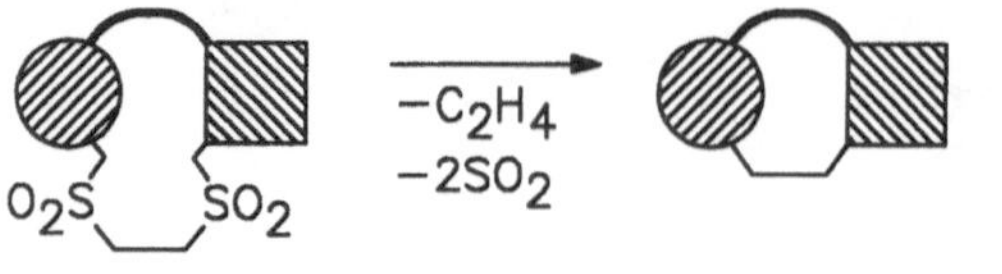

Scheme 9. Pyrolysis under simultaneous loss of fragments

This loss of fragments offers an alternative to pyrolysis of a highly strained, hardly accessible, both-sided benzylic monosulfone. A one-sided non-benzylic disulfone containing an ethano bridge in most cases is easier to synthesize in particular when incorporated into a highly strained skeleton. Two examples for this strategy were described by Vögtle et al. [26, 82]. The pentaphenylenicenes **278a/b** (P and M) and **17** were obtained in 46% and 9% yield, respectively. By incorporation of a bridge into the helix inversion of the helix is made more difficult and separation of the enantiomers should be possible.

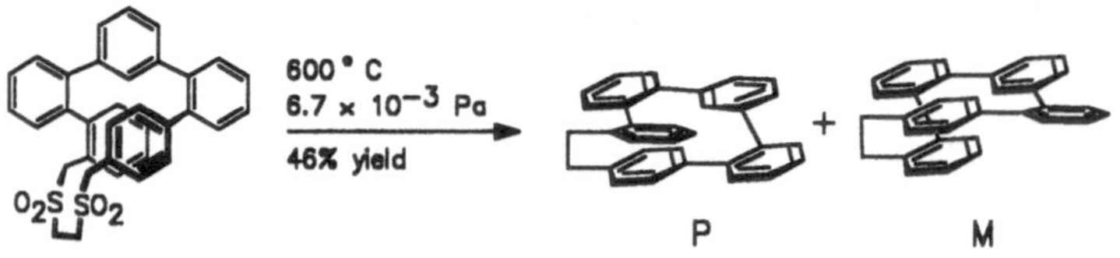

8 Concluding Remarks

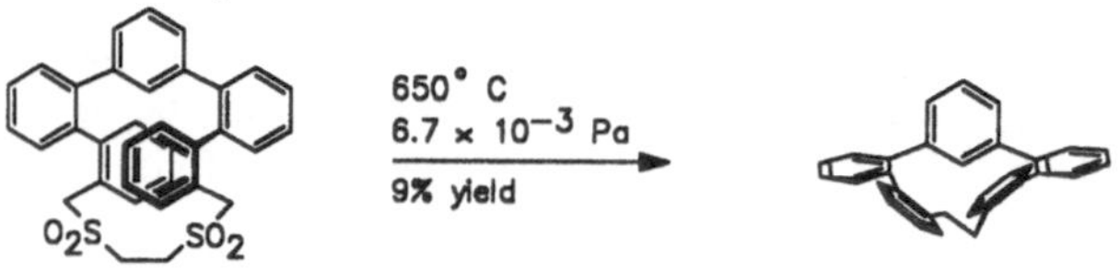

The large variety of examples presented in this review show the universal applicability of the sulfone pyrolysis as a unique synthetic method for the construction of complicated and strongly distorted molecular (hydrocarbon) skeletons bearing more-or-less rigidly (pre)organized aromatic and aliphatic structural units. Future synthetic efforts will focus on sulfone pyrolyses of oligosulfones, thus allowing for formation of several bonds in one step. So far, no case of six-fold sulfone pyrolysis has been described, but it is to expect, that in the light of successful four- and five-fold pyrolyses there will be successful six-fold pyrolyses in the future as well.

Pyrolysis of one-sided non-benzylic monosulfones still waits for application for the synthesis of strained unsymmetrical cyclophanes.

As some of the examples show, extremely strained molecules are easily accessible and fascinating distortions will be achieved by introducing strain by sulfone pyrolysis.

9 Acknowledgements

The authors kindly thank Dipl.-Chem. M. Frank, R. Friedrich, R. Güther, D. Karbach, A. Schröder, and C. Seel for computer drawings. J. D. thanks the Fonds der Chemischen Industrie for a doctoral scholarship.

10 References

1. a. Vögtle F (1969) Chem Ber 102: 1449;
   b. Vögtle F (1969) Chem Ber 102: 3077;
   c. Vögtle F (1969) Angew Chem 81: 258; Angew Chem Int Ed Engl 8: 274
2. Vögtle F, Rossa L (1979) Angew Chem 91: 534; Angew Chem, Int Ed Engl 18: 514
3. Rossa L, Vögtle F (1983) Top Curr Chem 113: 1
4. Müller E, Röscheisen G (1957) Chem Ber 90: 543
5. See, eg, Allinger NL, Da Rooge MA, Hermann RB (1961) J Am Chem Soc 83: 1974
6. Vögtle F, Knops P, Sendhoff N, Mekelburger HB (1990), Top Curr Chem (in press)
7. Ostrowicki A, Koepp E, Vögtle F (1990) Top Curr Chem (in press) [Beitrag 239]
8. Nokami J, Nishiuchi K, Wakabayashi S, Okawara R (1980) Tetrahedron Lett 21: 4455
9. See, eg, Givens RS, Olsen RJ, Wylie PL (1979) J Org Chem 44: 1608
10. See, eg, Bruhin J, Jenny W (1973) Tetrahedron Lett 1215
11. Bieber W, Vögtle F (1978) Chem Ber 111: 1653
12. Mitchell RH, Otsubo T, Boekelheide V (1983) Tetrahedron Lett 1975: 219
13. Mitchell RH, Boekelheide V (1974) J Am Chem Soc 96: 1547
14. Koray AR (1983) J Organomet Chem 243: 191
15. Higuchi H, Tani K, Otsubo T, Sakata Y, Misumi S (1987) Bull Chem Soc Jpn 60: 4027
16. Vögtle F, Dohm J, Rissanen K (1990) Angew Chem 102 943; Angew Chem, Int Ed Engl 29: 902
17. Staab HA, Reibel WRK, Krieger C (1985) Chem Ber 118: 1230
18. Staab HA, Wahl P, Kay KY (1987) Chem Ber 120: 541
19. Staab HA, Diederich F (1983) Chem Ber 116: 3487
20. Vögtle F, Fornell P, Löhr W (1979) Chem Ind (London) 416
21. Yamato T, Sakamoto H, Kobayashi K, Tashiro M (1986) J Chem Res (S) 352
22. Blank NE, Haenel MW (1981) Chem Ber 114: 1531
23. Hammerschmidt E, Vögtle F (1980) Chem Ber 113: 1125
24. Wittek M, Vögtle F, (1982) Chem Ber 115: 1363
25. Vögtle F, Palmer M, Fritz E, Lehmann U, Meurer K, Mannschreck A, Kastner F, Irngartinger H, Huber-Patz U, Puff H, Friedrichs E (1983) Chem Ber 116: 3112
26. Hammerschmidt E, Vögtle F (1980) Chem Ber 113: 3550
27. Vögtle F, Wittek M (1982) Chem Ber 115: 2533
28. Wingen R, Vögtle F (1980) Chem Ber 113: 676
29. Bodwell G, Ernst L, Haenel MW, Hopf H (1989) Angew Chem 101: 509; Angew Chem, Int Ed Engl 28: 455
30. Vögtle F, Neumann P (1970) Tetrahedron 26: 5847
31. Hopf H, Bodwell G, Ernst L (1989) Chem Ber 122: 1013

32. Sato T, Torizuka K, Komaki R, Atobe H (1980) J Chem Soc, Perkin Trans 2: 561
33. Torizuka K, Sato T (1980) Bull Chem Soc Jpn 53: 2411
34. Staab HA, Jörns M, Krieger C, Rentzea M (1985) Chem Ber 118: 796; Staab HA, Jörns M, Krieger C (1979) Tetrahedron Lett 2513
35. Staab HA, Schanne L, Krieger C, Taglieber V (1985) Chem Ber 118: 1204
36. Staab HA, Reimann-Haas R, Ulrich P, Krieger C (1983) Chem Ber 116: 2808
37. Tashiro M, Koya K, Yamato T (1983) J Am Chem Soc 105: 6650
38. Tashiro M, Yamato T (1985) J Org Chem 50: 2939
39. Tashiro M, Yamato T (1981) J Org Chem 46: 1543
40. Tashiro M, Fujimoto H, Tsuge A, Mataka S, Kobayashi H (1989) J Org Chem 54: 2012
41. Kawashima T, Tohda Y, Ariga M, Mori Y, Misumi S (1985) Heterocycles 23: 180
42. Kawashima T, Kurioka S, Tohda Y, Ariga M, Mori Y, Misumi S (1985) Chem Lett 1289
43. Haenel MW (1982) Chem Ber 115: 1425; Blank NE, Haenel MW, Krüger C, Tsay YH, Wientges H (1988) Angew Chem 100: 1096; Angew Chem, Int Ed Engl 27: 1064
44. Blank NE, Haenel MW (1983) Chem Ber 116: 827
45. Boekelheide V, Tsai CH (1976) Tetrahedron 32: 423
46. Haenel MW, Lintner B, Benn R, Rufinska A, Schroth G, Krüger C, Hirsch S, Irngartinger H, Schweitzer D (1985) Chem Ber 118: 4884
47. Haenel MW, Lintner B, Schweitzer D (1986) Z Naturforsch 41b: 223
48. Lintner B, Schweitzer B, Benn R, Rufinska A, Haenel MW (1985) Chem Ber 118: 4907
49. Staab HA, Kirrstetter RGH (1979) Liebigs Ann Chem 886
50. Kirrstetter RGH, Staab HA (1984) Liebigs Ann Chem 899
51. Sauer M, Staab HA (1984) Liebigs Ann Chem 615
52. Staab HA, Sauer M (1984) Liebigs Ann Chem 742
53. Staab HA, Diederich F, Caplar V (1983) Liebigs Ann Chem 2262
54. Leach DA, Reiss JA (1979) Tetrahedron Lett 4501
55. Banciu R, Pogany I, Mosara D, Rusu P, Stanciulescu R, Dezsö M (1985) Rev Roum Chim 30: 703
56. Böckmann K, Vögtle F (1981) Liebigs Ann Chem 467
57. Böckmann K, Vögtle F (1981) Chem Ber 114: 1048
58. Haenel MW, Irngartinger H, Krieger C (1985) Chem Ber 118: 144
59. Hatta T, Mataka S, Tashiro M (1986) J Heterocycl Chem 23: 813
60. Kawamata A, Fukazawa Y, Fujise Y, Ito S (1982) Tetrahedron Lett 1083
61. Vögtle F, Klieser B (1982) Angew Chem 94: 922; Angew Chem, Int Ed Engl 21: 928
62. Saitmacher K (1989) Ph D Thesis, Univ Bonn
63. Vinod TK, Hart H (1990) J Org Chem 55: 881; Vinod TK, Hart H (1988) J Am Chem Soc 110: 6574
64. Higuchi H, Kobayashi E, Sakata Y, Misumi S (1986) Tetrahedron 42: 1731
65. Staab HA, Alt R (1984) Chem Ber 117: 850
66. Staab HA, Ruland A, Kuo-chen C (1982) Chem Ber 115: 1755
67. Saito H, Fujise Y, Ito S (1983) Tetrahedron Lett 24: 3879
68. Kawamata A, Fukazawa Y, Fujise Y, Ito S (1982) Tetrahedron Lett 23: 4955
69. Haenel MW, Flatow A (1979) Chem Ber 112: 249
70. Otsubo T, Kitasawa M, Misumi S (1979) Bull Chem Soc Jpn 52: 1515
71. Otsubo T, Kohda T, Misumi S (1980) Bull Chem Soc Jpn 53: 512
72. Staab HA, Herz CP, Döhling A (1980) Chem Ber 113: 233
73. Staab HA, Herz CP, Krieger C, Rentea M (1983) Chem Ber 116: 3813
74. Staab HA, Hinz R, Knaus GH, Krieger C (1983) Chem Ber 116: 2835
75. Staab HA, Knaus GH, Henke HE, Krieger C (1983) Chem Ber 116: 2785; Staab HA, Knaus GH (1979) Tetrahedron Lett 4261
76. Staab HA, Döhling A, Krieger C (1981) Liebigs Ann Chem 1052
77. Staab HA, Starker B, Krieger C (1983) Chem Ber 116: 3831
78. Blank NE, Haenel MW (1981) Chem Ber 114: 1520
79. Yoshinaga M, Otsubo T, Sakata Y, Misumi S (1979) Bull Chem Soc Jpn 52: 3759
80. Staab HA, Riegler N, Diederich F, Krieger S, Schweitzer D (1984) Chem Ber 117: 246
81. Pascal Jr. RA, Grossmann RB, van Engen D (1987) J Am Chem Soc 109: 6878
82. Hammerschmidt E, Vögtle F (1979) Chem Ber 112: 1785

Ring Closure Methods in the Synthesis of Macrocyclic Natural Products

Qingchang Meng[1] and Manfred Hesse

Institute of Organic Chemistry, University of Zurich, Winterthurerstrasse 190, CH-8057 Zurich, Switzerland

Table of Contents

---

[1] Part of the Ph. D. thesis of Q. M., Zurich 1991

Topics in Current Chemistry, Vol. 161
© Springer-Verlag Berlin Heidelberg 1991

Macrocyclization methods by means of ring closure from linear bifunctional precursors are critically reviewed. The scope of this article is limited to macrocyclic natural products and, to some extent, their model compounds. However, not only macrolides, but all kinds of macrocyclic natural products which have been synthesized by ring closure reactions are covered. The commonly used ring closure methods in recent literature include those involving lactonization, lactamization, C−C bond formation, C=C bond formation, ether formation, amine formation, and exo ring formation. The activation of one or both interacting sites of the bifunctional linear precursor is the central issue of some ring closure methods. Stereochemistry plays an important role in ring closure reactions. Only when the stereochemistry of the linear precursor allows its two interacting sites to reach each other occurs a successful ring closure. If the adopted conformation of a linear precursor sufficiently resembles that of the corresponding macrocycle a ring closure can be very facile.

1 Introduction

The structural elucidation of *civetone* and *muscone* as large-ring ketones by Ruzicka [1] in 1926 earmarks the origin of macrocyclic chemistry. Generally, the term *macrocycle* refers to medium(8- to 11-membered)- and large(12-membered or bigger)-ring compounds and the term *macrolide*, coined by Woodward [2], refers to that subset of macrocycles which incorporates a lactone moiety [3]. Macrocyclic compounds constitute a large spectrum of natural products, e.g. macrolide antibiotics [4], alkaloids [5], and terpenes [6].

In recent years, many efforts have been focused on the synthesis of macrocyclic natural products, especially macrolide antibiotics [7]. The central issue in this area is the construction of macrocyclic rings. Three fundamental strategies are available for this purpose. They are ring enlargements [8], e.g. cleavage of the bridged bond in a bicycle (Scheme 1) [9], ring contractions, e.g. base-induced intramolecular acyl transfer (Scheme 2) [10], and ring closure reactions (Scheme 3)

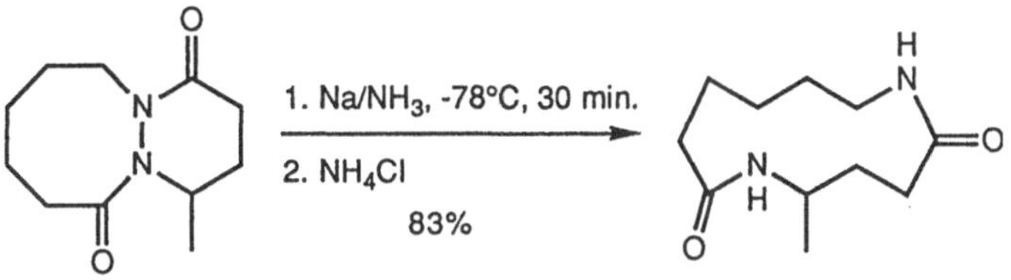

Scheme 1

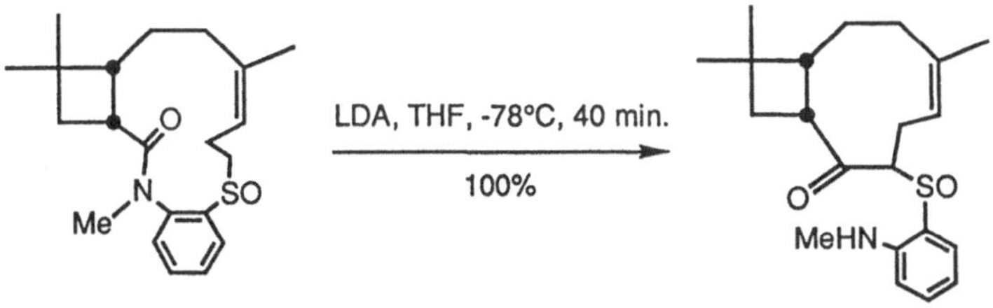

Scheme 2

Scheme 3

However, ring closure of bifunctional acyclic precursors is the most direct and general method, though such a cyclization is usually disfavored because of the loss of entropy and the gain of ring strain associated with the ring formation and requires high dilution techniques. Therefore, specially efficient methodology has to be devised in order to reach ring closure: not every intermolecular reaction can be employed intramolecularly! For instance, Scheme 4 illustrates such a difficulty. Before a Reformatsky-like aldol condensation was found successful, all attempts to effect direct aldol ring closure of linear precursor 7 to macrocycle 8, using 35 different conditions had proved fruitless (cf. Section 4.1) [11].

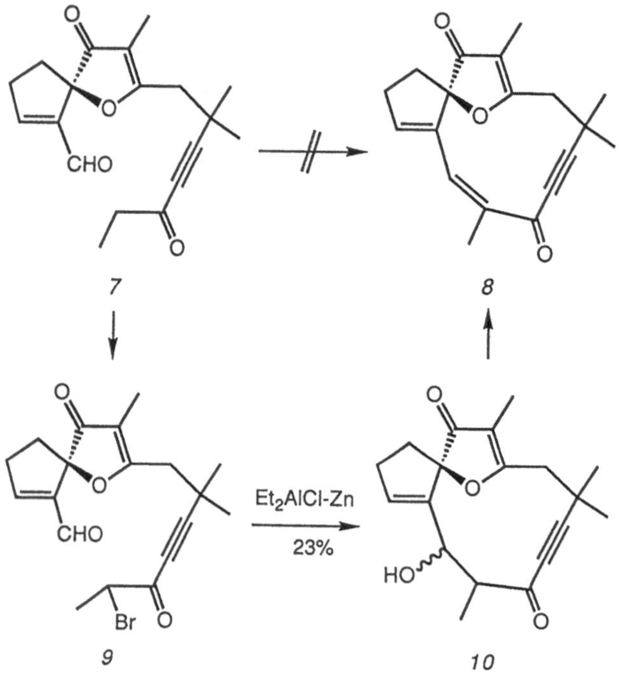

Scheme 4

Macrolide synthesis has been extensively reviewed by several authors in the last few years [7]. Synthesis of macrocyclic compounds, most being unnatural, via ring closure reactions by high dilution techniques was discussed by Rossa and Vögtle [12] in volume 113 of this series. And, Mandolini and coworkers [13] discussed the mechanistic aspects of macrocyclic ring closure reactions. However, all the macrocyclic natural products other than macrolides have not been equally treated. This contribution aims to cover all kinds of macrocyclic natural products and concentrates exclusively on the key cyclization step. In the following sections, efficient methods of ring closure from linear bifunctional precursors 5 to macrocycles 6 as delineated in Scheme 3, which have developed recently in the synthesis of macrocyclic natural products and are generally applicable in further synthetic pursuits, are dicussed according to the type of bond formed in the ring closure step.

2 Methods Involving Lactonization

To make intramolecular esterification feasible it is always necessary to activate one or both interacting sites of a hydroxy acid precursor. All the methods of macrocyclic ring closure involving lactonization come from this principal idea.

2.1 The Corey Method

In 1974, Corey and Nicolaou [14] found that hydroxy acid *11* can be efficiently activated by 2,2'-dipyridyl disulfide (DPDS).[1] As shown in Scheme 5, in the presence of triphenylphosphine, *11* reacts with DPDS yielding 2-pyridinethiol ester *12*. The proton transfer from hydroxyl group to carbonyl in 2-pyridinethiol ester *12* is facilitated by the basic nitrogen of the pyridine nucleus and a dipolar intermediate *13* is formed. Then a facile, electrostatically driven cyclization occurs.

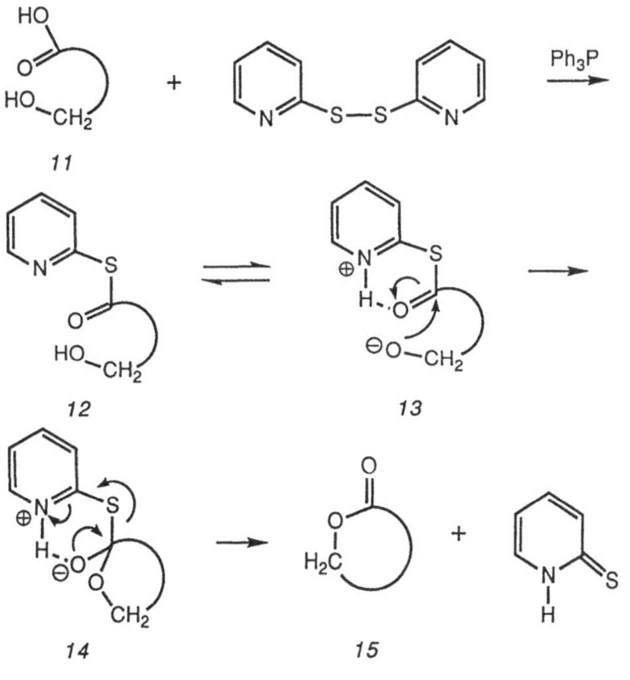

Scheme 5

---

[1] A list of symbols and abbreviations is given in chapter 10.

The 2-pyridinethiol ester *12* can also be prepared from hydroxy acid and 2-thiopyridyl chloroformate in the presence of triethylamine [15].

This double activation method has been successfully used in numerous syntheses of complex natural products [7]. Plata and Kallmerten [16] claimed that this procedure was the most reliable one examined for effecting macrocyclization in the synthesis of the naturally occuring antibiotic *(+)-18-deoxynargenicin A₁ (18)*. As shown in Scheme 6, treatment of the hydroxy acid *16* with DPDS in the presence of triphenylphosphine led to a thiolester, which upon slow addition to refluxing xylene afforded macrocycle *17* in 38% yield. The latter was converted to *18*.

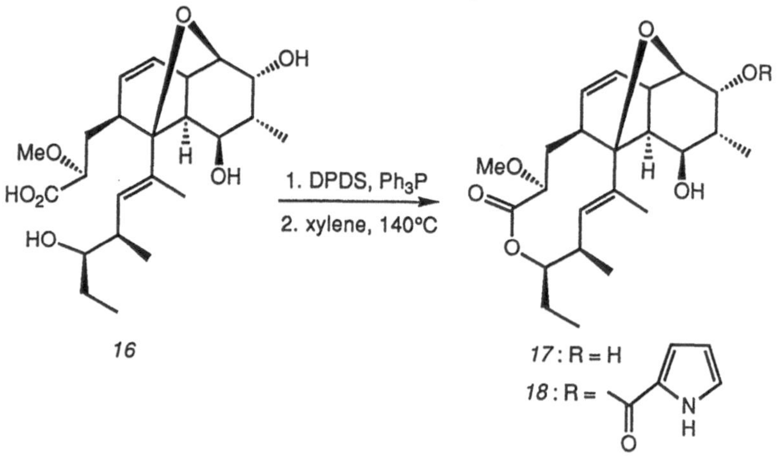

Scheme 6

An improtant modification of the Corey method was reported by Gerlach and Thalmann [17], who used silver ions ($AgClO_4$ or $AgBF_4$) to activate the 2-pyridinethiol ester by complexation as indicated in Scheme 7.

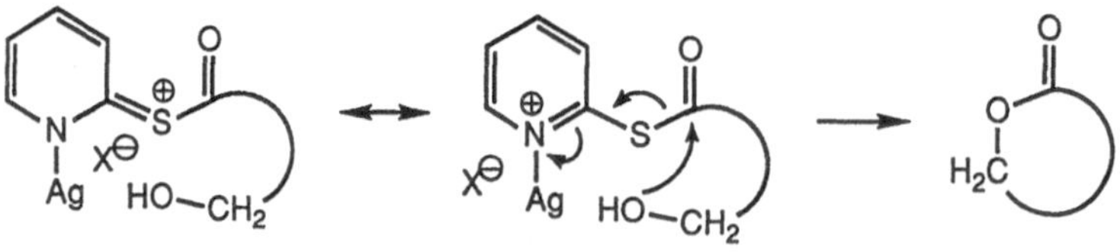

Scheme 7

This modification has been applied to the synthesis of the macrotetrolide antibiotic *nonactin (22)* by Gerlach et al. [18]. As shown in Scheme 8, the linear

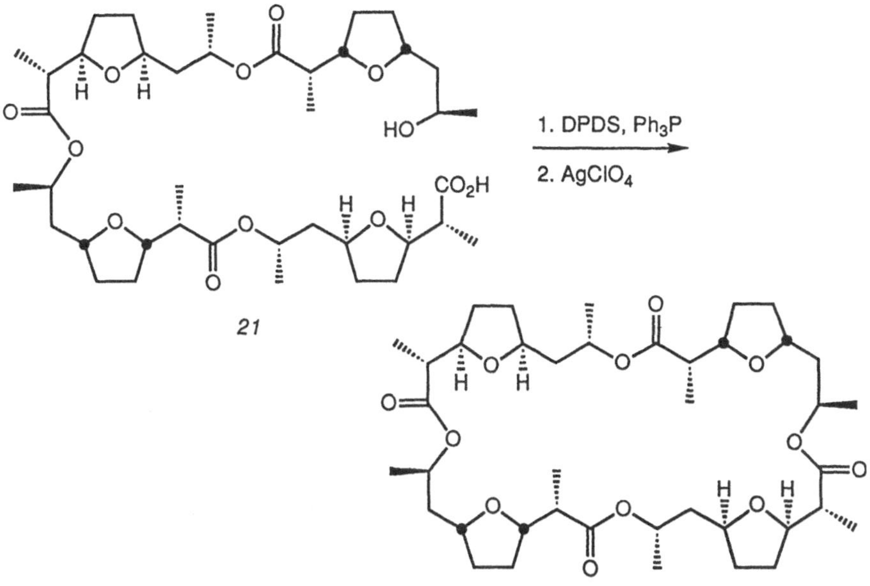

1. DPDS, Ph₃P

2. AgClO₄

21

22

Scheme 8

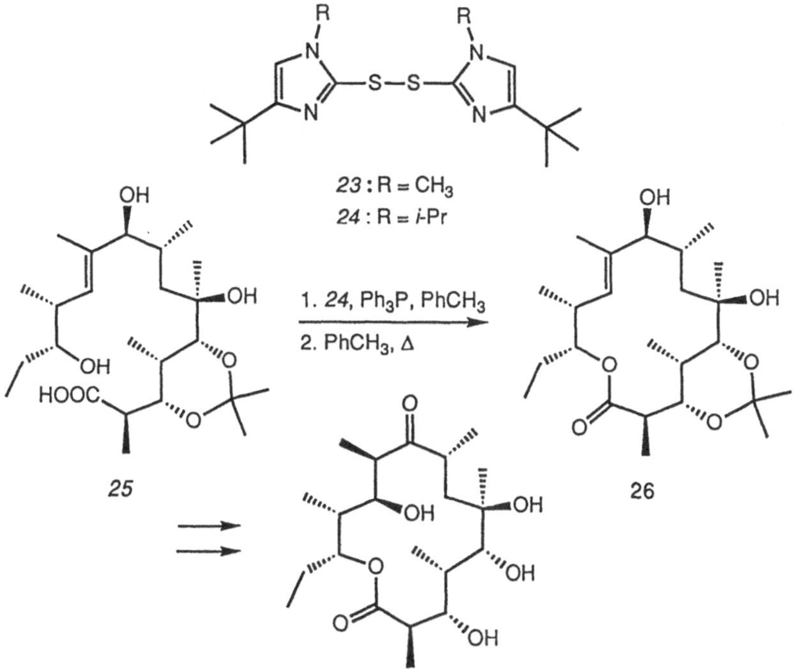

23 : R = CH₃
24 : R = i-Pr

1. 24, Ph₃P, PhCH₃

2. PhCH₃, Δ

25

26

27

Scheme 9

hydroxy acid *21* was first converted to 2-pyridinethiol ester and then treated with silver perchlorate. In benzene solution at 25 °C (0.5 h) a 20% yield of the cyclic tetramers was obtained, whereas the yield rose to 40% in acetonitrile at 80 °C (1 h). From the four possible tetrameric diastereoisomers (starting with racemic nonactic acid) only three were observed. *Nonactin* (*22*) composed 25% of the mixture and was isolated by chromatography (cf. ref. 7a).

Two superior, alternative reagents for the Corey method are the disulfides *23* and *24* [19]. For example, in the first synthesis [20] of *erythronolide B* (*27*), the aglycone of the important antibiotic *erythromycin B*, cyclization of the hydroxy acid *25* to the 14-membered lactone *26* was effected in 50% yield via the thiol ester of 4-*t*-butyl-*N*-isopropyl-2-mercaptoimidazole by heating in dry toluene under reflux (Scheme 9).

2.2 The Mukaiyama Method

In 1976, Mukaiyama et al. [21] developed an efficient macrocyclization method mediated by 1-methyl-2-chloropyridinium iodide (*28*). As shown in Scheme 10, the mechanism of this method is similar to that of Corey. It was found that triethylamine is the most suitable base and optimal yields were obtained in refluxing acetonitrile or dichloromethane.

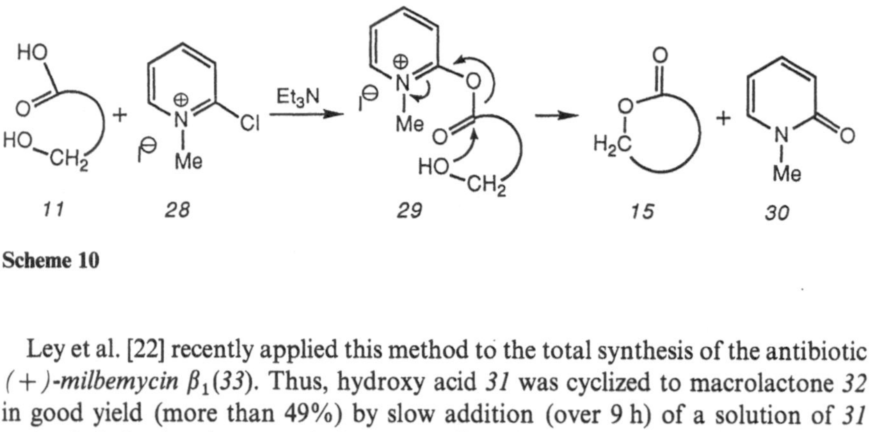

Scheme 10

Ley et al. [22] recently applied this method to the total synthesis of the antibiotic *(+)-milbemycin* β_1(*33*). Thus, hydroxy acid *31* was cyclized to macrolactone *32* in good yield (more than 49%) by slow addition (over 9 h) of a solution of *31* and triethylamine in acetonitrile to a refluxing solution of *28* in acetonitrile (Scheme 11). Another recent application of the Mukaiyama method is due to White and Bolton [23].

Mukaiyama and coworkers [24] found that 2-chloro-3-methoxymethyl-1-methylpyridinium iodide is also suitable for effecting macrolactonization. Furthermore [25], the cyclization mediated by the 2-chloropyridinium salts described above sometimes gives no satisfactory yields. It is mainly due to the decomposition of the pyridinium salts under the cyclization conditions by the attack of triethylamine to either the 1-methyl group or the 2-position of the pyridinium ring to form 2-chloropyridine or 2-ammoniopyridinium salts. To solve this

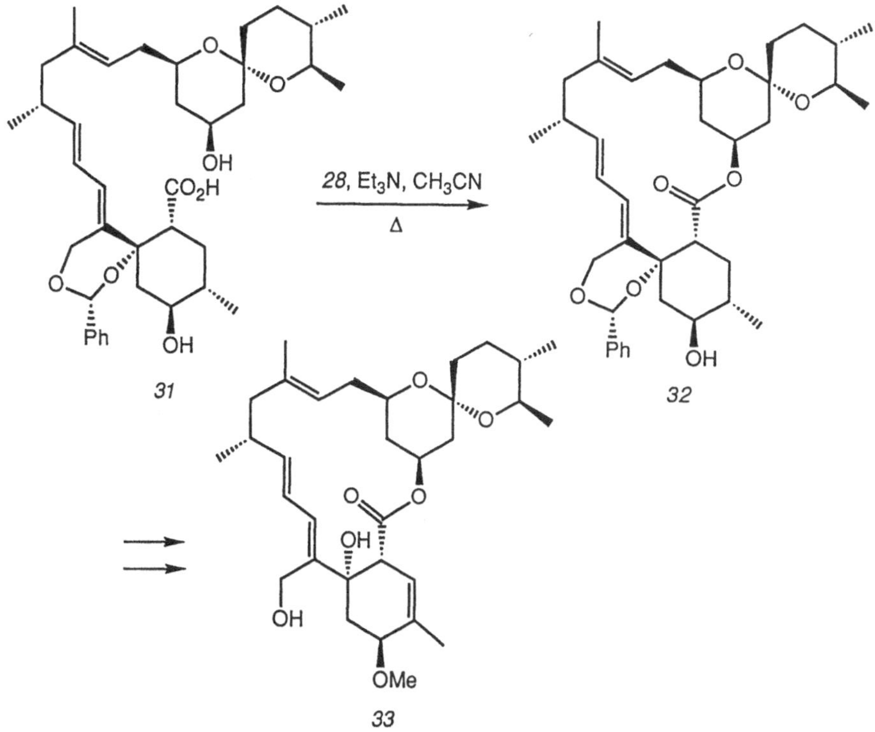

28, Et₃N, CH₃CN
Δ

Scheme 11

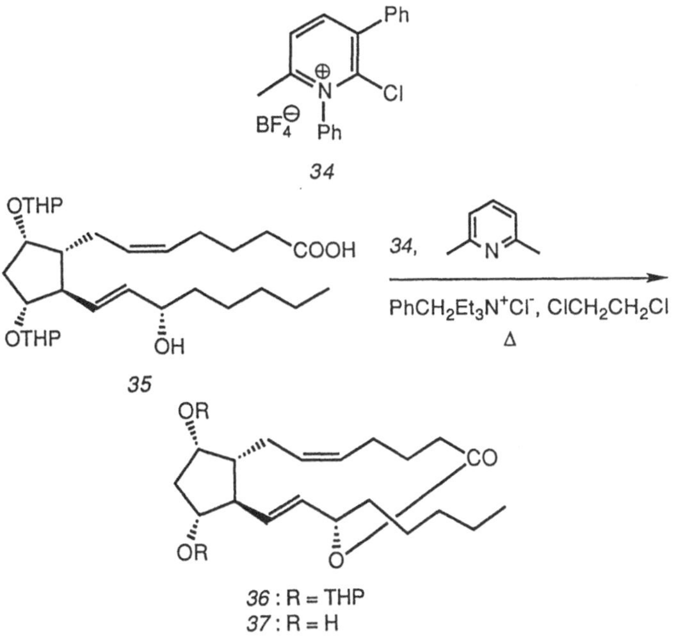

34

Scheme 12

36 : R = THP
37 : R = H

problem, a stable pyridinium salt, 2-chloro-6-methyl-1,3-diphenylpyridinium tetrafluoroborate (*34*) has been developed as an efficient reagent for lactonization. 2,4,6-Triphenylpyridine or 2,6-dimethylpyridine, instead of triethylamine, is used as a base and in addition benzyltriethylammonium chloride is necessary to achieve lactonization. Prostaglandin $F_{2\alpha}1,15$-*lactone* (*37*) was synthesized by this method [25]. Thus, as shown in Scheme 12, a 1,2-dichloroethane solution of the hydroxy acid *35* was added to a 1,2-dichloroethane solution of *34*, 2,6-dimethylpyridine and benzyltriethylammonium chloride under reflux over 4.25 h. After reflux for another 1 h, the lactone *36* was obtained in 91% yield.

Meanwhile, Mukaiyama et al. [26] developed another similar procedure, by the use of 6-phenyl-2-pyridyl esters, for the synthesis of macrocyclic lactones. This method has been used in the synthesis of (+)-*ricinelaidic acid lactone* (*40*) [27]. Thus, as shown in Scheme 13, a mixture of 6-phenyl-2-pyridone, 2-chloro-1-methylpyridinium iodide (*28*), and triethylamine in dichloromethane was stirred at room temperature for 1 h. To this solution was added a dichloromethane solution of the hydroxy acid *38* and triethylamine under reflux over 6 h to give the activated ester *39* in 99% yield. A dichloromethane solution of *39* was added to a *p*-toluenesulfonic acid solution in dichloromethane under reflux over 11 h. Acid-induced lactonization led to the macrocycle *40* in 96% yield!

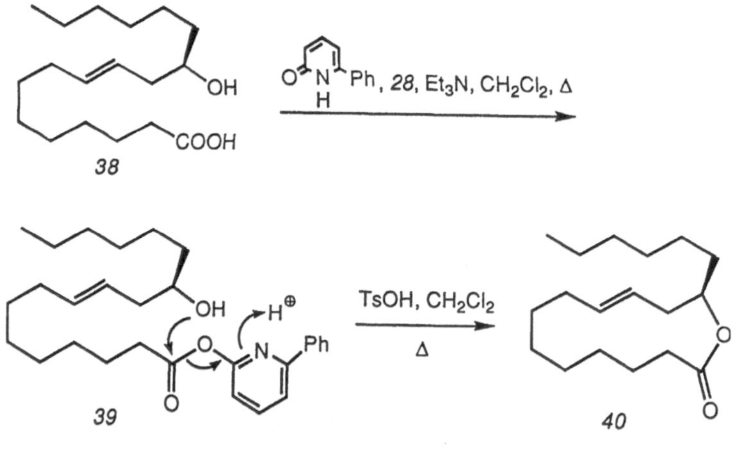

Scheme 13

2.3 The Masamune Method

In the total synthesis of the macrolide antibiotic *methymycin* (*49*), Masamune and coworkers [28] developed a new macrolactonization method which makes use of the electrophilicity of Hg(II) toward bivalent sulfur. It involves the *S-t*-butyl thiolester *44* of the hydroxy acid *41* and employs mercuric trifluoroacetate as an activating agent. The required *S-t*-butyl thiolester *44* can be prepared in high yields from the corresponding hydroxy acid *41* and thallous 2-methylpropane-2-thioate (TlSBu^t) via the acid chloride *42* or the mixed phosphoric anhydride *43* (Scheme 14). The question of whether the reaction proceeds via the suggested mercury complex *45* or through the intermediacy of a mixed trifluoroacetic anhydride *46*, or both, has not been fully clarified [7a]. An advantage of this method is that the S-*t*-butyl thiol ester group can serve as a protecting group and it can be introduced at an early stage of synthesis.

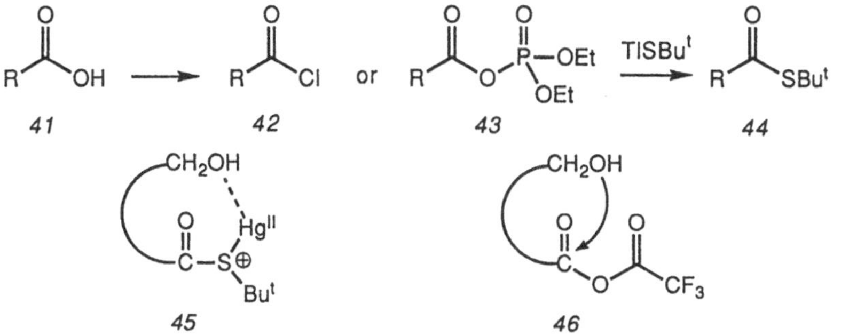

Scheme 14

In Masamune's *methymycin* synthesis [28b], the S-*t*-butyl thioester *47* upon treatment with 2 equiv of mercuric trifluoroacetate in acetonitrile solution (0.01 M) at room temperature for 1 h afforded the lactone *48* in up to 30% yield (Scheme 15).

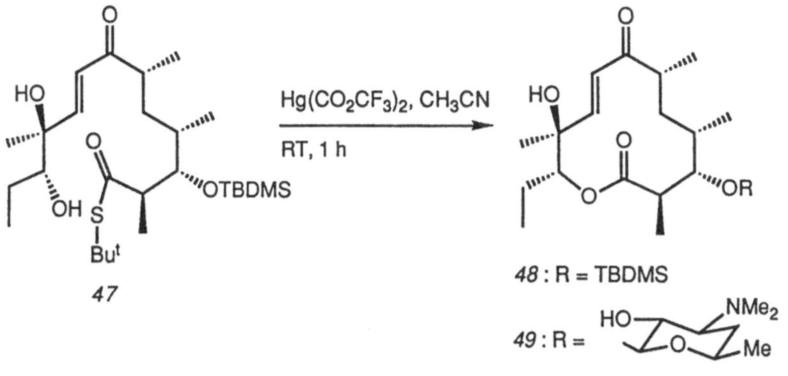

Scheme 15

Huang and Meinwald applied the Masamune method to the synthesis of the 11-membered lactonic pyrrolizidine alkaloid *crobarbatine acetate*. However, mercuric trifluoroacetate, a mixture of mercuric chloride and cadmium carbonate, and copper(I) trifluoroacetate all failed to lactonize the thiol ester. In all cases the starting material was quantitatively recovered. Finally copper(I) trifluorome- thanesulfonate-benzene complex was found to effect the crucial lactonization in reasonable yield [29].

Masamune et al. [30] found that benzenethiol ester is also suitable for ion lactonization. Tatsuta et al. [31] applied this method to the synthesis of the antibiotic A26771B (53). Thus, seco acid 50 was treated with diethylphos- phorochloridate and triethylamine in THF for 3 h to give the corresponding mixed phosphoric anhydride, which was in turn converted to the benzenethiol ester with thallium benzenethioate in 93% yield from 50. After deacetonation with di- fluoroacetic acid, the resultant triol 51 was treated with Na_2HPO_4 and $AgCO_2CF_3$ in benzene at 70 °C for 3 days and the 16-membered lactone 52 was obtained in 10% yield (Scheme 16)! The medicinally important antibiotics, carbomycin B and josamycin, have also been synthesized using this cyclization procedure [32].

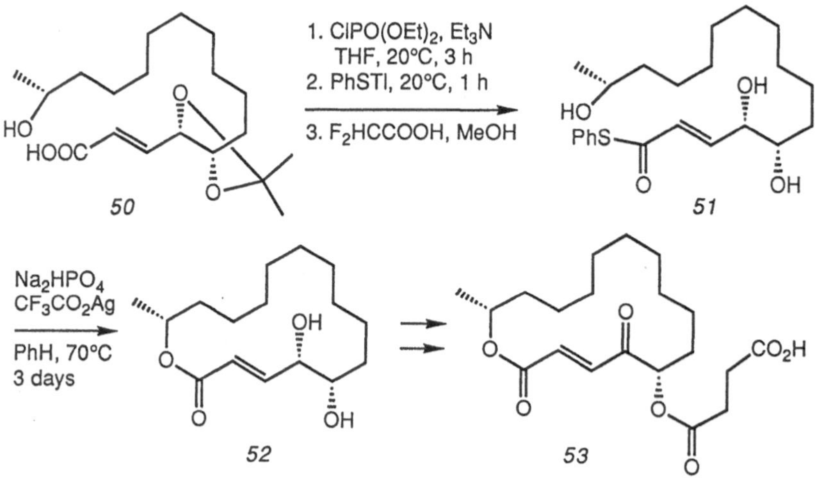

Scheme 16

2.4 Mixed Anhydride Methods

2.4.1 Mixed Pivalic Anhydride

This procedure has been used by Roush and Blizzard [33] in the synthesis of the macrocyclic mycotoxin verrucarin J (55). Thus, seco acid 54 was treated with 2 equiv of pivaloyl chloride and 3 equiv of triethylamine in dichloromethane (0.01 M) and the resultant mixed anhydride was treated in situ with 4-pyrrolidinopyridine (4-PP) to effect the ring closure at 23 °C. Verrucarin J (55) was obtained in up to 60% yield (Scheme 17). The mixed pivalic anhydride method has also been applied, e.g. to the synthesis of verrucarin B [34] and 4-epiverrucarin A [35] as the key cyclization step.

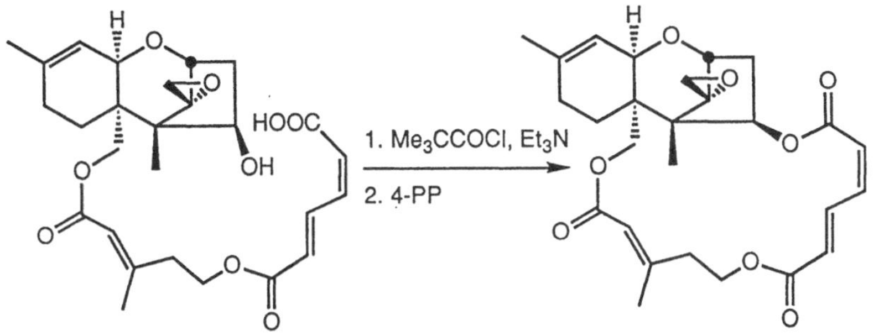

54

55

Scheme 17

2.4.2 Mixed 2,4,6-Trichlorobenzoic Anhydride

This procedure was developed by Yamaguchi and coworkers [36]. Yamada and coworkers [37] applied it to the synthesis of the macrocyclic pyrrolizidine alkaloids *(−)-integerrimine* and *(−)-monocrotaline*. For instance, in the synthesis of *(−)-integerrimine* (*58*) [37a], treatment of seco acid *56* with 1.1 equiv of 2,4,6-trichlorobenzoyl chloride and 4 equiv of triethylamine in THF at room temperature for 2 h afforded a mixed anhydride which was slowly added to refluxing toluene containing 6 equiv of 4-dimethylaminopyridine (DMAP) over 1.5 h followed by refluxing for 2 h. The cyclized product *57* was obtained in 75% yield (Scheme 18).

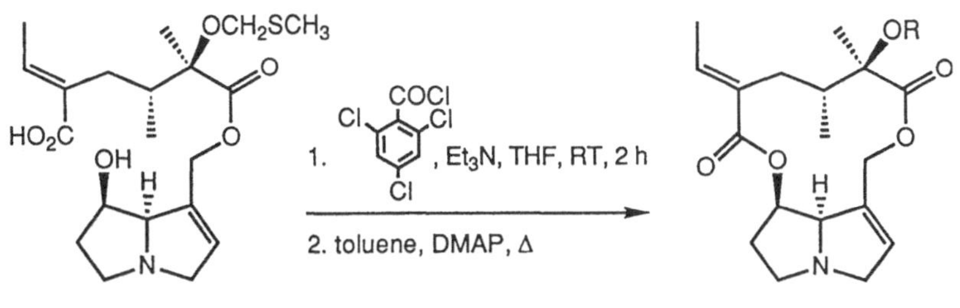

56

57: R = CH$_2$SCH$_3$
58: R = H

Scheme 18

2.4.3 Mixed 2,6-Chlorobenzoic Anhydride

Zwanenburg and coworkers [38] modified the Yamaguchi procedure in the synthesis of the 12-membered macrolide *patulolide C* (*61*). Mixed 2,6-chlorobenzoic anhydride, instead of mixed 2,4,6-trichlorobenzoic anhydride, was used.

Thus, as shown in Scheme 19, treatment of the hydroxy acid *59* with the commercially available 2,6-dichlorobenzoyl chloride, triethylamine, and DMAP in toluene at 100 °C gave the cyclized product *60* in 67% yield.

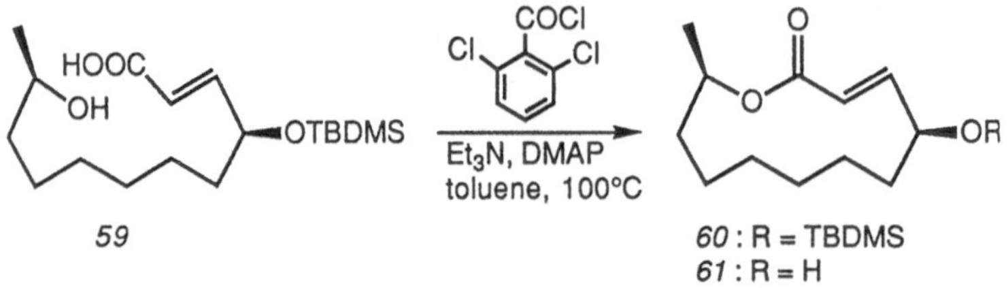

Scheme 19

2.4.4 Mixed Phosphoric Anhydride

In the particular case of the synthesis of the 14-membered polyoxomacrolide *narbonolide*, direct cyclization through the activation of thiol ester failed and then the mixed phosphoric anhydride derived from diphenylphosphorochloridate was developed to effect cyclization by Masamune and coworkers [39]. Fukumoto et al. [40] applied this procedure to the synthesis of the macrocyclic alkaloid *vertaline* (*63*). Thus, seco acid *62* was treated with diphenyl phosphorochloridate and triethylamine followed by refluxing in benzene in the presence of DMAP at high dilution afforded *63* in 54% yield (Scheme 20).

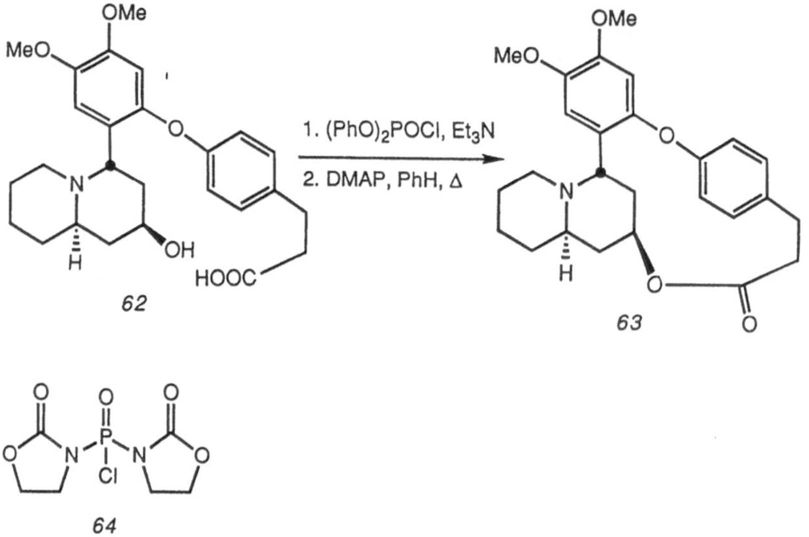

Scheme 20

2.4.5 N,N-Bis(2-oxo-3-oxazolidinyl)phosphordiamidic Chloride

Another useful reagent for macrolactonization via mixed phosphoric anhydride is N,N- bis(2-oxo-3-oxazolidinyl)phosphordiamidic chloride (BOP-Cl, 64), which was used by Corey and coworkers [41] in the first synthesis of *aplasmomycin* (67), a novel boron-containing macrocyclic antibiotic. As shown in Scheme 21, the linear precursor 65 was treated with 3 equiv of BOP-Cl (64) and 7 equiv of triethylamine in dichloromethane at 23 °C for 6 h to give the dilactone 66 in 71% yield.

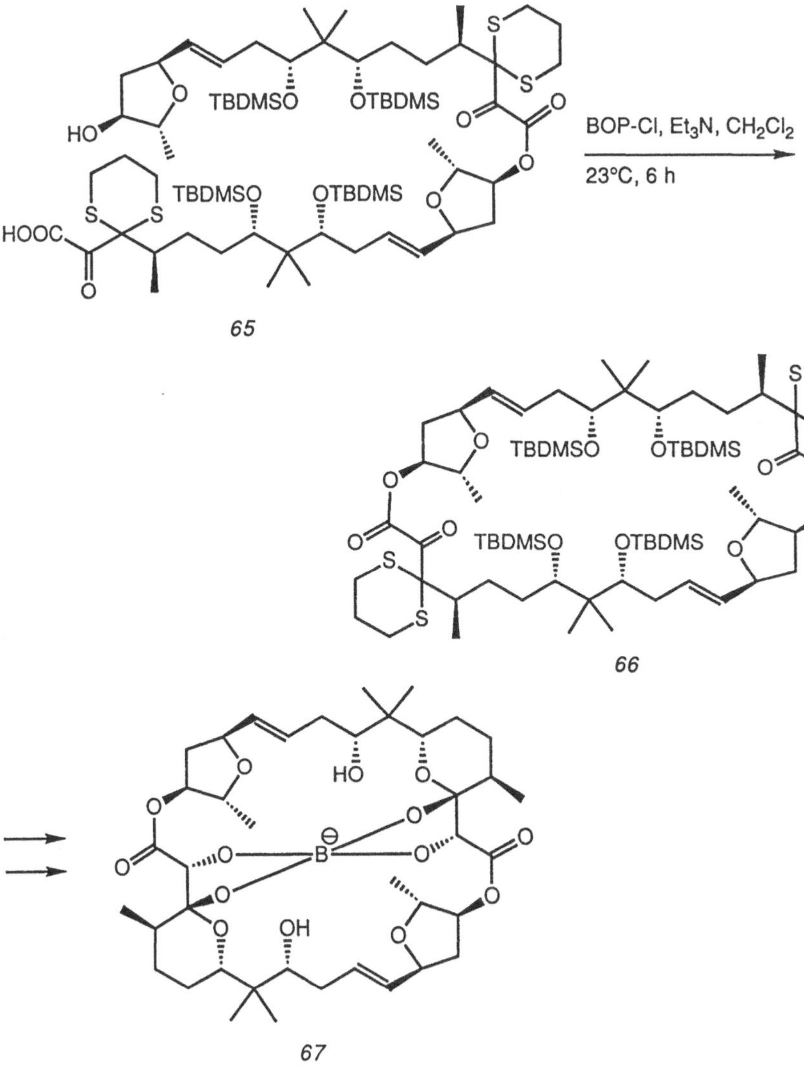

Scheme 21

2.4.6 Mixed Trifluoroacetic Anhydride

The mixed trifluoroacetic anhydride method played an important role in the first synthesis of the 14-membered macrolide *zearalenone* (*70*) by Taub et al. [42]. Hydroxy acid *68* was treated with trifluoroacetic anhydride (TFAA) and lactone *69* was obtained in 15% yield (Scheme 22).

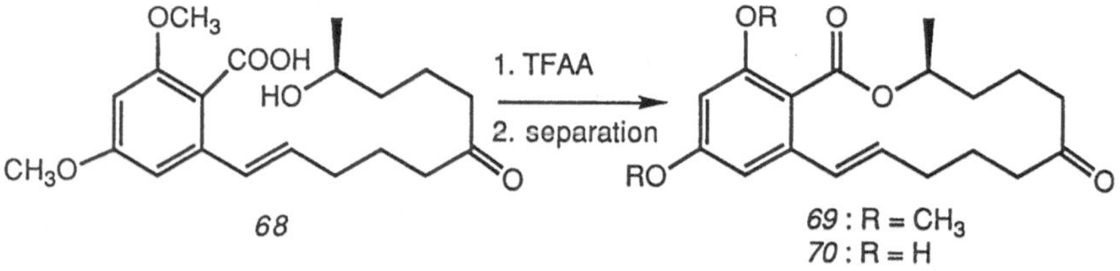

Scheme 22

2.4.7 Mixed Sulfonic Anhydride

In a model study for *erythronolide B*, White et al. [43] synthesized lactone *72* from hydroxy acid *71* via a mixed sulfonic anhydride. Thus, as shown in Scheme 23, exposure of *71* to *p*-toluenesulfonyl chloride and triethylamine in benzene followed by chromatography afforded *72* in 52% yield.

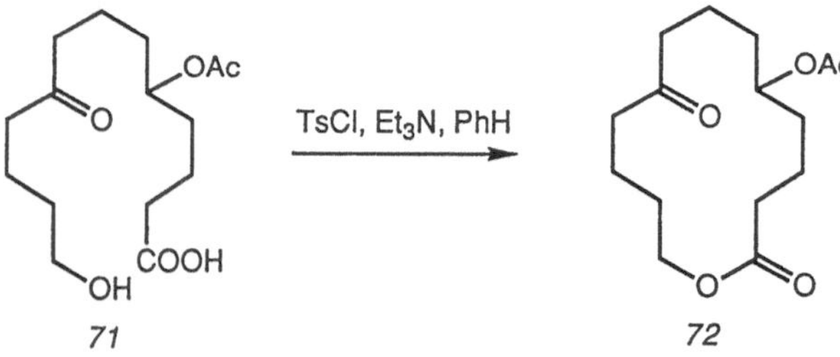

Scheme 23

2.5 Cyclization Mediated by the Mitsunobu Reaction

Macrolactonization can also be achieved by the Mitsunobu reaction [44] with inversion of the configuration of the alcohol. The reaction principle and mechanism are demonstrated in Scheme 24. Addition of triphenylphosphine to diethyl azodicarboxylate (DEAD, *73*) forms a quaternary phosphonium salt *74*, which is protonated by hydroxy acid *11*, followed by phosphorus transfer from nitrogen to oxygen yielding the alkoxyphosphonium salt *76* and diethyl hydrazinedicarboxylate *75*. Then, an intramolecular S_N2 displacement of the important intermediate *76* results in the formation of the lactone *15* and triphenylphosphine oxide.

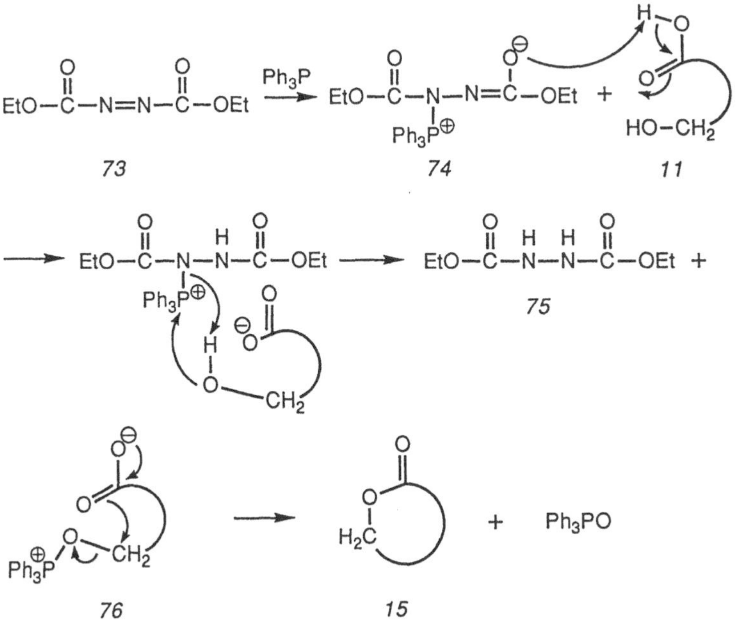

Scheme 24

Seebach et al. [45] employed this procedure in the synthesis of the germination self-inhibitor *gloeosporone* (*79*). Treatment of seco acid *77* with 2 equiv of DEAD and 2 equiv of triphenylphosphine in benzene (0.005 M) for 10 min afforded the

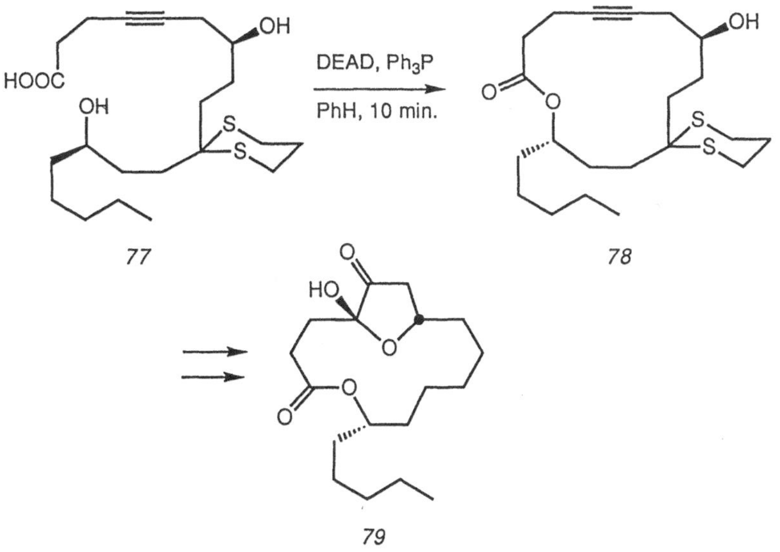

Scheme 25

macrolactone *78* in 67% yield (Scheme 25). The Mitsunobu reaction has been widely used for macrolactonization. Another example is the synthesis of the macrolide toxin *(+)-latrunculin B* [46].

A further reaction mechanistically similar to the Mitsunobu reaction as shown in Scheme 26, with the use of *N,N*-dimethylformamide dineopentylacetal (*80*), can also be employed for macrolactonization [47]. Takei and coworkers [48] applied it to the synthesis of the macrocyclic antibiotic *A26771B* (*53*). As shown in Scheme 27, treatment of the linear precursor *82* with *80* in refluxing dichloromethane for 7 h afforded the lactone *83* (39% yield).

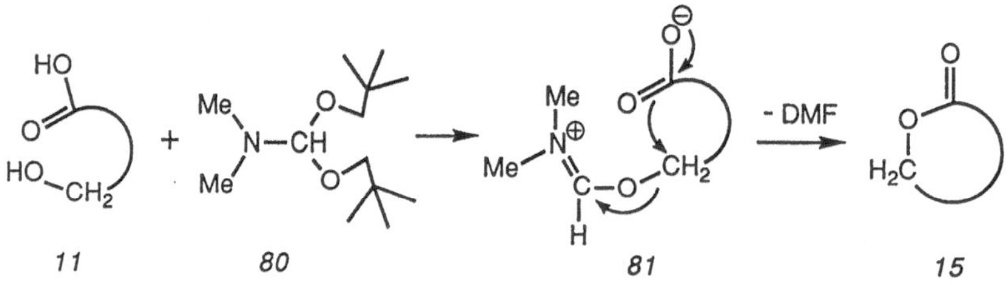

Scheme 26

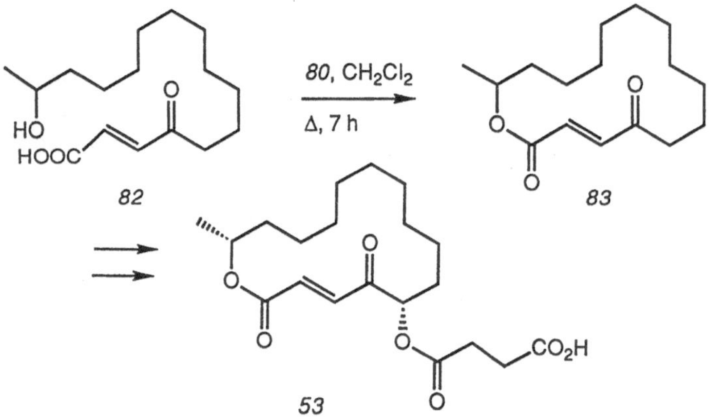

Scheme 27

2.6 Cyclization Mediated by Carbodiimide

Dicyclohexylcarbodiimide (DCC) is a well known reagent for peptide bond formation. It is also a useful activator for macrolactonization. However, DCC alone does not give satisfactory results. Boden and Keck [49] found that the combination of DCC and a proper base, normally DMAP, works perfectly. Hanessian et al. [50] used the DCC-DMAP system in the synthesis of the antibiotic *(+)-avermectin B*$_{1a}$ (*86*). As shown in Scheme 28, macrolactone *85* was obtained in 30% yield from the linear hydroxy acid *84*. A couple of other natural products [51], such as the macrolide *milbemycin E* and the cyclodepsipeptide *geodiamolide A* have been recently synthesized using this procedure as the key cyclization step.

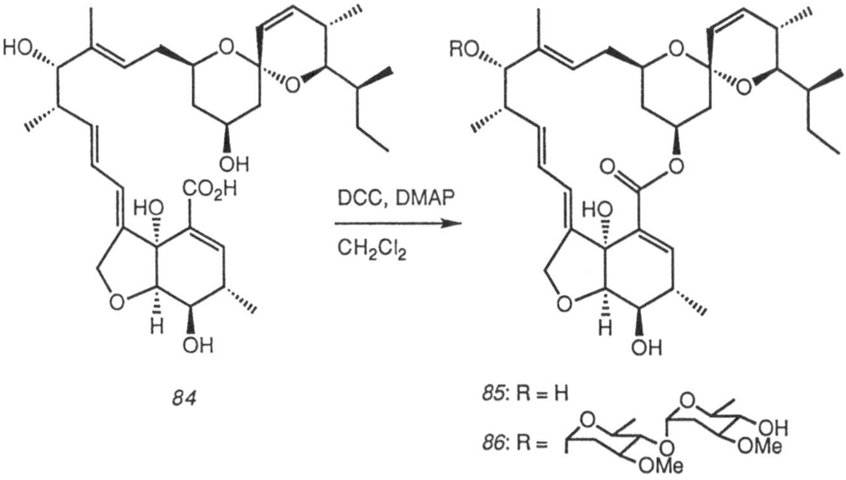

Scheme 28

Another carbodiimide, 1-cyclohexyl-3-(2-morpholinoethyl)carbodiimide methyl *p*-toluenesulfonate, has been used for macrocyclization in the synthesis of *milbemycin* β_3 [52].

2.7 Activated Ester Methods

In fact, many methods discussed above, such as that of Corey and Mukaiyama, in the final analysis, fall into the activated ester class. Several other "real" activated ester methods are given below.

2.7.1 Imidazole Ester

In the first synthesis of the dilactonic antibiotic *pyrenophorin* (*89*) by Raphael and coworkers [53], imidazole ester was used to achieve cyclization. As shown in Scheme 29, the linear precursor *87* was converted to the imidazolide by the action of diimidazol-1-yl ketone (Im$_2$CO) and followed by 1,5-diazabicyclo[4.3.0]non-5-ene (DBN) induced cyclization to give the dilactone *88* in 60% yield.

2.7.2 "Push-pull" Acetylene Derived Ester

In the synthesis of *brefeldin A* (*92*) by Gais and Lied [54], seco acid *90* was lactonized to *91* with the "push-pull" acetylene *93* (Scheme 31) in 71% yield (Scheme 30). The mechanism is outlined in Scheme 31.

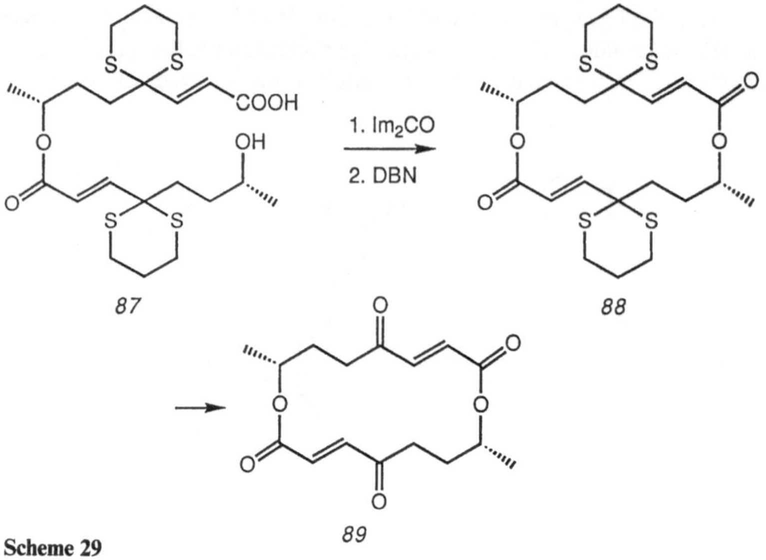

Scheme 29

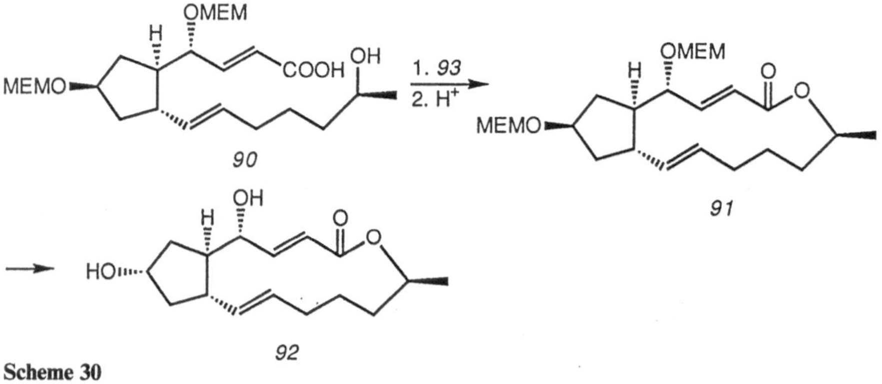

Scheme 30

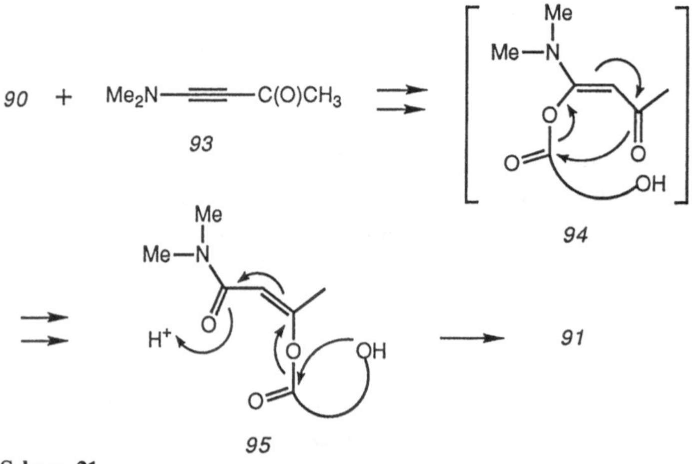

90 + Me$_2$N—≡≡—C(O)CH$_3$

93

Scheme 31

2.7.3 (Methylsulfonyl)methyl Ester

Narasaka et al. [55] developed a macrocyclization method utilizing (methyl-thio)methyl (MTM) group as an activable protecting group of carboxylic acid. As shown in Scheme 32, the macrocyclic pyrrolizidine alkaloid *integerrimine* (*58*) has been synthesized using this method as a key step. Thus, hydroxy MTM ester *96* was oxidized with hydrogen peroxide-Mo(VI) to (methylsulfonyl)methyl ester *97*, which was treated with 1 equiv of *n*-butyllithium or (triphenylmethyl)lithium (TPMLi) in THF, yielding the lactone *98* in 40% yield.

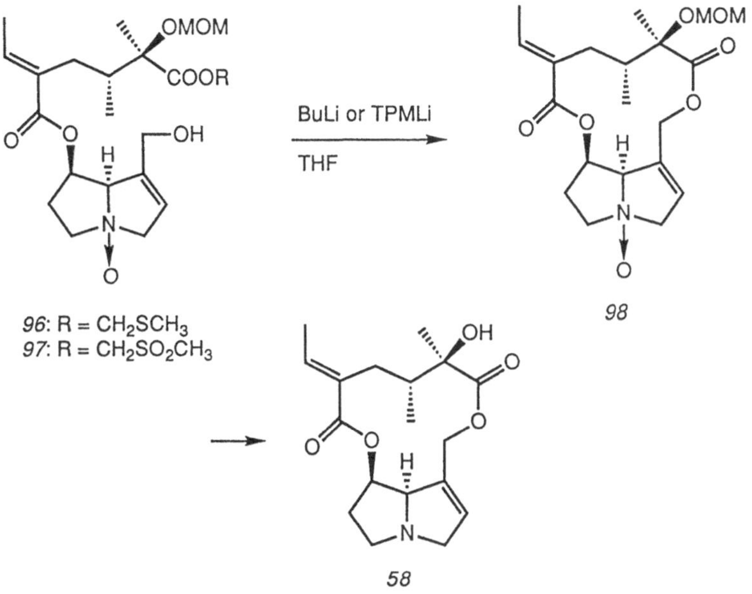

96: R = CH$_2$SCH$_3$
97: R = CH$_2$SO$_2$CH$_3$

Scheme 32

2.8 Cyclization Mediated by Organotin Reagents

In 1980, Hanessian and coworkers [56] developed an organostannyl oxide-induced lactonization method which can be used in the synthesis of macrocycles. For instance, 15-hydroxypentadecanoic acid has been converted to the corresponding lactone by treating with 0.1 equiv of *n*-Bu$_2$SnO in refluxing mesitylene for 20 h in 60% yield. As demonstrated in Scheme 33, the mechanism of this reaction is envisioned to involve complexation, loss of water, and cleavage of the catalyst *n*-Bu$_2$SnO with incident lactone formation. This procedure is promising but the reaction is reversible, heavily depending on the nature of the substrate, the concentration, and the type of organotin oxide used. A similar method, which also depends on the coordinative ability of organotins, has been reported by another group simultaneously [57].

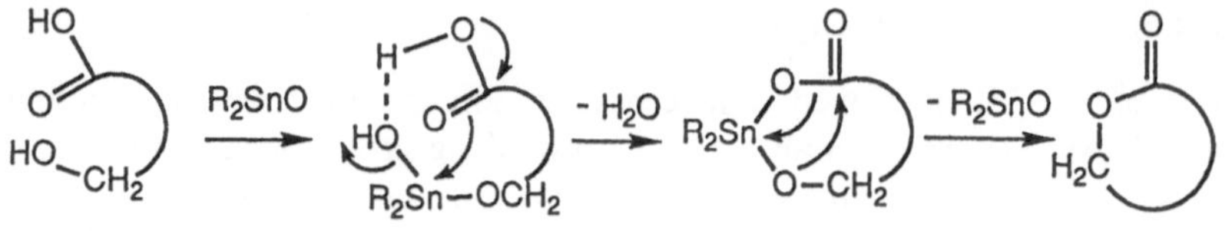

Q. Meng and M. Hesse

Scheme 33

2.9 Other Methods

2.9.1 Intramolecular Ketene Trapping

Recently, Boeckman and Pruitt [58] demonstrated the use of dioxolenones as precursors of β-acyl ketenes, which can be thermally generated under mild neutral conditions in the absence of other nucleophiles and trapped intramolecularly by the hydroxy group to afford good yields of macrocyclic lactones. The 16-membered macrolide (−)-kromycin (102) has been synthesized in this way by thermolysis of the β-acyl ketene precursor 101 in toluene at high dilution (0.0001 M) in 70% yield (Scheme 34).

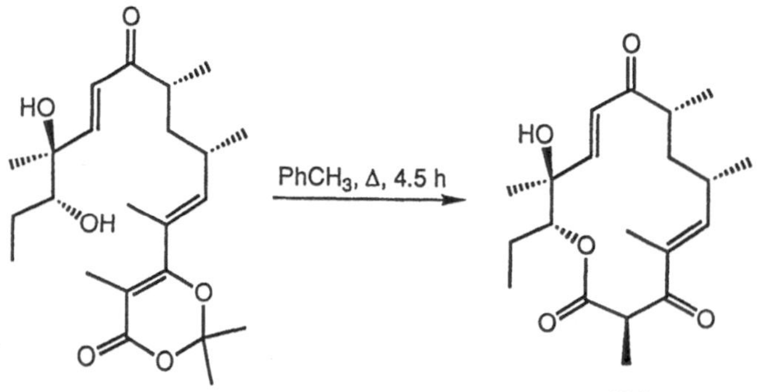

Scheme 34

2.9.2 Oxazole as Masked Activated Ester

Oxazoles may be used as masked forms of activated carboxylic esters since they readily form tertiary amides on reaction with singlet oxygen. Wasserman et al. [59] applied this principle to the synthesis of *di-O-methylcurvalarin* (105). As shown in Scheme 35, the oxazole containing precursor 103 was readily converted in 31% yield by photooxygenation to the corresponding tertiary amide 104, which underwent acid-catalyzed cyclization in refluxing benzene to yield 105 (30% yield).

128

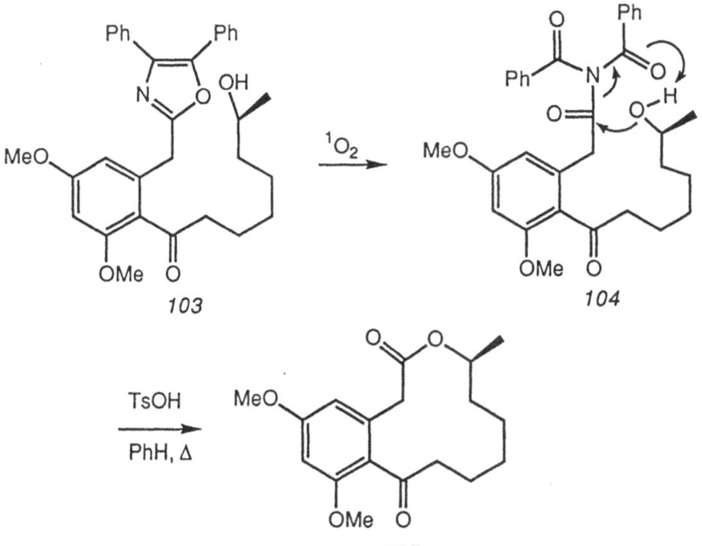

103 104 105

Scheme 35

2.9.3 Activation of Hydroxy Group

The hydroxy part of a hydroxy acid can also be activated for macrolactonization. Vedejs et al. [60] applied such a strategy to the synthesis of the macrocyclic pyrrolizidine alkaloid *monocrotaline* (*108*). Thus, the seco acid derivative *106* was first mesylated with MsCl/Et_3N in dichloromethane, and the crude product was added over 3 h to an excess of tetrabutylammonium fluoride trihydrate in acetonitrile at 34 °C to effect ring carboxy deprotection and ring closure to give *107* in 71% yield (Scheme 36). It has been noted that the active intermediate of this kind of lactonization may be an allylic chloride rather than a mesylate [61 a]. In addition, an intramolecular nucleophilic displacement process of chloride from an α-chloro ketone moiety by a remote carboxylate has been recently reported as an efficient approach to macrocyclic keto lactones [61 b].

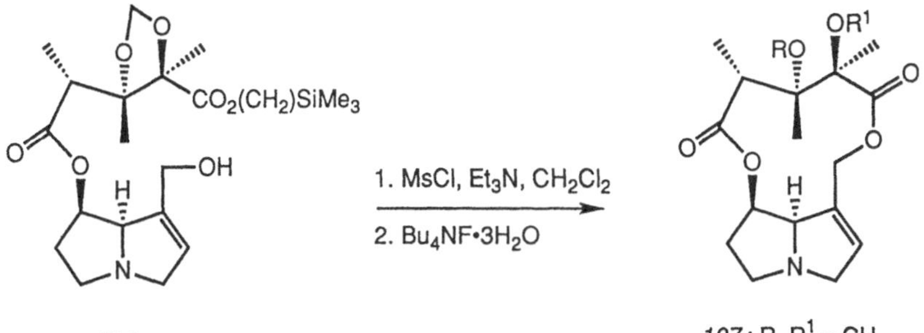

106

107: R, R^1 = CH_2
108: R = R^1 = H

Scheme 36

2.9.4 Oxidation of Hemiacetal

In the synthesis of *verrucarin J* (*55*) by Fraser-Reid and coworkers [62], triol *109* was treated with pyridinium dichromate (PDC) for 3 days, resulting in oxidative cleavage of the adjacent diol to the corresponding aldehyde and further oxidation of the presumed cyclic hemiacetal intermediate gave *55* in 50% yield (Scheme 37).

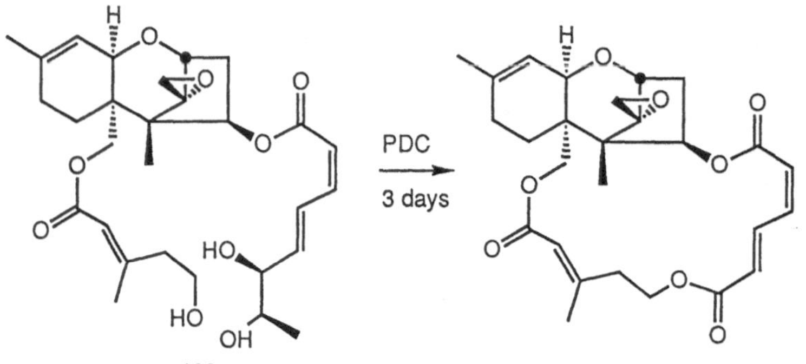

Scheme 37

2.10 Stereochemistry, Another Important Factor

The stereoelectronic theory of organic chemistry has been established on the basis of experimental evidence [63]. Most types of organic reactions depend upon the relative stereochemistry of the particular electron pairs concerned. Only when the electron pairs are properly oriented in space the reaction takes place. This principle also applies to ring closure reactions.

Woodward et al. [64] extensively investigated the structure/reactivity relationships of macrolactonization in the course of the synthesis of *erythromycin*, one of the biologically most important mold metabolites. They found that the proper functionalization of a linear precursor is critical for the successful lactonization. For example, all attempts to lactonize seco acids *110* and *111* using several of known methods including Corey's and Masamune's were uniformly unsuccessful. However, as shown in Scheme 38, subjection of seco acid *112* to the Corey method furnished lactone *113* in 70% yield. The conclusion is that certain structural features of the linear precursor are required for efficient lactonization: a) (*S*) configuration at C(9) and b) "cyclic" protecting groups at C(3)/C(5) and C(9)/C(11). These structural characteristics arise from conformational requirements for lactonization. In particular, the required pattern of cyclic protecting groups in a (9*S*) precursor may assist in adopting a conformation sufficiently resembling that of the corresponding lactone to facilitate ring closure.

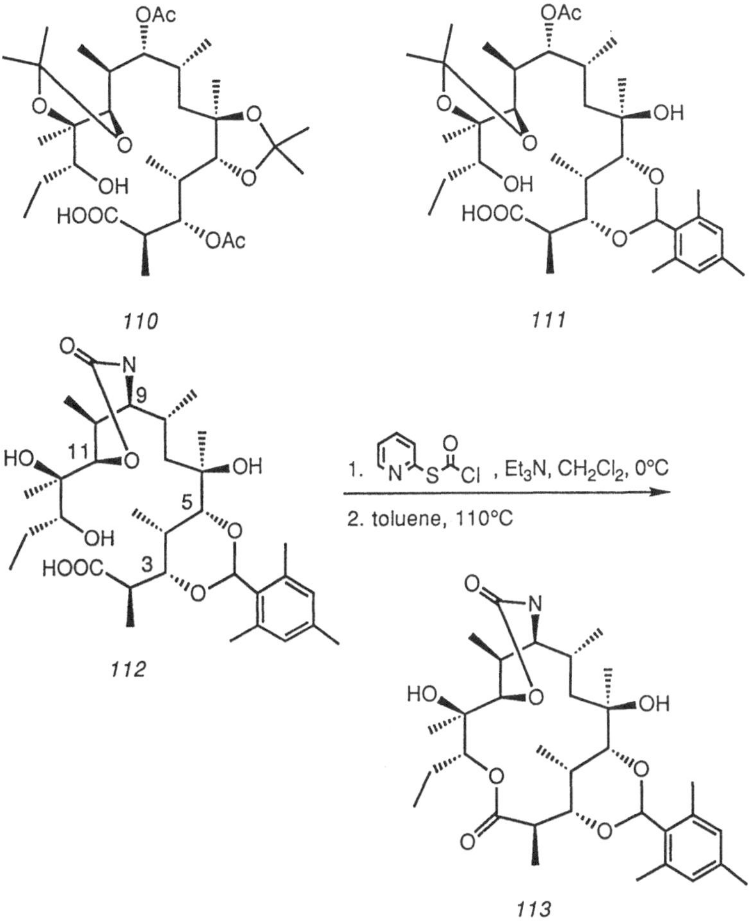

Scheme 38

The same structural requirements have been recently recognized by Stork and Rychovsky [65] in the course of the total synthesis of *(+)-(9S)-dihydroery-thronolide A*, though a different ring closure method was employed. In addition, a further structural requirment for efficient lactonization has been found. Seco acids *114* and *115* failed to lactonize while seco acid *116* was cyclized using the DCC-DMAP method to lactone *117* in 64% yield (Scheme 39). The explanation is as follows: The conformation in solution of the C(8)–C(11) portion of *117* is shown in Fig. 1, in which a 1,3-diaxial interaction is present between R^2 and C(8). Therefore, when R^2 is an alkyl group, the resulting severe interaction should make cyclization of the seco acid very unfavorable.

Yonemitsu and coworkers [66] have observed the same phenomenon in the synthesis of *erythronolide A*. For the conformationally favorable precursor even the high dilution technique was not necessary and the yield of a 14-membered lactone was as high as 92%.

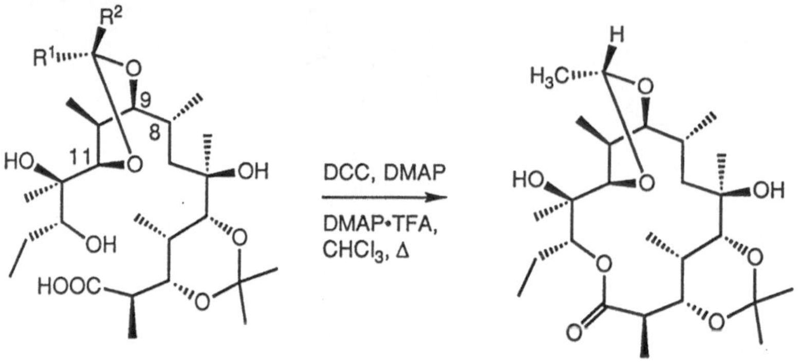

114 : R^1 = R^2 = CH$_3$
115 : R^1 = H, R^2 = CH$_3$
116 : R^1 = CH$_3$, R^2 = H

Scheme 39

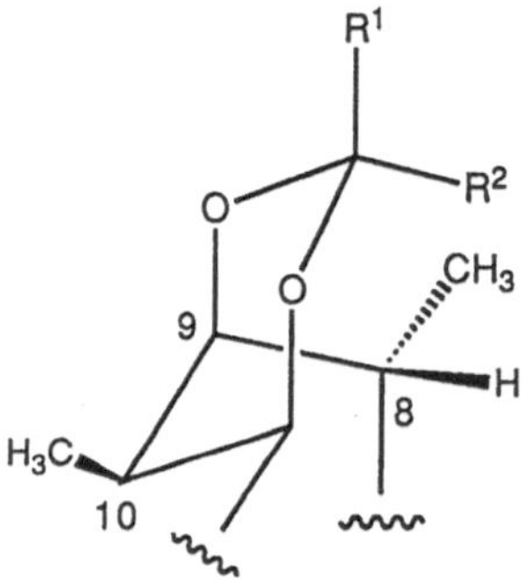

Fig. 1. The C(8)–C(11) portion of *117*

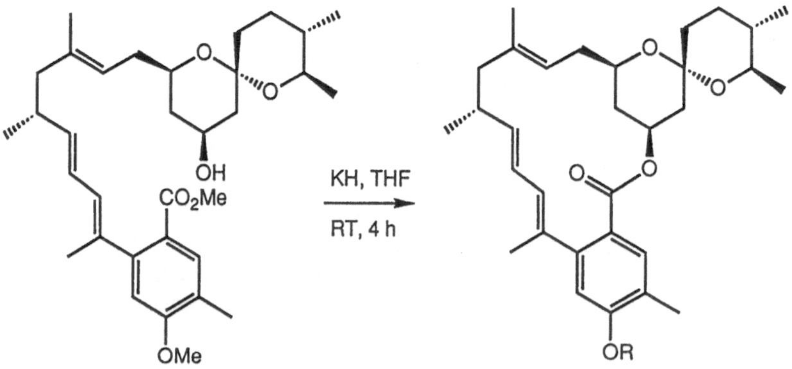

118

119 : R = CH$_3$
120 : R = H

Scheme 40

Favorable configuration and efficient activation complement with each other. If the configuration is favorable enough even a methyl ester is as good as an activated ester. For instance, as shown in Scheme 40, in the synthesis of the macrolide antibiotic *milbemycin* β_3 (*120*) by Smith et al. [67], alcohol ester *118* was lactonized by potassium hydride to *milbemycin* β_3 methyl ether (*119*) in high yield (more than 76%).

3 Methods Involving Lactamization

Amide or lactam bond formation is the major issue of peptide synthesis. The synthesis of regular macrocyclic peptides, which is beyond this review, can be easily achieved using well-known procedures [68]. However, in the case of irregular-peptidal and non-peptidal macrocyclic natural products, where no intramolecular hydrogen bonds are available to facilitate the ring closure, cyclization by lactam formation is still challenging. Though traditional peptide synthesis procedures [68] can sometimes be employed, usually more efficient activation methods are required. Some recent developments are dicussed below.

3.1 Mixed Anhydride Methods

3.1.1 Mixed Phosphoric Anhydrides

N,N-Bis(2-oxo-3-oxazolidinyl)phosphorodiamidic chloride (BOP-Cl, *64*) has been found to be an efficient macrolactamization reagent [69]. Tatsuta and coworkers [69c] used it in the synthesis of the ansamycin antibiotic *rifamycin W* (*123*). As shown in Scheme 41, amino acid *121* was cyclized to *122* using 4 equiv of BOP-Cl and 10 equiv of diisopropylethylamine in toluene at 85 °C for 3 h. Upon deprotection and oxidation *123* was obtained in 30% yield from *121*.

More than a decade ago, Yamada and coworkers [70] developed two efficient reagents, diphenylphosphoryl azide (DPPA) and diethyl phosphorocyanidate (DEPC), for peptide done construction. The exact mechanism still remains speculative. For example, in the case of DPPA, the intermediacy of the carboxylic acid azide is tentative, but it seems attractive to consider a concerted process as shown in Figure 2 [70a].

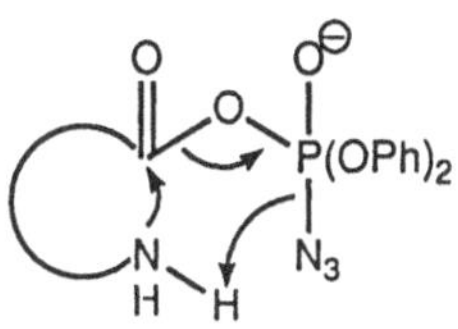

Fig. 2. A tentative mechanism of the DPPA-mediated ring closure

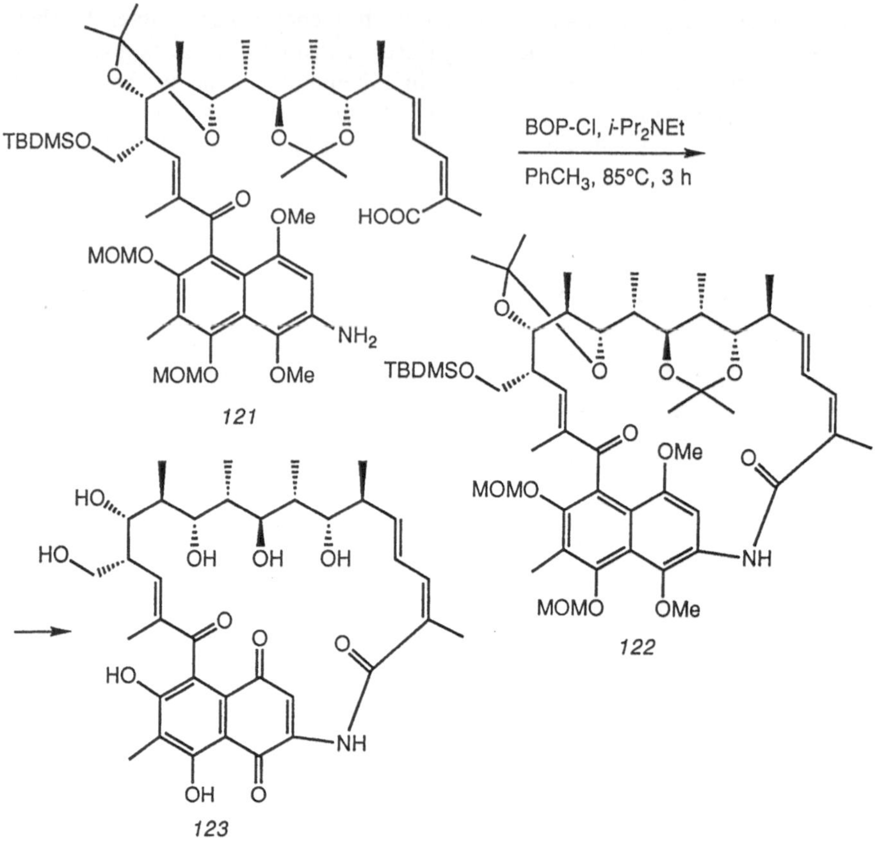

Scheme 41

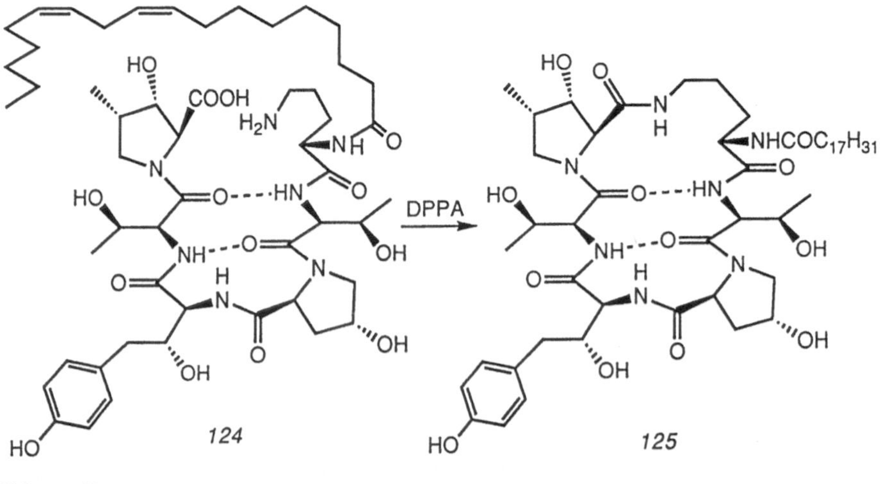

Scheme 42

Kurokawa and Ohfune [71] employed DPPA in the synthesis of the hexapeptide *echinocandin D* (*125*). As shown in Scheme 42, the cyclization of the linear peptide *124* was accomplished by DPPA to give *125* in 50% yield. There are more applications of DPPA in the synthesis of macrocyclic natural products [72]. DEPC is relatively less commonly used than DPPA. Kishi and coworkers [73] have successfully achieved the synthesis of the macrocyclic antibiotic *rifamycin* S using DEPC as a macrolactamization promoter.

3.1.2 Mixed Phosphonic Anhydride

Durette et al. [74] have achieved the total synthesis of the hexadepsipeptide antibiotic *L-156,602* (*128*) using the mixed phosphonic anhydride method as key macrolactamization step. As shown in Scheme 43, treatment of the linear depsipeptide *126* with *n*-propylphosphonic anhydride and DMAP in dichloromethane at high dilution afforded the macrocycle *127* in more than 57% yield.

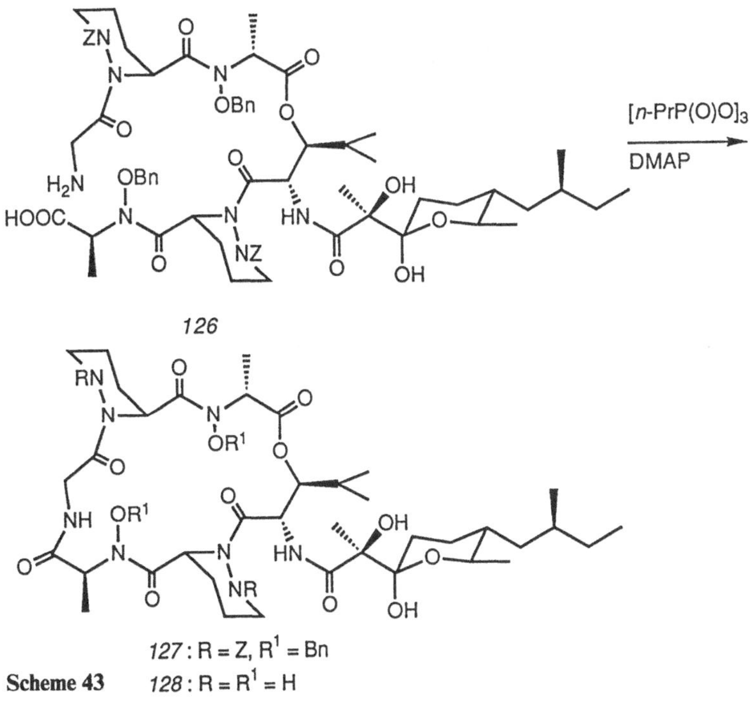
ZN
OBn
H2N
OBn
HOOC
NZ
OH
OH
[n-PrP(O)O]3
DMAP
126
RN
OR1
NH
OR1
NR
OH
OH
127: R = Z, R1 = Bn
128: R = R1 = H

Scheme 43

3.1.3 Mixed Sulfonic Anhydride

In the first synthesis of the potent antitumor agent *maytansine* (*131*) by Corey et al. [75], linear amino acid *129* was first converted to the soluble tetra-*n*-butylammonium salt and then slowly added to a solution of excess mesitylenesulfonyl chloride and diisopropylethylamine in benzene at 40 °C for 28 h to afford macrolactam *130* in 71% yield (Scheme 44).

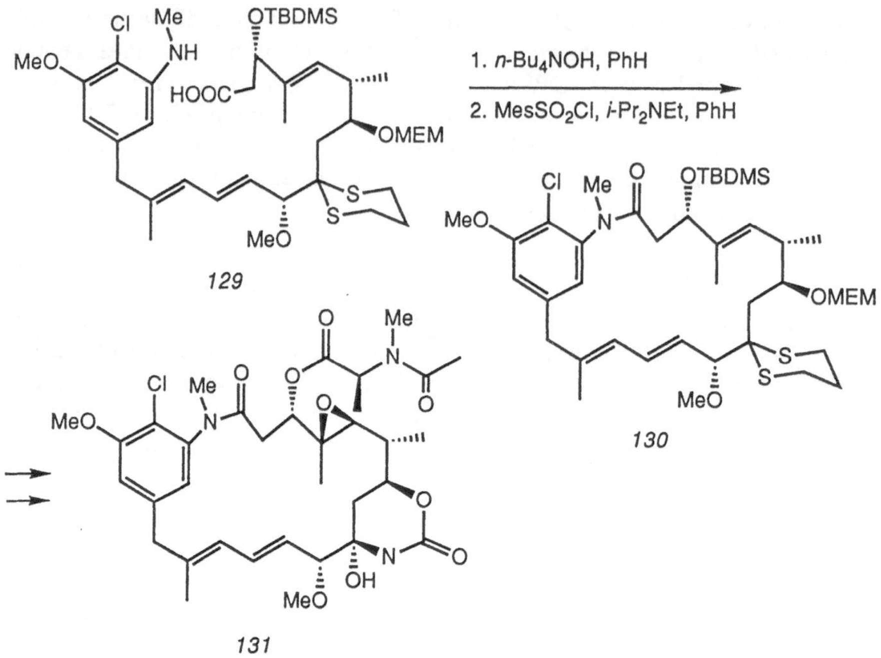

Scheme 44

Baker and Castro [69e] reported that DCC, DPPA, or DPEC did not induce the macrocyclic lactam closure in the synthesis of the antitumor antibiotic *(+)-macbecin I*, but it was achieved by mesitylenesulfonyl chloride.

3.1.4 Mixed Carbonic Anhydride

In the studies of the synthesis of the ansamycin antibiotic *rifamycin S (135)*, Corey and Clark [76] found numerous attempts to effect the lactam closure of the linear precursor *132* to *134* uniformly unsuccessful under a variety of experimental conditions, e.g. via activated ester with imidazole and mixed benzoic anhydride. The crux of the problem was associated with the quinone system which so deactivates the amino group to prevent its attachment to mildly activated carboxylic derivatives. Cyclization was achieved after conversion of the quinone system to the hydroquinone system. Thus, as shown in Scheme 45, treatment of *132* with 10 equiv of isobutyl chloroformate and 1 equiv of triethylamine at 23 °C produced the corresponding mixed carbonic anhydride in 95% yield. The quinone C=C bond was reduced by hydrogenation with Lindlar catalyst at low temperature. A cold solution of the hydroquinone was added over 2 h to THF at 50 °C and stirred for an additional 12 h at the same temperature. Oxidation with aqueous potassium ferricyanide afforded the cyclic product *134* in 80% yield. Kishi and coworkers [73] gained a similar result by using mixed ethyl carbonic anhydride.

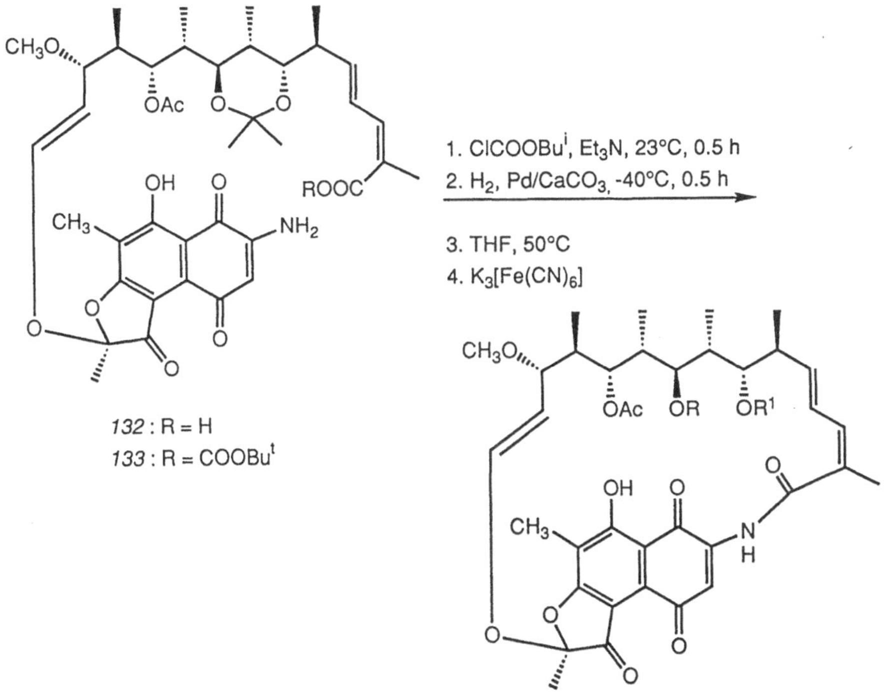

1. ClCOOBu^i, Et$_3$N, 23°C, 0.5 h
2. H$_2$, Pd/CaCO$_3$, -40°C, 0.5 h

3. THF, 50°C
4. K$_3$[Fe(CN)$_6$]

132 : R = H
133 : R = COOBu^t

134 : R, R^1 = C(CH$_3$)$_2$
135 : R = R^1 = H

Scheme 45

3.2 Activated Ester Methods

3.2.1 N-Hydroxysuccinimide Ester

This procedure has been long known in the peptide chemistry. Sakai and coworkers [77] used it in the synthesis of *indolactam V* (*139*), an active fragment of a group of potent tumor promotors. As shown in Scheme 46, precursor *136* was hydrolyzed and treated with N-hydroxysuccinimide (HOSu) and DCC in acetonitrile to produce the activated ester *137* in 57% yield. After deprotection the amino ester was treated with a weak base to afford lactam *138* in 64% yield. Another similar example is due to Nakatsuta et al. [78].

3.2.2 p-Nitrophenyl Ester

This is also a traditional peptide synthesis procedure. Joullié and coworkers [79] used it in the synthesis of the cyclopeptide alkaloid *dihydromauritine A* (*142*). As shown in Scheme 47, the linear precursor *140* was treated with p-nitrophenyl trifluoroacetate in pyridine to give the p-nitrophenyl ester. After cleavage of the Boc group, the amino ester was subjected to cyclization in dilute DMF in the presence of hydroxybenztriazole (HOBt) and diisopropylethylamine at 25 °C for 5 days. The cyclic product *141* was obtained in 10% yield only.

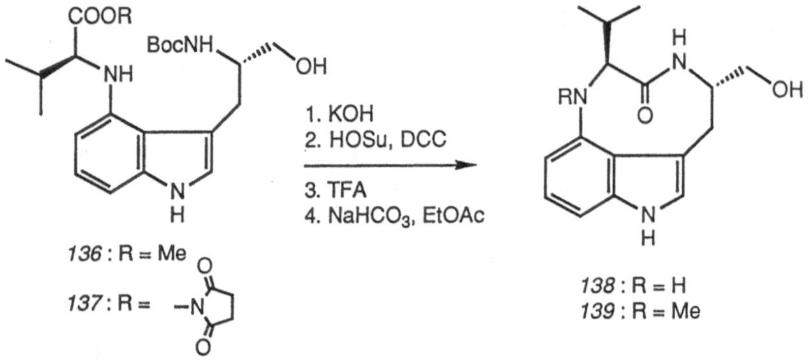

Scheme 46

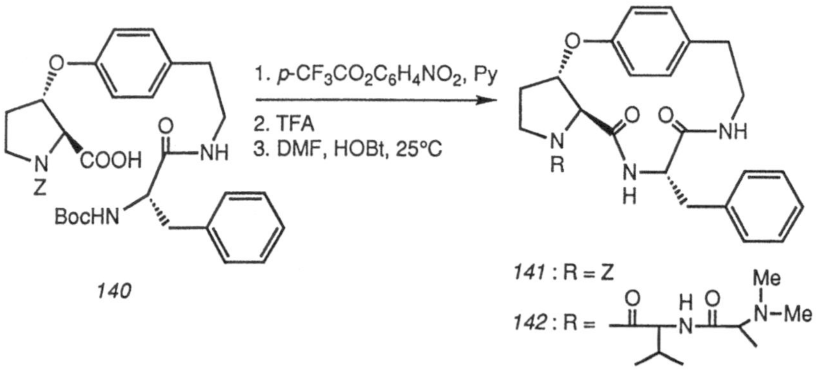

Scheme 47

3.2.3 Pentafluorophenyl Ester

Schmidt and coworkers [80] developed the pentafluorophenyl ester method for synthesizing cyclopeptides, particularly applicable for 13- and 14-membered ansa peptides. This procedure is superior to the *p*-nitrophenyl ester method in respect to short reaction time and easy separation of products. Evans and Ellman [81] applied this method to the synthesis of the cyclic tripeptide *K-13* (*146*). As shown in Scheme 48, the reaction of the linear precursor *143* with pentafluorophenol and DCC afforded pentafluorophenyl ester *144* in 87–93% yield. Then, under catalytic hydrogenation condition in the presence of a mild base and ethanol, *144* was cyclized to *145* in up to 70% yield. There are more applications of the pentafluorophenyl ester procedure in recent literature [82].

3.2.4 Imidazole Ester

In 1962, Wieland and Vogeler [83] found the aminolysis of carboxylic alkyl ester in a melt with imidazole to be a good method for peptide bond construction.

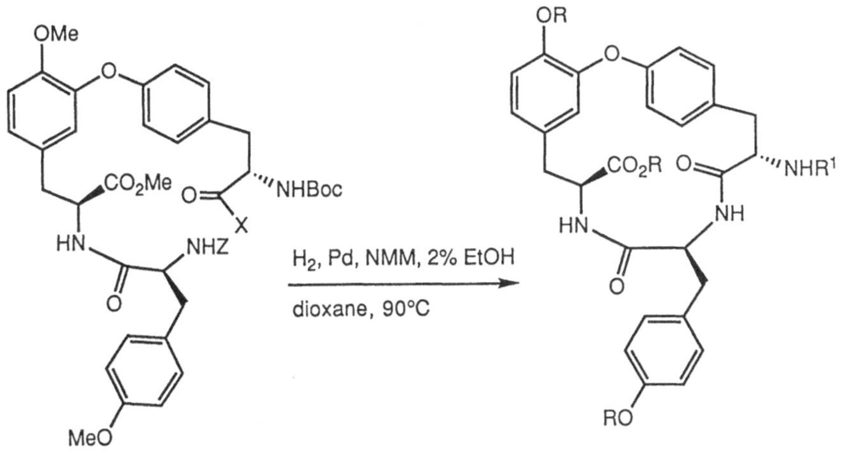

143 : X = OH
144 : X = OC_6F_5

145 : R = Me, R^1 = Boc
146 : R = H, R^1 = Ac

Scheme 48

Wälchli-Schaer and Eugster [84] applied this method to the synthesis of the spermidine alkaloid *dihydropalustrine* (*148*). Thus, as shown in Scheme 49, amino ester *147* was heated in imidazole at 120 °C for 2.5 h to achieve lactam closure. After detosylation *148* was obtained in low yield.

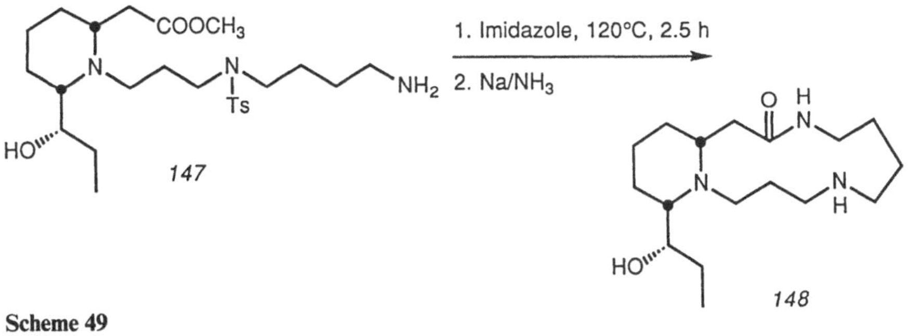

147

148

Scheme 49

3.3 The Mukaiyama Method

The Mukaiyama reagent, 1-methyl-2-chloropyridinium iodide (*28*), is also suitable for macrolactamization [85]. Jones et al. [86] achieved the first total synthesis of the important immunosupressant *(−)-FK-506* (*151*) using the Mukaiyama method as a key cyclization step. As shown in Scheme 50, the unstable amino acid *149* was treated with *28* under high dilution to give the macrocycle *150* in

81–85% yield. Schreiber and coworkers [87] also finished a total synthesis of *FK-506* using the Mukaiyama lactamization method. Danishefsky and coworkers [88] tried to synthesize *FK-506* alternatively by lactonization with little success up till now.

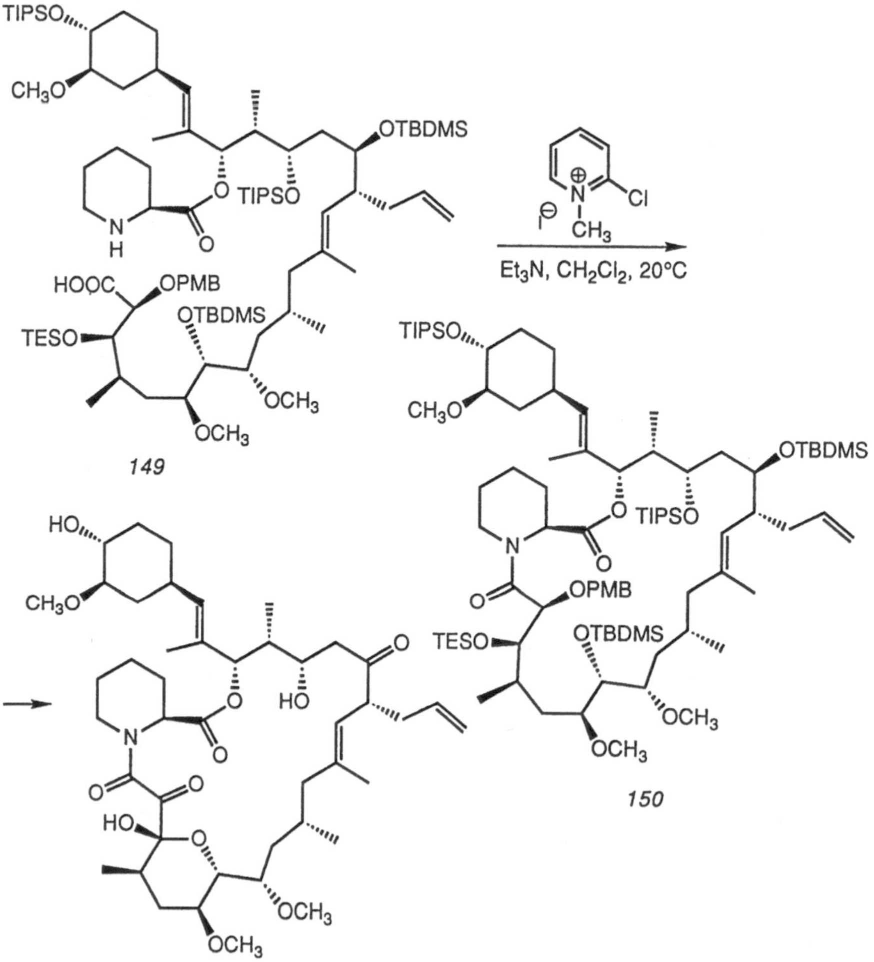

149

150

151

Scheme 50

3.4 Ketene Trapping Method

This method has been used for lactone closure (cf. Sect. 2.9). It is also suitable for lactam closure. The tetrasmic acid antibiotic *(+)-ikarugamycin* (*154*) was synthesized by two groups, both using intramolecular trapping of β-acyl ketene to achieve lactam formation [89]. For instance, in Boeckman's synthesis [89a], ketene precursor 152 was heated in toluene at 105 °C for 8–10 h to give the macrolactam *153* in 77% yield (Scheme 51).

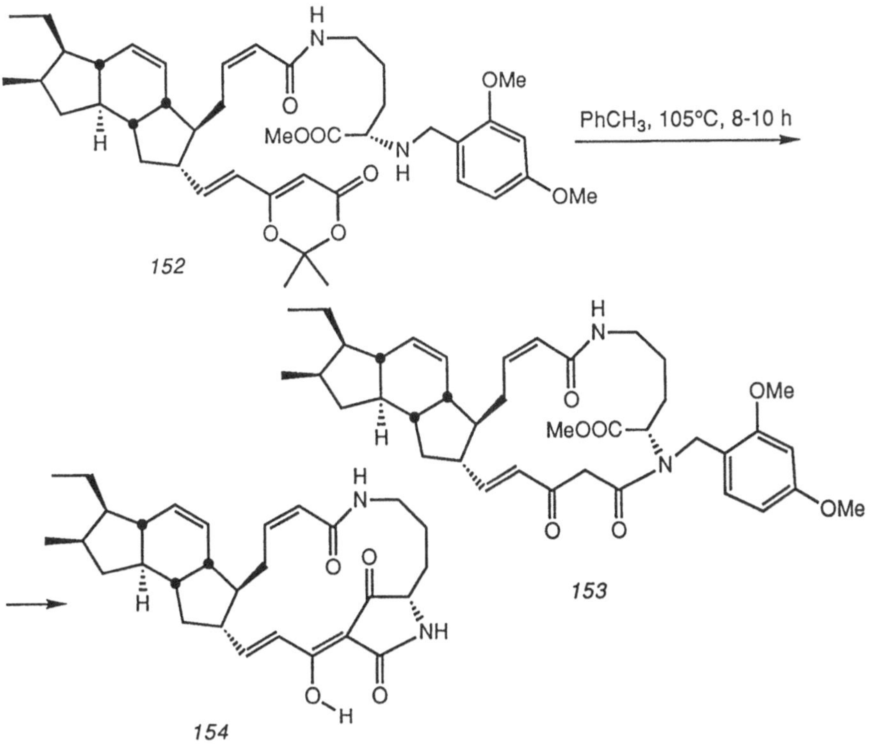

Scheme 51

3.5 Boron-Template Method

Yamamoto and Maruoka [90] found that the complexation of triamino ester with tris(dimethylamino)borane is a highly efficient lactam closure method. *Celacinnine* (*159*) and other similar spermidine alkaloids were synthesized in this manner. As shown in Scheme 52, the cyclization of the linear precursor *155* was effected with tris(dimethylamino)borane in xylene under reflux to furnish the lactam *158* via the intermediates *156* and *157* in 90% yield.

Earlier, Ganem and coworkers [91] reported lactam closure reactions mediated by catecholborane. However, they are likely to involve boron-containing activated esters rather than triaminoborane intermediates.

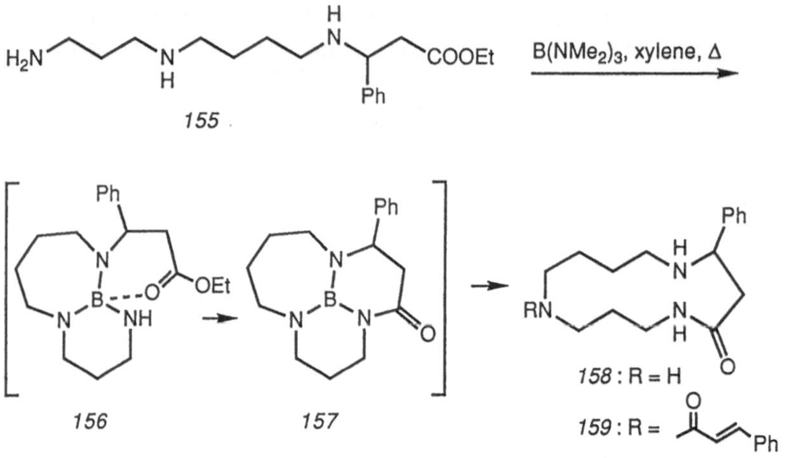

Scheme 52

3.6 Other Methods

An acyl chloride was used in the synthesis of the spermidine alkaloid *cannabisativine* (*162*) by Natsume and coworkers [92]. As shown in Scheme 53, the precursor *160* was hydrolyzed with Ba(OH)$_2$ and the resulting amino acid was directly converted to acyl chloride HCl salt, a cold solution of which was added dropwise to a potassium carbonate suspension in acetonitrile to achieve the lactam closure. Deprotection of the methoxymethyl groups with acid afforded macrolactam *161* in 43% yield from *160*.

Hanessian and coworkers [56] reported that lactam closure could also be achieved by the organotin-induced reaction (cf. Sect. 2.9).

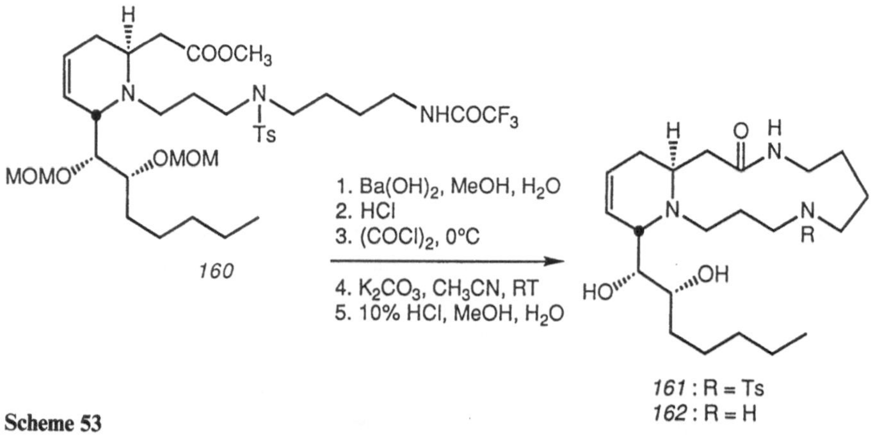

Scheme 53

4. Methods Involving C−C Bond Formation

4.1 Nucleophilic Addition to Carbonyls

4.1.1 Aldol Reaction

In the studies on the synthesis of the antitumor agents *esperamicin* A_1 and *calicheamicin* γ_1 by Magnus et al. [93], an aldol reaction was found suitable for macrocyclization after a number of unsuccessful attempts. Thus, as shown in Scheme 54, the diynene core structure (*165*) of the two antitumor agents was synthesized from the dicobalt hexacarbonyl derivative *163*. When *163* was treated with n-Bu$_2$BOTf/DABCO/Et$_3$N in CH$_2$Cl$_2$-THF the aldol product *164* was isolated as a single stereoisomer in 45% yield. Although alkynyl aldehydes undergo similar crossed aldol condensation, their dicobalt hexacarbonyl derivatives react with moderate to excellent *syn* diastereoselectivity [94].

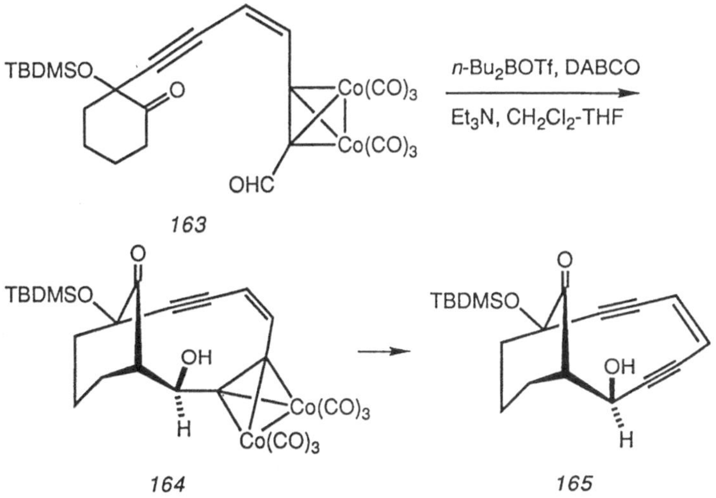

Scheme 54

Although direct aldol condensations can be used for macrocyclization, the indirect one is more suitable. Smith et al. [95] reported that all attempts to effect a direct cyclization of precursor *7* to the desired 11-membered ring system *8* employing a wide variety of different acidic and basic reagents met with complete failure (cf. Scheme 4). The events required for the successful cyclization are chemospecific generation of the enolate of the ketone side instead of the aldehyde side and irreversible addition of the enolate to the aldehyde. The Mukaiyama acetal-aldol condensation [96] fulfilled these requirements and finally the long sought after cyclization was successfully effected. The conditions derived here have been used in the synthesis of the antileukemic diterpene *(+)-hydroxyjatro-phone B* (*168*) to construct the 11-membered ring [97]. As shown in Scheme 55, keto acetal *166* was first changed to enol silyl ether with LDA and TMSCl and then treated with $TiCl_4$ to provide β-alkoxy ketone *167* in 65% yield.

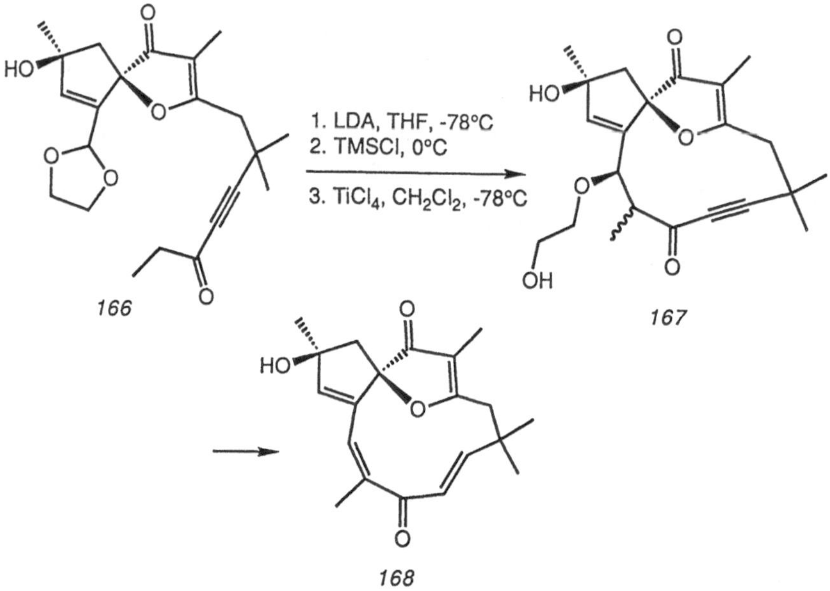

1. LDA, THF, -78°C
2. TMSCl, 0°C

3. $TiCl_4$, CH_2Cl_2, -78°C

166

167

168

Scheme 55

Kocienski and coworkers [98] have reported the synthesis of 8-membered cyclic ketones by intramolecular aldol reaction of enol silanes and acetals mediated by Lewis acid.

4.1.2 Sulfone-Carbonyl Coupling

Reactions [99] of active methylene nucleophiles other than aldehydes and ketones with carbonyl compounds are also used for cyclizations. An example is the coupling reaction between sulfone and carbonyl. A sulfonyl group is easy to introduce and easy to remove. Yoshii and coworkers [100] used a sulfone-mediated macrocycliza-tion in the synthesis of *(±)-O(26)-methyl-28,29-bisnor-kijanolide* (*171*), the aglycone of the antitumor antibiotic *kijanimicin*. As shown in Scheme 56, treatment of the linear precursor *169* with 1 equiv of sodium *t*-amylate at room temperature for 10 min. afforded the macrocycle *170* in 82% yield. On the other hand, a diastereoisomer of *169* with opposite configuration at all the stereocenters in the octalin system resisted cyclization under the same conditions. This failure comes from a severe steric hindrance in the transation state of cyclization. Once again, stereochemistry is also an important factor for macrocyclization (cf. Section 2.10).

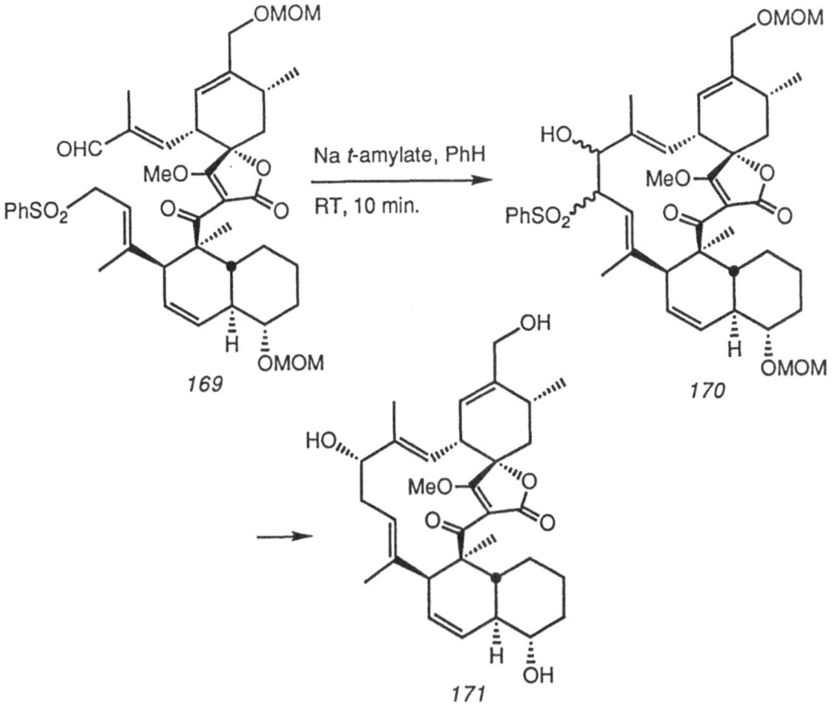

Scheme 56

4.1.3 Addition of Alkenylchromium to Aldehyde

Takai et al. [101] found that alkenyl halide is readily reduced with $CrCl_2$ to give the corresponding organochromium species which adds selectively to an aldehyde

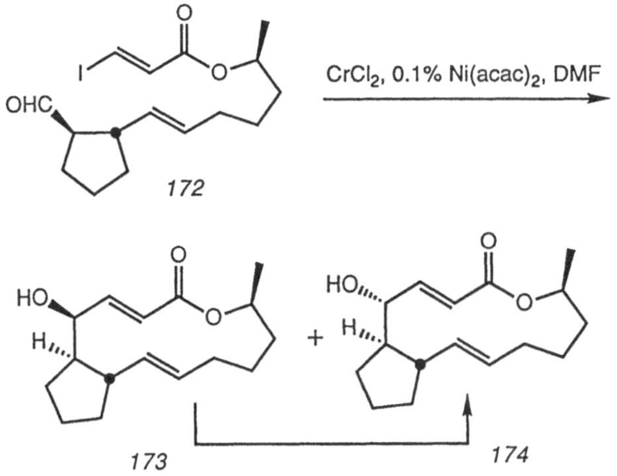

Scheme 57

145

moiety in the presence of ketone or cyano groups. This reaction is suited for macrocyclization. Rowley and Kishi [102] applied this procedure with a modification to the construction of the 8-membered ring in the studies toward the tricyclic sesterterpenes *ophiobolins* and *ceroplastols*. Schreiber and Meyers [103] applied the Kishi's modification to the synthesis of the macrolide antibiotic *(+)-brefeldin C (174)*. As shown in Scheme 57, subjection of the cyclization substrate *172* with CrCl$_2$ in the presence of Ni(acac)$_2$ in DMF produced a 4:1 mixture of *4-epi-brefeldin C (173)* and *174* in high yield.

4.1.4 Addition of Allylchromium to Aldehyde

Hiyama and coworkers [104] reported that chromic chloride is easily reduced by a half molar equivalent of lithium aluminium hydride in THF and the resulting salt, presumably Cr(II), reduces allylic halides to produce allylchromium species which add efficiently to carbonyls with high degree of stereo- and chemoselectivity. Kitagawa and coworkers [105] successfully applied this procedure to the synthesis of the 10-membered antitumor germacranolides *costunolide* and *dihydrocostunolide*. Still and Mobilio [106] synthesized the cembranoid antitumor agent *asperdiol* (*178*) using this method. As shown in Scheme 58, after a number of other methods based on homoallylic alcohol preparations failed, the Hiyama-Heathcock reaction using 5 equiv of CrCl$_2$ in THF was effective at cyclizing *175* and gave in 64% yield a 4:1 mixture of the desired isomer *176* and its diastereoisomer *177*.

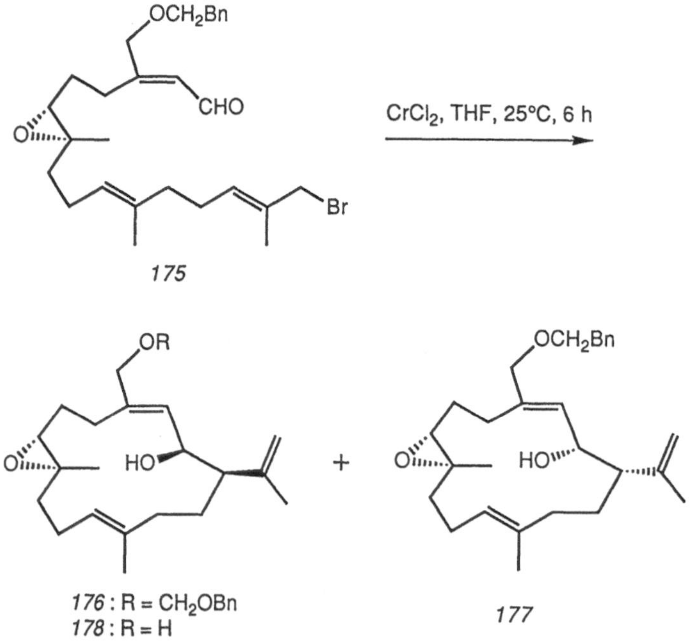

Scheme 58

4.1.5 Addition of Allylstannane to Aldehyde

Marshall and coworkers [107] sucessfully applied this reaction to the synthesis of some 14-membered isoprenoids. For example, *cembranolide 181* (unnamed) was synthesized as shown in Scheme 59 [107b]. Thus, treatment of (α-alkoxyallyl)-stannane aldehyde *179* with $BF_3 \cdot Et_2O$ at $-78\,°C$ in dichloromethane at high dilution afforded a $(Z) : (E)$ $(88 : 12)$ mixture of macrocycle *180* in 88% yield.

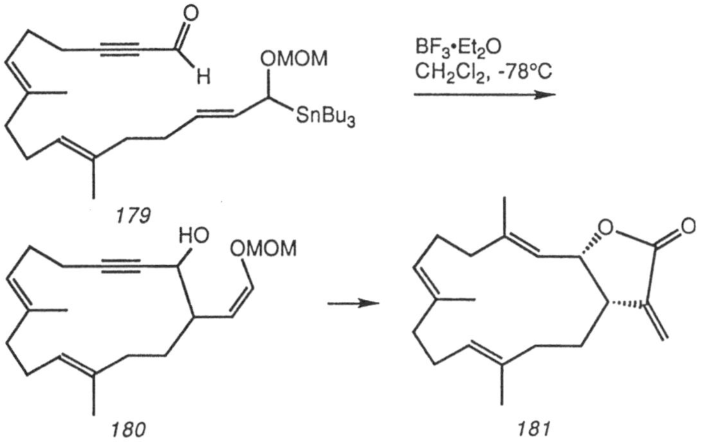

Scheme 59

4.1.6 Acetylene-Aldehyde Coupling

In the synthesis of the core structure of the diynene antitumor antibiotics *esperamicins* and *calicheamicins* by Danishefsky et al. [108], an acetylene-aldehyde coupling reaction was used to achieve the cyclization. As shown in Scheme 60, reaction of a toluene solution of acetylene aldehyde *182* with potassium hexame-thyldisilazide at $-78\,°C$ for 20 min afforded a 52% yield of a 10 : 1 ratio of *183* and *184*.

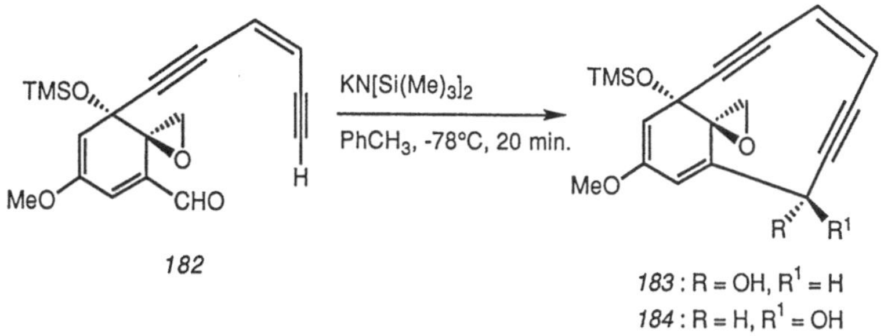

Scheme 60

147

4.1.7 Reformatsky Reaction

Inanaga and coworkers [109] used a modified intramolecular Reformatsky reaction in the synthesis of the beetle aggregate phermone *ferrulactone I* (*187*). As shown in Scheme 61, precursor *185* was cyclized with SmI_2 followed by acylation of the resulting unstable β-hydroxydecadienolide to afford the 11-membered lactone benzoate *186* in 47% yield. There are more applications of the Reformatsky reaction for macrocyclization [110].

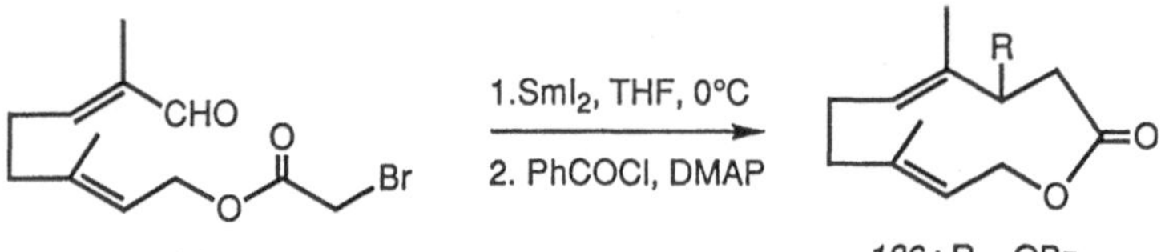

Scheme 61

4.2 Alkene-Acetal Coupling

As discussed at the beginning of Sect. 4.1, a masked carbonyl, such as acetal, is more suitable for macrocyclization than the carbonyl itself because the nucleophilic substitution reaction on the former is irreversible. Overman and coworkers [111] developed a C−C bond-forming cyclization approach to 8-membered ethers by Lewis acid-promoted alkene substitution of mixed acetals. Thus, oxocenes with 4,5-unsaturation (3,6,7,8-tetrahydro-2*H*-oxocins) can be accessed with moderate to excellent efficiency and with perfect regiochemical fidelity. The yields of

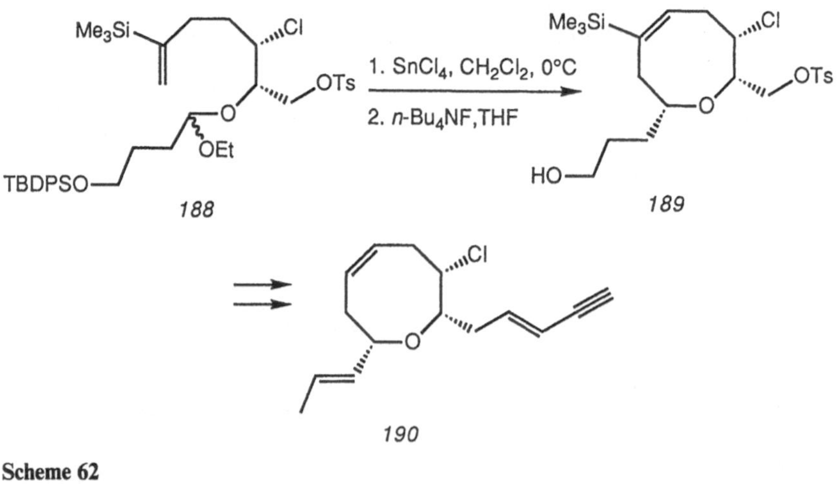

Scheme 62

Δ^4-oxocenes increase as the 5-substituent of a 5-hexenyl acetal is varied from H to SiMe$_3$ to SPh. As shown in Scheme 62, the C_{15}-nonisoprenoid metabolite $(-)$-laurenyne (190) was synthesized using this method [111b]. Thus, mixed acetal precursor 188 was treated with 2 equiv of stannic chloride in dichloromethane at 0 °C for 1.5 h followed by O-desilylation to produce oxocene 189 as the sole cyclic ether product in 37% yield.

Trost and Lee [112] reported a similar approach, using vinylcyclopropanols as cyclization terminators, to construct 8-membered rings. The composite of an olefin and a hydroxyl group mediated through a cyclopropane in the form of a 1-vinylcyclopropanol permits polarization of the double bond to enhance its nucleophilicity. For example, treatment of alkene acetal 191 with 1 equiv of trimethylsilyl triflate and 0.7 equiv of pyridine at 0.001 M in dichloromethane at -40 to -20 °C afforded a mixture of bicyclo[6.3.0]undecane ring systems in 96% yield. The major isomer 192 was obtained in 85% yield (Scheme 63).

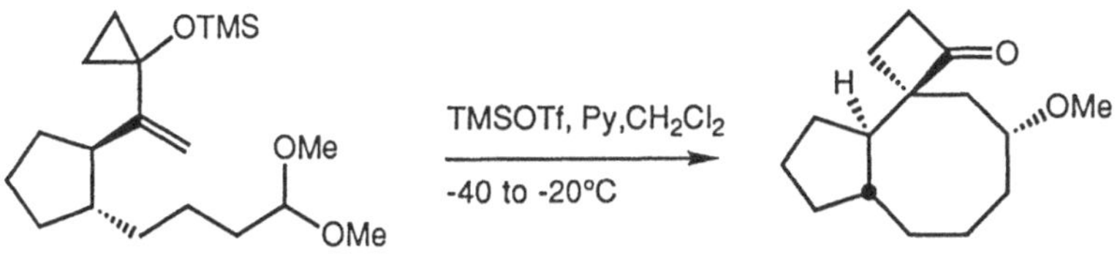

Scheme 63

4.3 Alkylation of Carbanions or Enols

4.3.1 Alkylation of Cyanohydrins

Tsuji and coworkers [113] reported a simple synthetic method for macrocyclic ketones based on intramolecular alkylation of the carbanion generated from protected cyanohydrins. The reaction is rapid and irreversible. It has been used in the synthesis of a couple of natural products [114]. For example, the macrolide zeralenone (70) was synthesized using this method [114a]. As shown in Scheme 64, linear procursor 193 was converted to cyanohydrin with sodium bisulfite and sodium cyanide at 0 °C. Protection of the resulting hydroxyl group gave 194 in 90% overall yield. The cyclization was carried out by adding 194 in THF over 1 h to sodium hexamethyldisilazane in THF at 45 °C and stirring the mixture at 60 °C for 30 min. The cyclized product 195 was obtained in 85% yield.

4.3.2 Alkylation of Sulfones

Marshall and Cleary [115] synthesized 7(8)-desoxyasperdiol (198), a precursor of the cembranoid asperdiol by using sulfonyl stabilized anion displacement of an allylic halide as a key macrocyclization step. As shown in Scheme 65, iodo sulfone *196* was slowly added to an excess of KN(TMS)$_2$ in THF in the presence of 18-crown-6 at 0 °C to effect cyclization to *197* in 53% yield. It was proposed, though not pursued, that the cyclization proceeds via a dianion intermediate.

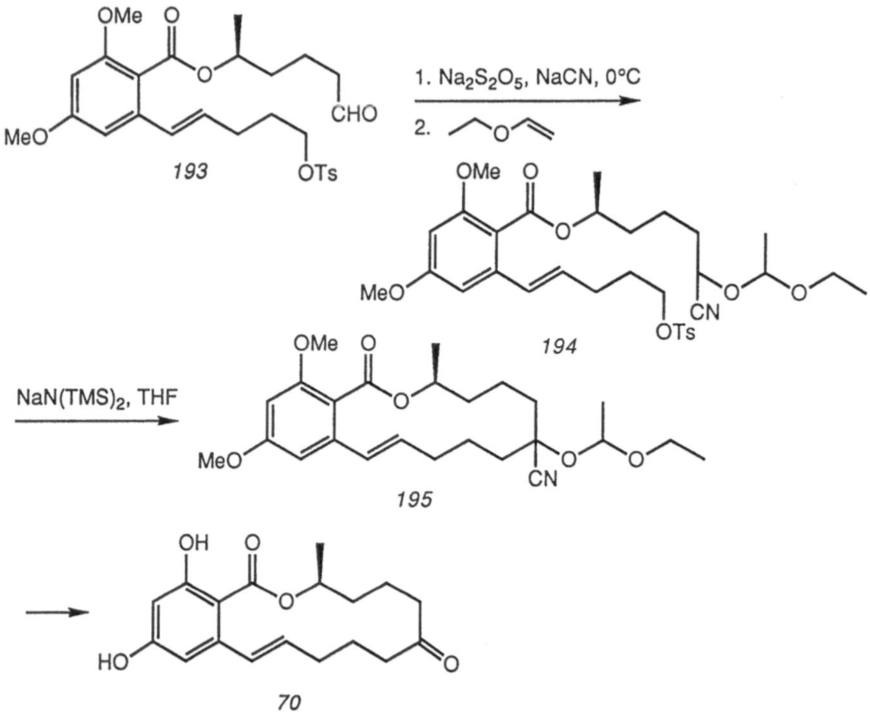

Scheme 64

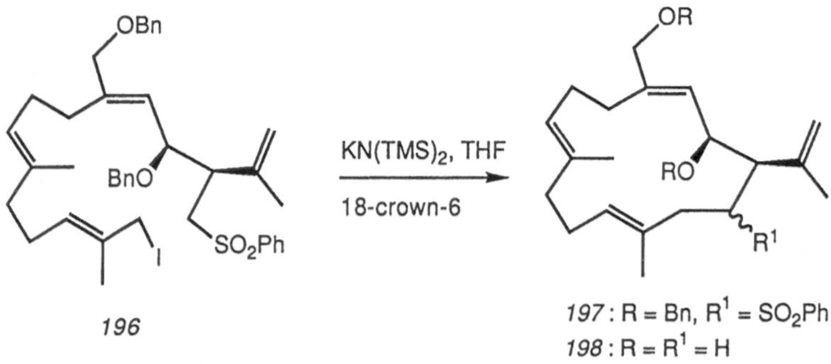

197 : R = Bn, R^1 = SO$_2$Ph
198 : R = R^1 = H

Scheme 65

4.3.3 Palladium-Catalyzed Allylic Alkylation

Trost and coworkers [116] found that palladium-catalyzed intramolecular allylic alkylation to α-sulfonyl ketones is a good means of performing macrocyclization. This reaction involves the intermediacy of a π-allylpalladium complex as an enolonium equivalent to initiate cyclization. For instance, this method was used in the synthesis of the cytochalasin (−)-aspochalasin B (201) [117]. As shown in Scheme 66, cyclization of the linear precursor 199 using 10 mol% (Ph$_3$P)$_4$Pd in the presence of 10 mol% DPPP in THF created the 11-membered carbocycle 200 as a single diastereoisomer in 49% yield.

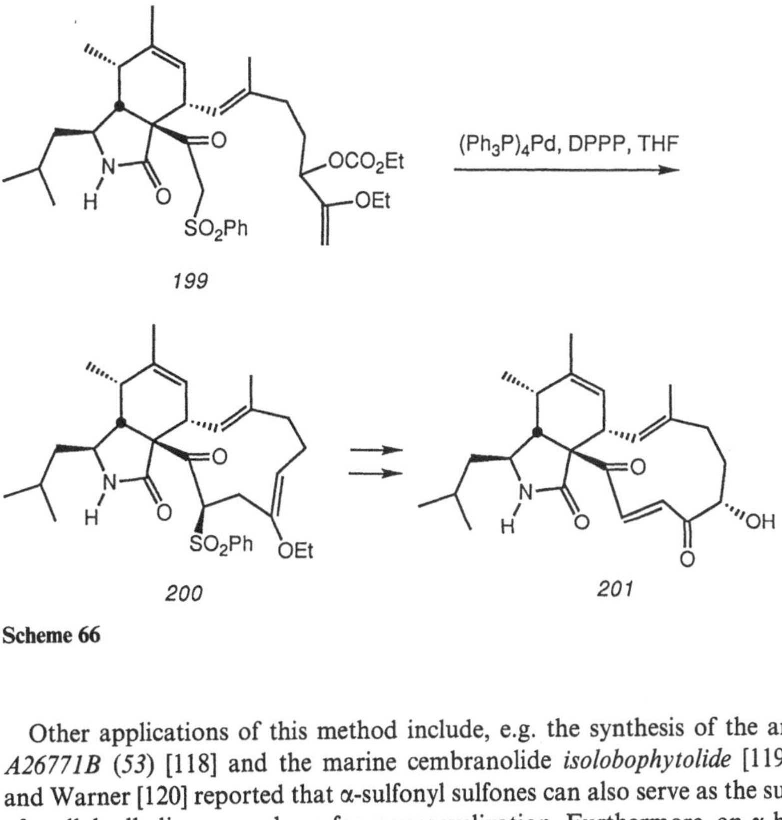

Scheme 66

Other applications of this method include, e.g. the synthesis of the antibiotic A26771B (53) [118] and the marine cembranolide isolobophytolide [119]. Trost and Warner [120] reported that α-sulfonyl sulfones can also serve as the substrates of π-allylpalladium complexes for macrocyclization. Furthermore, an α-hydroxy-carbonyl ketone was alkylated intramolecularly by a π-allylpalladium complex in the total synthesis of the macrocyclic sesquiterpene humulene [121].

4.3.4 Alkylation of Sulfides

Ito and coworkers [122] found that allylsulfides can be alkylated intramolecularly by epoxides in the presence of a proper base to form macrocycles. Tsuji and coworkers [123] reported that the alkylation of phenylsulfide anions is also suitable for macrocyclization. Both methods have been used in the synthesis of a number of natural products [122–124]. For instance, *(±)-cubitene* (*204*), a diterpene component of the defence secretion of termites, has been synthesized by Kodama et al. [124b]. Treatment of sulfenyl epoxide *202* with *n*-butyllithium in the presence of 1,4-diazabicyclo[2.2.2]octane (DABCO) afforded the 12-membered cycle *203* in 73% yield (Scheme 67).

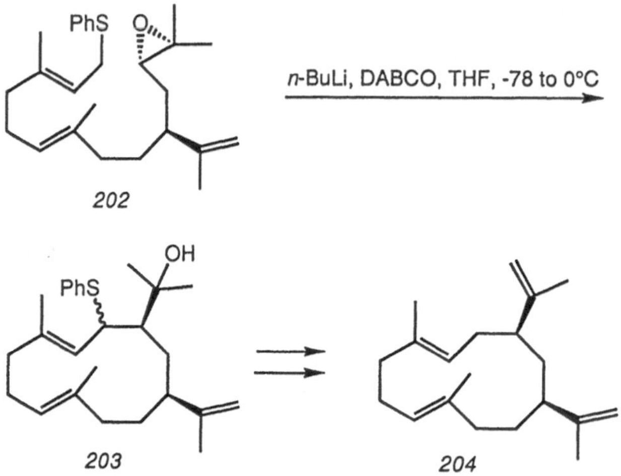
PhS
O
n-BuLi, DABCO, THF, -78 to 0°C
202
OH
PhS
203
204

Scheme 67

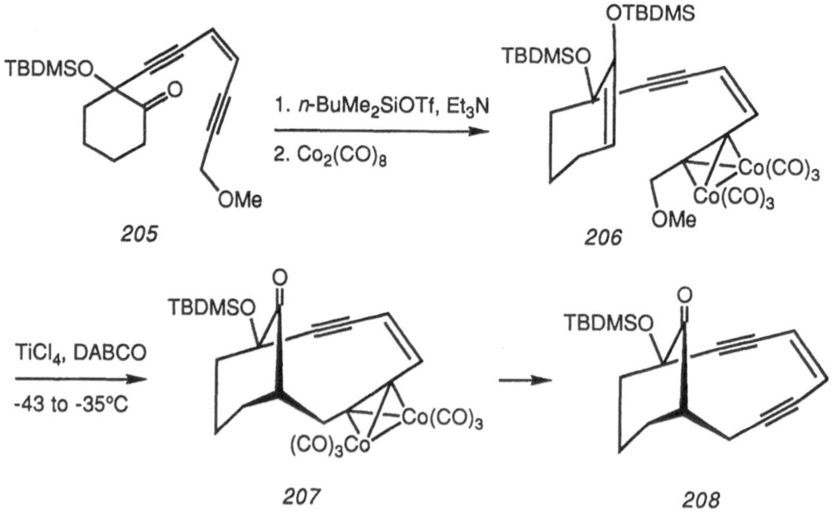
TBDMSO
O
OMe
205
1. n-BuMe2SiOTf, Et3N
2. Co2(CO)8
OTBDMS
TBDMSO
Co(CO)3
Co(CO)3
OMe
206
TiCl4, DABCO
-43 to -35°C
O
TBDMSO
Co(CO)3
(CO)3Co
207
O
TBDMSO
208

Scheme 68

4.3.5 Alkylation by Dicobalt Hexacarbonyl Propargyl Cations

The stability of carbocations flanked by π-coordinated organic moieties is dramatically enhanced so that they react easily with nucleophiles [125]. Magnus et al. [126] applied this principle to the synthesis of the core structure (*208*) of the diynene antibiotics *esperamicins* and *calicheamicins*. As shown in Scheme 68, diynene *205* was converted to enol silyl ether which was treated with $Co_2(CO)_8$ to give dicobalt hexacarbonyl adduct *206*. Exposure of *206* to 3 equiv of $TiCl_4$ and 1 equiv of DABCO at -43 to $-35\,°C$ gave macrocycle *207* in 50% yield.

4.4 Pinacol and Related Reactions

4.4.1 Aldehyde-Aldehyde Coupling

McMurry and coworkers [127] found that the titanium-induced pinacol coupling reaction is a general and effective means of preparing carbocyclic rings of all sizes. For instance, the anticancer cembranoid *sarcophytol B* (*210*) was synthesized using this method [128]. As shown in Scheme 69, dialdehyde *209* in DME was added over 30 h at $-40\,°C$ to a stirred slurry of a low-valent titanium reagent prepared by reduction of $TiCl_3(DME)_2$ with Zn-Cu in DME. Hydrolysis of the product followed by chromatography afforded *sarcophytol B* (*210*) in 46% yield. The diterpenoid *crassin* has also been synthesized by using such an intramolecular pinacol reaction as a cyclization step [129].

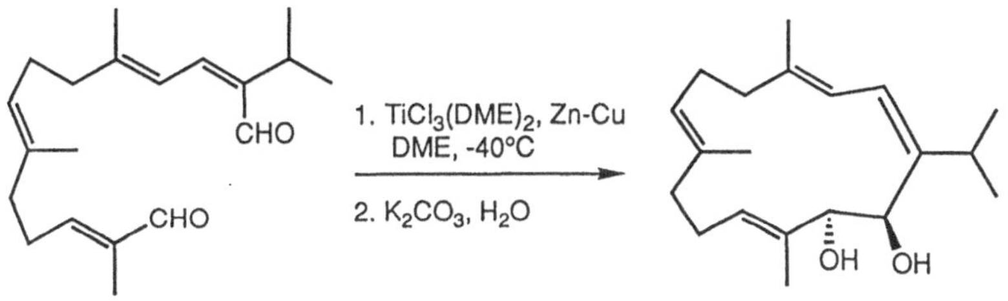

Scheme 69

4.4.2 Aldehyde-Ester Coupling

McMurry and Miller [130] found that a similar coupling reaction with higher oxidation state, carbonyl ester rather than dicarbonyl, also occurs with synthetically acceptable yields for macrocyclization. This method has been applied to the synthesis of the sesquiterpene *isocaryphyllene* (*213*) [131]. As shown in Scheme 70, keto ester *211* in DME was slowly added to the refluxing low-valent titanium slurry prepared from $TiCl_3$ and $LiAlH_4$ over 16.5 h, followed by an additional 3 h period of reflux. After quenching, the 9-membered cycle *212* was obtained in 38% yield. An unexpected double bond isomerization occured during the cyclization.

211

212 : R = O
213 : R = CH_2

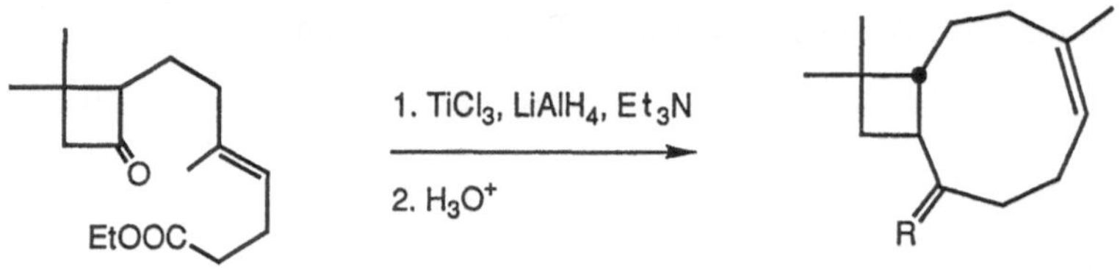

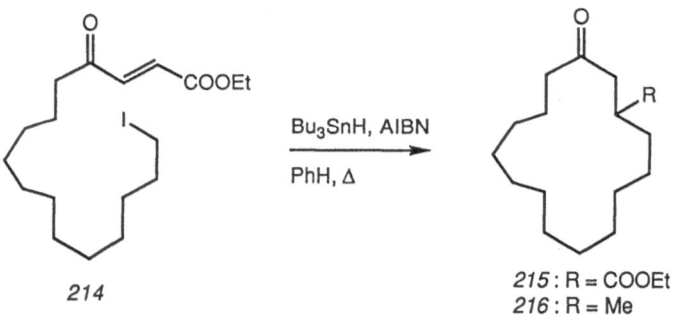

Scheme 70

4.5 Free-Radical Macrocyclization

4.5.1 Alkyl Radical

The construction of 5- and 6-membered rings by free-radical reactions has received much attention. However, preparations of macrocycles by radical cyclization have evolved only in recent years. Porter and coworkers [132] have investigated intramolecular alkyl radical addition to olefins substituted with electronwithdrawing groups. Macrocycles of 12 to 20 members are generated by this procedure in good yields. For example, *muscone* (*216*) was synthesized as shown in Scheme 71 [132c]. Iodide *214* in refluxing benzene was reacted with 1.1 equiv of Bu_3SnH and 0.1 equiv of the azo initiator AIBN for 3 h. The endocyclic product *215* was obtained in 57% yield.

214

215 : R = COOEt
216 : R = Me

Scheme 71

4.5.2 Allylic Radical

Pattenden and coworkers [133] have described the use of allylic radical intermediates in macrocyclizations leading to several natural products. For example, the cembranoid *mukulol* (*220*) [133a] was synthesized as shown in Scheme 72. Iodotetraenone *217* was heated under reflux for 3 h in the presence of Bu_3SnH and AIBN in benzene. A 1 : 4 mixture of *218* and *219* was obtained together in 40% yield and *219* was converted into *mukulol* (*220*).

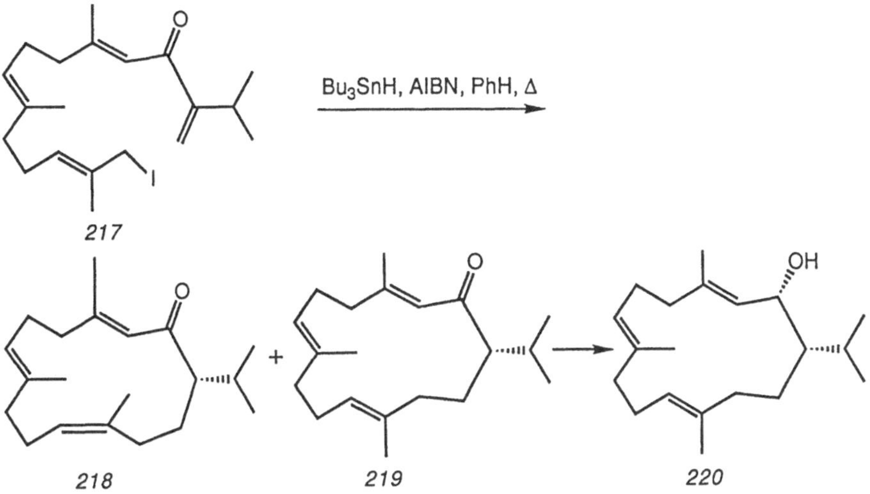

Scheme 72

4.5.3 Acyl Radical

Boger and Mathvink [134] recently reported the generation of acyl radicals from phenyl selenoesters and their participation in macrocyclization through free-radical alkene addition reactions. As shown in Scheme 73, the 16-membered macrocycle 222 was obtained from phenyl selenoester 221 in 68% yield.

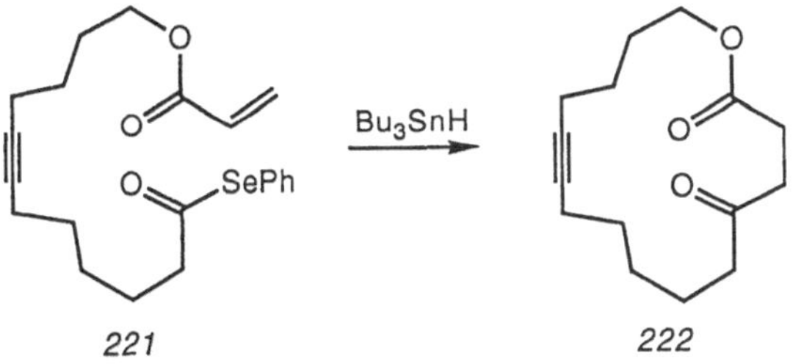

Scheme 73

4.6 Nickel-Mediated Cyclization

4.6.1 Bis-Allylic Bromide

The method for forming cycloolefins from bis-allylic halides and nickel carbonyl was first reported by Corey and coworkers [135] and has been applied to the synthesis of several macrocyclic natural products [136]. The 14-membered diterpene *cembrene* (225) was synthesized by Dauben et al. [136b], for example. As shown in Scheme 74, dibromide *223* was reacted with nickel carbonyl in *N*-methylpyrrolidone (NMP) at 52 °C to give a (*E*, *Z*)-mixture of monomer *224* in 25% yield, from which *cembrene* (*225*) was prepared.

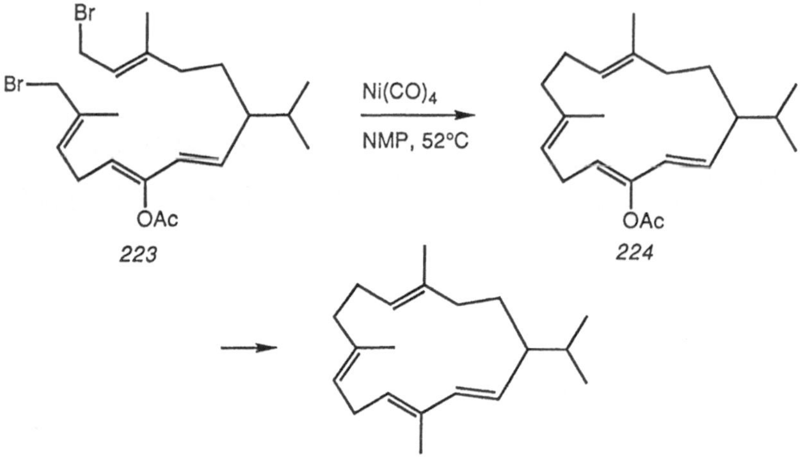

Scheme 74

4.6.2 Alkyl-Allylic Dihalide Coupling

Sato and coworkers [137] reported the synthesis of macrocyclic lactones via intramolecular alkylation of π-allylic nickel complexes, based on the facile reactivity of π-allylic nickel complexes toward alkyl halides. As shown in Scheme 75, this method was applied to the synthesis of the macrolide *recifeiolide* (*227*). The linear precursor *226* in benzene was added slowly over 1 h to nickel carbonyl in benzene at 50 °C. After additional 1.5 h at 50 °C, *227* was obtained in 32% yield.

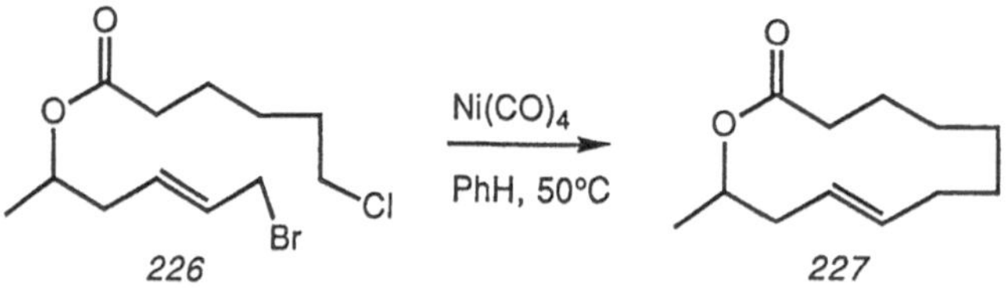

Scheme 75

4.7 Bis-Aryl Halide Coupling

Though many traditional reactions are available for couplings between aryl rings [138], only a few are suited for intramolecular reactions due to the strain of the ring formed thereof. Semmelhack and coworkers [139] reported that zero-valent nickel reacts rapidly with aryl halides to produce the symmetrical products. The reaction is a low-temperature analog of the Ullmann reaction and proceeds through oxidative addition of the organic halide to Ni(0). The resulting aryl nickel complex decomposes rapidly to biaryl in DMF and an intramolecular process is suitable for macrocyclization. For example, the 13-membered *meta*-bridged cyclic biphenyl, *alnusone*, was prepared efficiently with the crucial aryl halide coupling to form the ring proceeding in 50% yield [139b]. Hart and coworkers [140] applied the Semmelhack modification of Ullmann reaction to the synthesis of the *Lythraceae* alkaloid *lythrancepines II* and *III*. Thus, in the synthesis of *lythrancepine II* (*230*), diiodide *228* was treated with an excess of tetrakis(triphenylphosphine)nickel(0) in DMF to afford biaryl *229* in 20% yield (Scheme 76). Another application of the Semmelhack method is due to Helquist and coworkers [141].

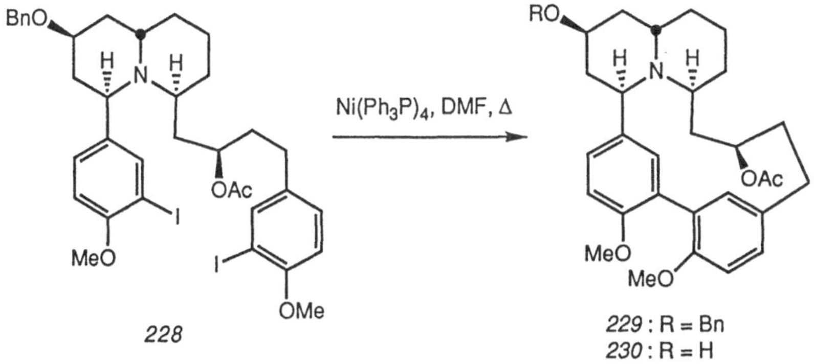

Scheme 76

4.8 Other Methods

4.8.1 Nucleophilic Addition to Carboxylic Ester

This method was employed by Meyers et al. [142] in the synthesis of (±)-*maytansinol* (*233*), the common precursor to the ansa macrocyclic antitumor agents, maytansinoids, to construct the macrocycle. As shown in Scheme 77, the cyclization of amide ester *231* to *232* was accomplished in 58% yield by using 4 equiv of lithium(hexamethylsilyl)amide at −78 °C for 4 h.

4.8.2 Acylation of Alkene and Alkyne by Acid Halide

Kitahara and coworkers [143] found that a terminal double bond in acyclic precursors of terpenes is selectively acylated when treated with acid chlorides in the presence of $SnCl_4$ or $AlCl_3$ and the intramolecular acylation of polyenyl acid chlorides is suitable for macrocyclization. For example, the *cembrene* skeleton *235* was efficiently constructed by intramolecular acylation of *trans*-geranylgeranic

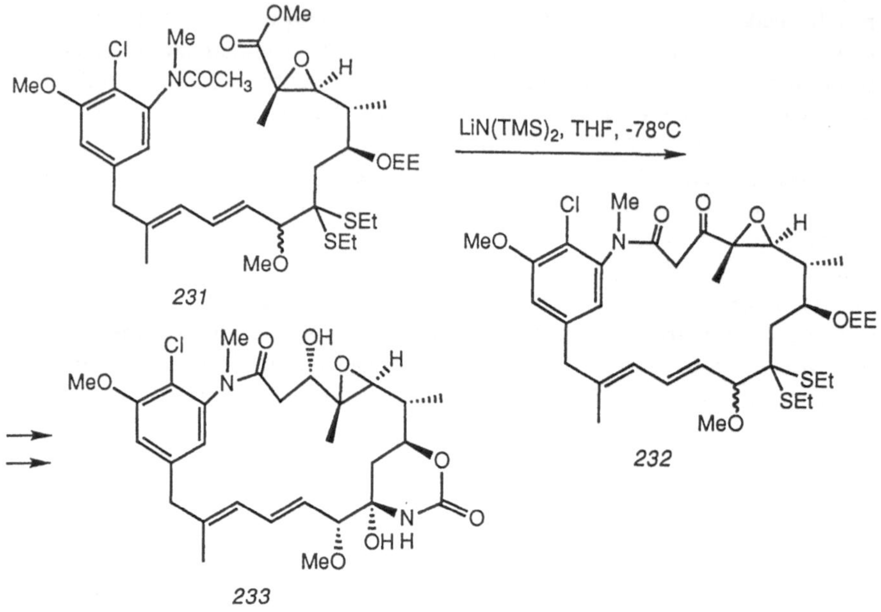

Scheme 77

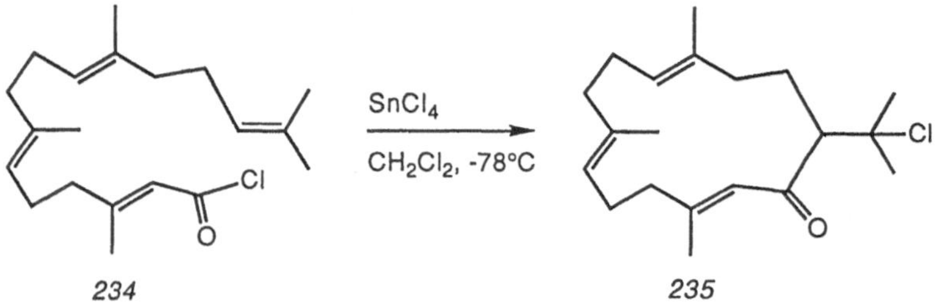

Scheme 78

acid chloride (*234*) [143c]. Thus, as shown in Scheme 78, *234* was treated with 1 equiv of SnCl$_4$ in CH$_2$Cl$_2$ at $-78\,°C$ for 1.5 h to produce *235* in 71% yield.

More interestingly [144], as shown in Scheme 79, under the same condition geranylfarnesonic acid chloride (*236*) gave exclusively the corresponding 14-membered product *237*, although the inspection of a Dreiding model suggests that the terminal double bond in *236* could interact with the acyl cation without any hindrance. The reason for this switch mode of ring closure is not yet entirely clear.

Utimoto et al. [145] reported a procedure for the synthesis of macrocyclic ynones by intramolecular acylation of ω-(trimethylsilyl)ethynylalkanoyl chlorides in the presence of Lewis acid. For example, in the synthesis of *(−)-muscone* (*216*), cyclization of alkyne acid chloride *238* gave the macrocycle *239* in 52% yield, which was hydrogenated to *(−)-muscone* (Scheme 80).

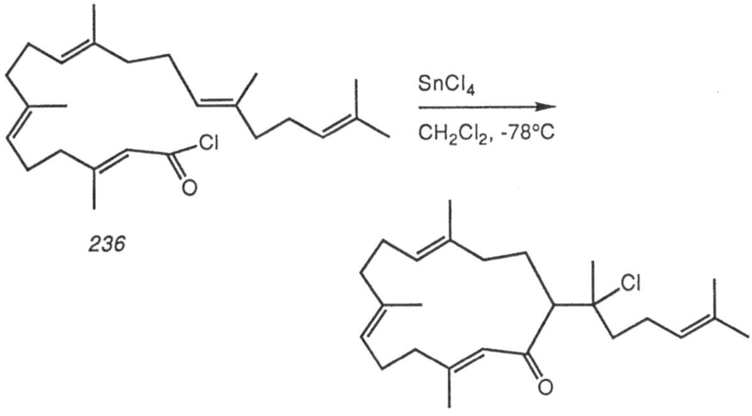

236

237

Scheme 79

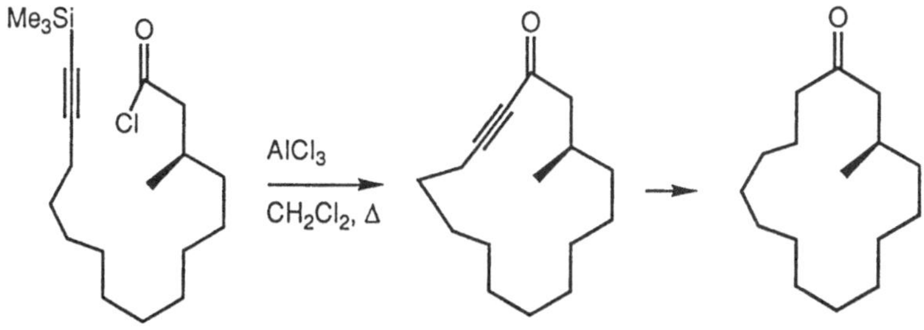

238

239

216

Scheme 80

4.8.3 Intramolecular Friedel-Crafts Reaction

This strategy has been used in the synthesis of the mould metabolite *curvularin* and its dimethyl ether derivative [146]. In the synthesis of *curvularin* (*242*) by Gerlach [146a], carboxylic acid *240* was exposed to a mixture of trifluoroacetic acid and its anhydride to give the macrocycle *241* via a Friedel-Crafts acylation.

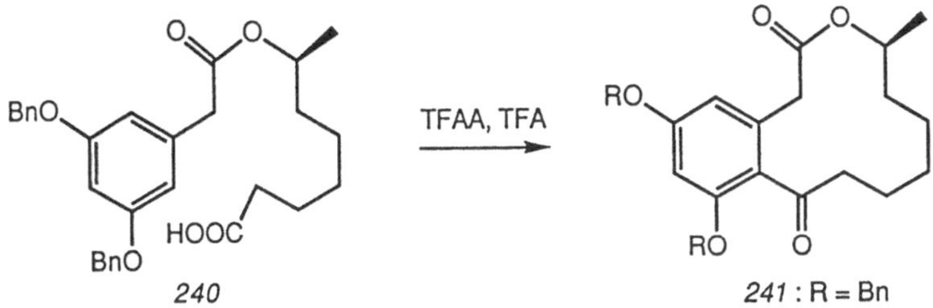

240

241 : R = Bn
242 : R = H

Scheme 81

159

4.8.4 Intramolecular Dieckmann Condensation

Hurd and Shah [147] found that the Dieckmann condensation is suitable for macrocyclization and superior to the parallel Thorpe-Ziegler condensation. For example, *zearalanone* (*246*) was prepared from diester *243* by using this method

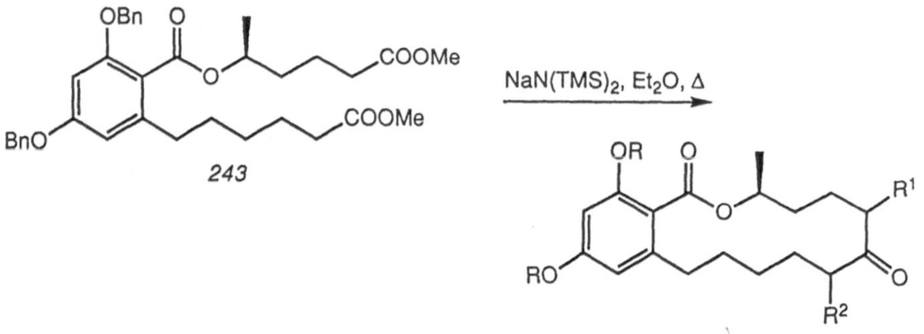

244 : R = Bn, R^1 = COOMe, R^2 = H
245 ; R = Bn, R^1 = H, R^2 = COOMe
246 : R = R^1 = R^2 = H

Scheme 82

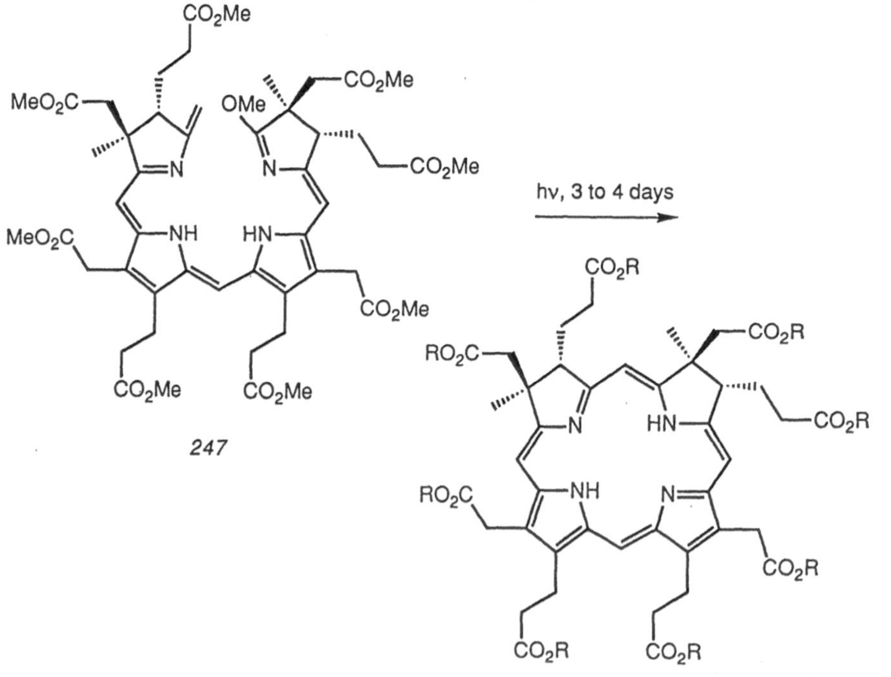

247

hv, 3 to 4 days

248 : R = Me
249 : R = H

Scheme 83

as ring closure step. Thus, as shown in Scheme 82, addition of *243* to a refluxing ether solution of NaN(TMS)$_2$ over 8 h resulted in a mixture of cyclic products *244* and *245* together in 77% yield, both equally usefull for the synthesis of *246*.

4.8.5 Photocyclization

The central macrocyclic rings of vitamin B$_{12}$ [148] and other similar natural products [149] have been constructed by using unique methodologies, e.g. electrochemical oxidation. As a special example of macrocyclization, here photocyclization is discussed. Battersby and coworkers [149b] efficiently synthesized *sirohydrochlorin* (*249*), an isobacteriochlorin isolated as the metal-free prosthetic group of sulphite reductase. Photochemical treatment of the seco precursor *247* for 3 to 4 days yielded *248* which was hydrolyzed to *249* (Scheme 83).

5 Methods Involving C=C Bond Formation

In fact, several methods described in Sect. 4, e.g. aldol reaction, can be used to construct C=C bond-containing macrocycles with modification of conditions or an additional step. Two methods involving "real" C=C bond formation are discussed below.

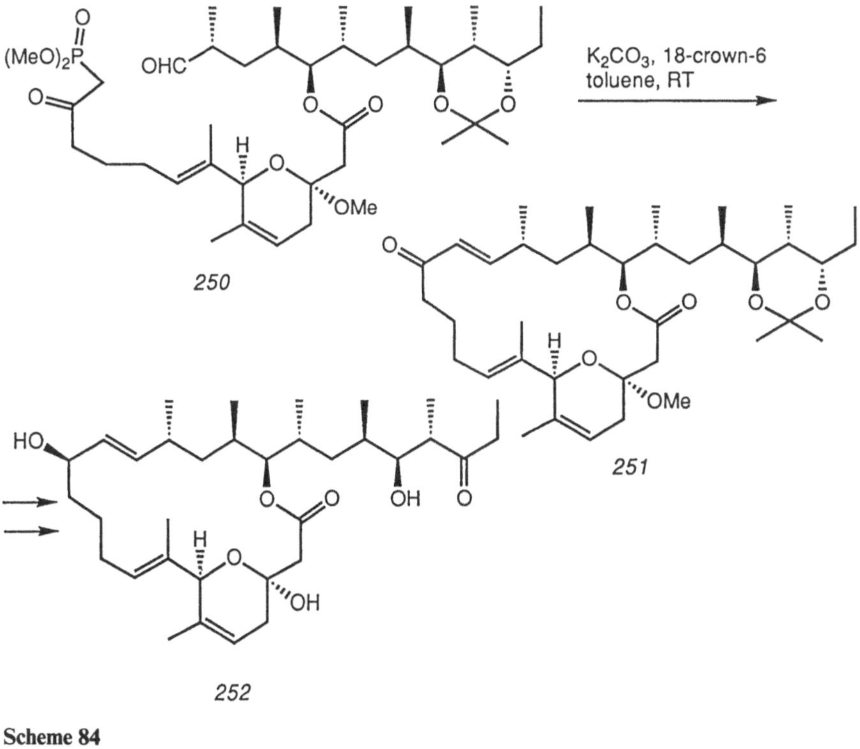

Scheme 84

5.1 Wittig-Like Reactions

The Wittig-like reactions are well known for C=C bond formation. An intramolecular process can serve as a ring closure and has been generally employed in the synthesis of numerous macrocyclic natural products [150]. Oishi and coworkers [151] applied this strategy to the synthesis of the aglycone (*252*) of the antibiotics *venturicidins A* and *B*. Thus, as shown in Scheme 84, the aldehyde phosphate *250* was subjected to the modified intramolecular Wittig-Horner condensation with a mild base, yielding the macrocycle *251* in 48% yield.

5.2 The McMurry Method

In 1977, McMurry and Kees [152] developed a titanium-induced intramolecular coupling procedure to form cycloalkenes from dicarbonyl compounds. Mechanistically, as shown in Scheme 85, the coupling reaction proceeds by an initial pinacol dimerization of the dicarbonyl *253* to *254*, followed by titanium-induced deoxygenation to afford alkene *255*.

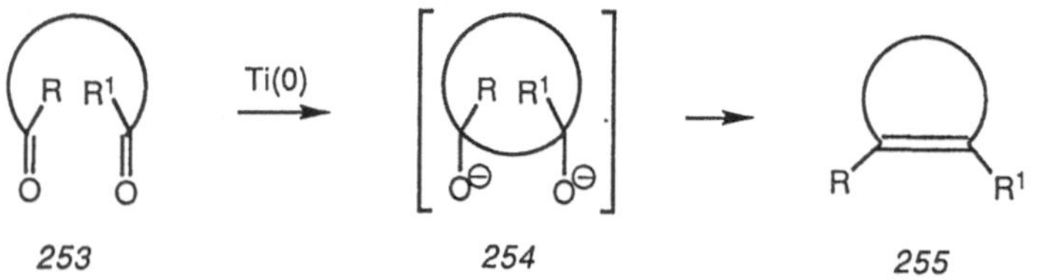

Scheme 85

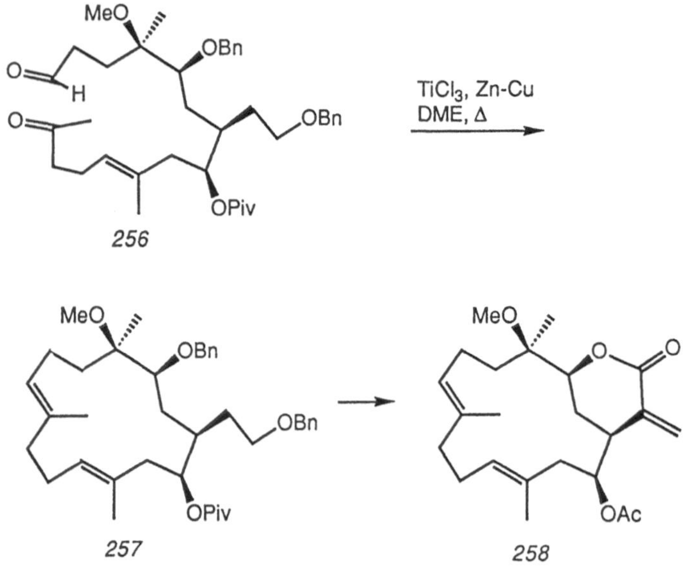

Scheme 86

Dauben et al. [153] successfully applied this procedure to the total synthesis of the cembranoid *crassin acetate methyl ether* (*258*). As shown in Scheme 86, the keto aldehyde *256* was treated with $TiCl_3/Zn$-Cu in refluxing DME to give the cyclic olefin *257* in 65% yield ($E : Z = 4 : 3$). The germacrane sesquiterpenes have also been synthesized using this method [154].

6 Methods Involving Ether Formation

6.1 Alcohol-Halide Coupling

In the synthetic studies on cembranoid natural products by Marshall et al. [155], an alcohol-halide coupling was used to construct the macrocycles. As shown in Scheme 87, the linear precursor *259* was cyclized to *260* in 71% yield by addition of 1 equiv of EtMgBr to a solution of the chloro alcohol *259* in HMPA-THF and stirring at reflux for 4 h. There are also other examples of this macrocyclization method [156].

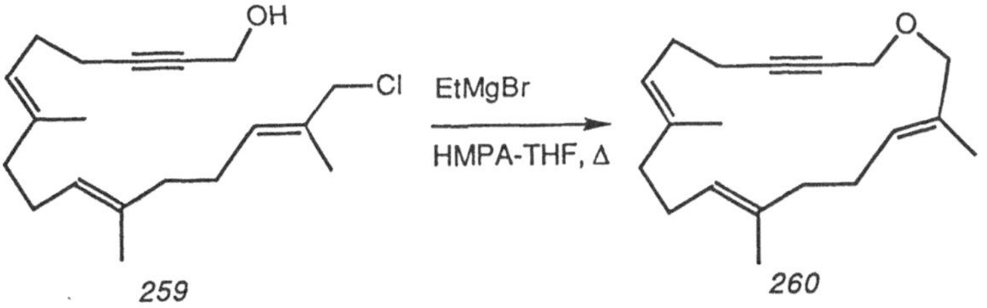

Scheme 87

6.2 Alcohol-Dithioketal Coupling

In the synthetic studies of the marine natural products *brevetoxins A* and *B*, Nicolaou and coworkers [157] developed a highly efficient cyclization reaction of hydroxy dithioketals leading to oxocenes. As shown in Scheme 88, exposure of the hydroxy dithioketal *261* to 1.1 equiv of *N*-chlorosuccinimide (NCS) in CH_3CN in the presence of 2 equiv of 2,6-lutidine, 1.1 equiv of $AgNO_3$, molecular sieves and silica gel at 25 °C for 5 min led to the oxocene *262* in 95% yield.

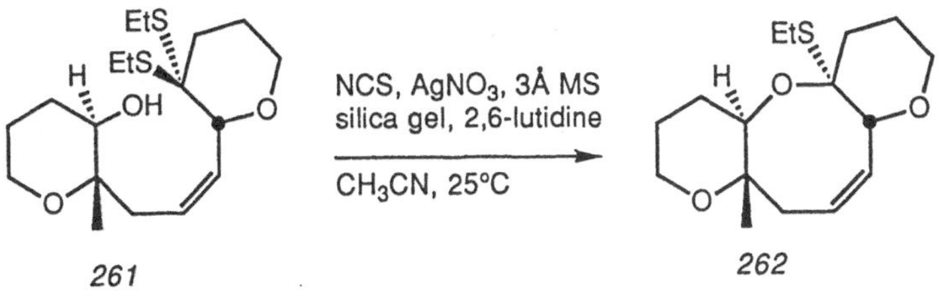

Scheme 88

163

6.3 Diphenol Coupling

Yamamura and coworkers [158] developed an oxidative cyclization method to construct biphenyl ether bonds by thallium trinitrate (TTN) oxidation of the corresponding *O,O'*-dihalophenols followed by zinc reduction. The antibiotic *piperazinomycin* (*266*) was synthesized using this method as a key cyclization step [159]. As shown in Scheme 89, the diketopiperazine *263* was subjected to TTN oxidation in MeOH to afford an inseparable mixture containing plausible intermediate *264*, which was directly reduced with zinc powder in AcOH-THF to give rise to the strained 14-membered biphenyl ether *265* in 19% yield together with two other isomers.

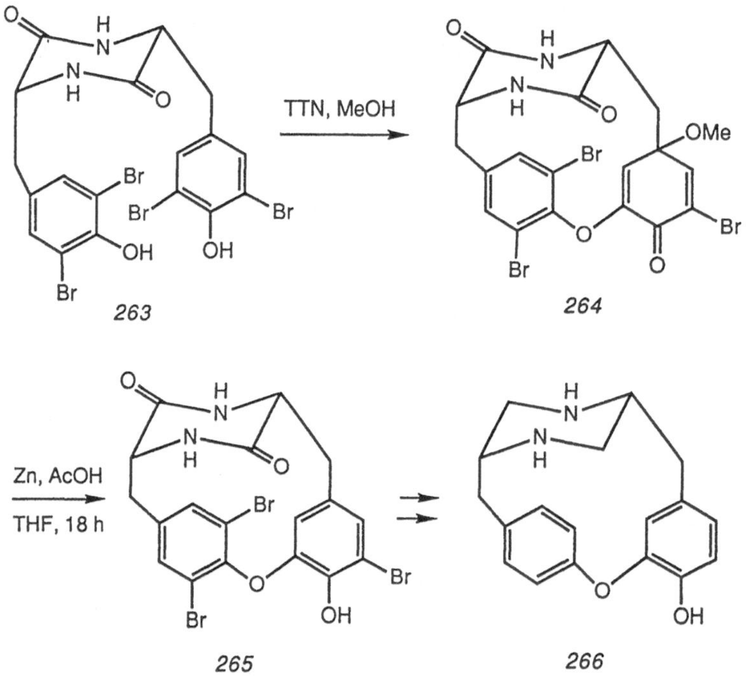

Scheme 89

The mechanism of the cyclization step above can be explained as indicated in Scheme 90. Thus, diphenol *263* is first oxidized by TTN to the Tl complex *267*, which collapses to intermediate *268*. Then the nucleophilic attack of MeOH and loss of HBr result in macrocycle *264*.

Ring closure direction can be controlled by employing different halogen substituents in both phenol sites [160]. For example, in the case where one phenol is flanked by two bromine atoms and the other by two chlorine atoms, bromine will be replaced by oxygen.

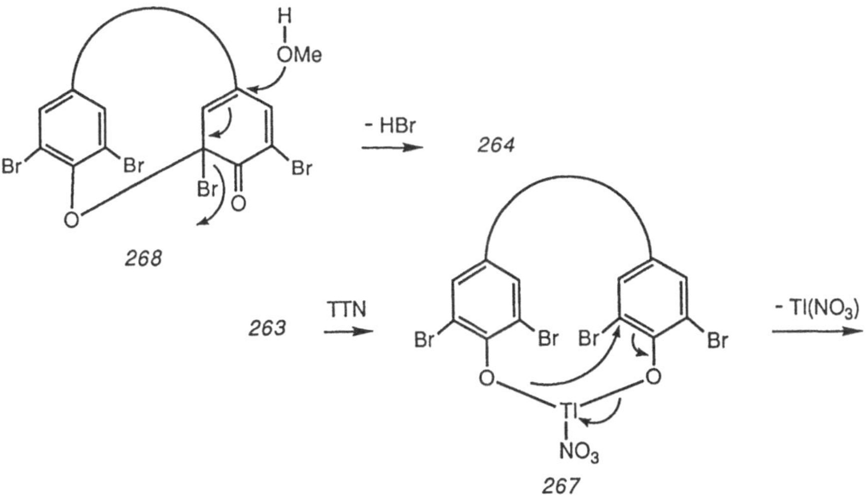

Scheme 90

Evans et al. [161] applied this strategy to the synthetic approach to the glycopeptide antibiotic *vancomycin* and modified the reduction step by using $CrCl_2$ instead of Zn/AcOH. In the synthetic studies on *vancomycin* by Yamamura and coworkers [162] even no reduction step was necessary. This macrocyclization method has also been used in the synthesis of *OF4949-III* and *K-13* [163].

7 Methods Involving Amine Formation

7.1 Iminium Cyclization

In the synthesis of the spermidine alkaloid *(+)-dihydroperiphylline* (*271*), Kibayashi and coworkers [164] employed intramolecular iminium cyclization to achieve the ring closure. As shown in Scheme 91, cleavage of the Boc group of aldehyde *269* resulted in *in situ* cyclization to the 13-membered ring via iminium formation and subsequent reduction by $NaBH_4$ afforded *270* in 61% yield.

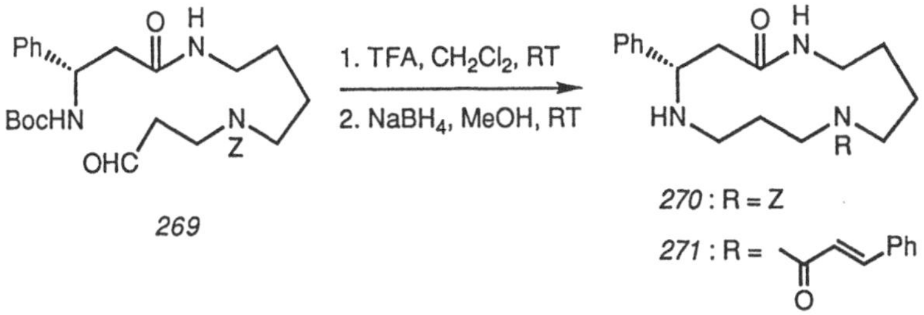

Scheme 91

165

7.2 Palladium-Catalyzed Macroheterocyclization

Trost and Cossy [165] developed this method to construct the macrocyclic ring of the spermidine alkaloid *inandenin-12-one* (*274*). As shown in Scheme 92, subjection of amino ketone *272* to 10 mol% of (Ph₃P)₄Pd and 8 mol% of 1,4-bis(diphenylphosphino)butane (DPPB) in THF at elevated temperature led, in high yield, to *273*, which was converted to *274*.

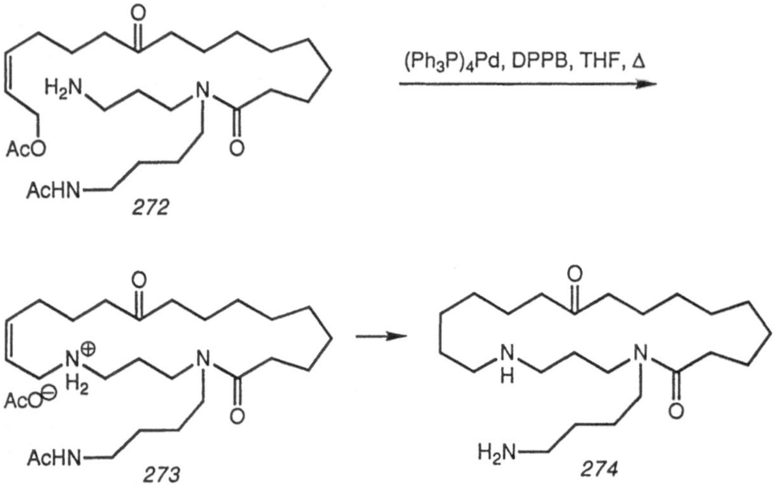

Scheme 92

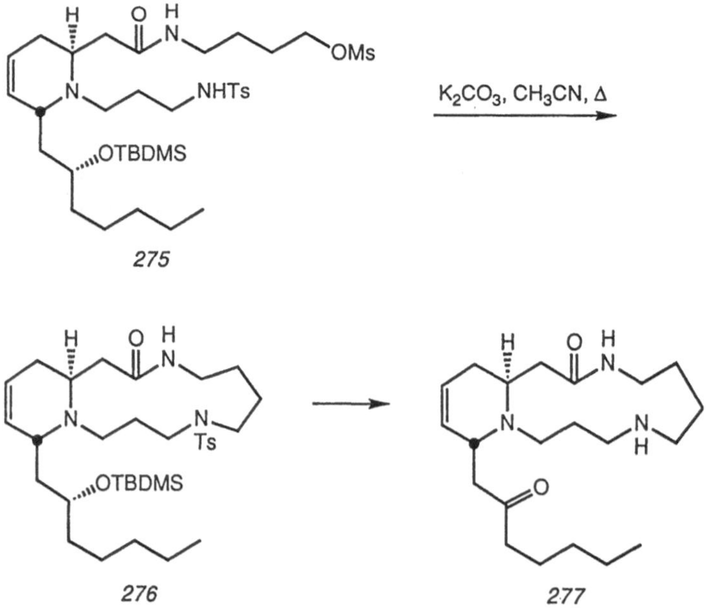

Scheme 93

7.3 Alkylation of Sulfonamide

Weinreb and coworkers [166] achieved the synthesis of the spermidine alkaloid *anhydrocannabistivene* (*277*) using alkylation of sulfonamide as a ring closure means. As shown in Scheme 93, the linear precursor *275* was treated with an excess of K_2CO_3 in refluxing CH_3CN to afford the desired lactam *276* in 58% yield.

8 Methods Involving *exo* Ring Formation

8.1 Diels-Alder Reaction

Intramolecular Diels-Alder reaction can be used as a macrocyclization means. Thomas and Whitehead [167] applied this approach to the synthesis of the 13-membered cytochalasan *proxiphomin* (*280*). As shown in Scheme 94, the long chain precursor *278* was heated in toluene at 100 °C for 5 h to give the 13-membered skeleton *279* and the *endo* adduct (52 : 48) in 52% yield. There are several other examples of the application of intramolecular Diels-Alder reaction to the synthesis of macrocyclic natural products [168].

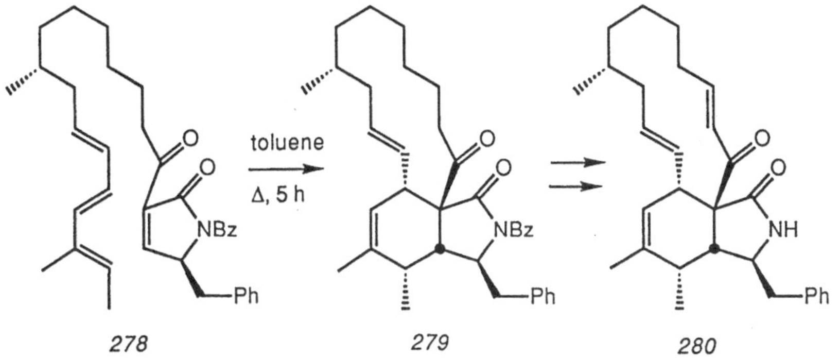

Scheme 94

8.2 [3 + 2] Dipolar Cycloaddition

In the studies toward the total synthesis of the antitumor substance *maytansine* (*131*) by Ko and Confalone [169], an intramolecular [3 + 2] dipolar cycloaddition route was developed to construct the macrocyclic ring. As shown in Scheme 95, model compound *283* was synthesized from precursor *281*. Treatment of *281* with *p*-chlorophenylisocyanate and Et_3N in toluene at 80 °C effected a smooth transformation to the macrocyclic isoazoline *283* in 68% yield, presumably via the nitrile oxide-olefin intermediate *282*.

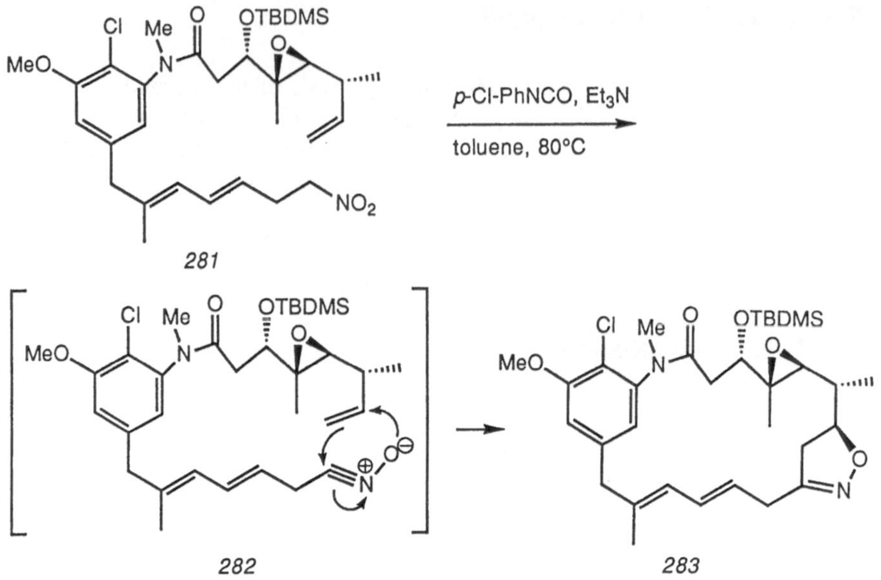

Scheme 95

8.3 Carbene-Olefin Coupling

Takahashi and coworkers [170] synthesized the antifungal diterpene *casbene* (*287*) by using carbene-olefin coupling for ring closure. As shown in Scheme 96, the

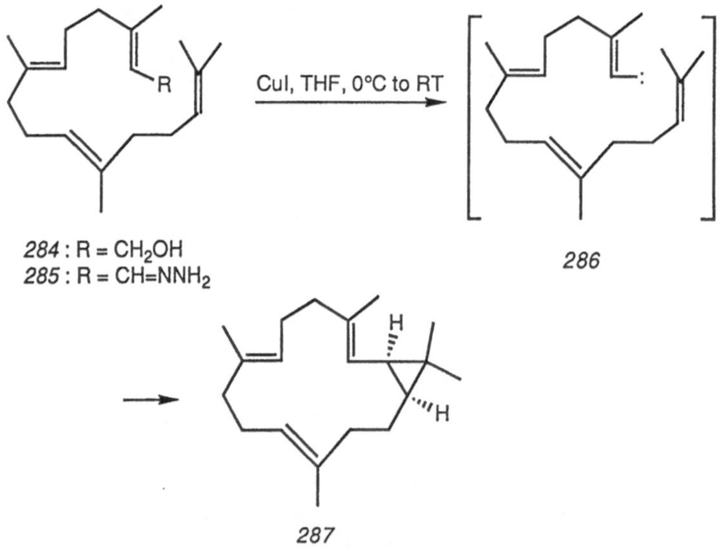

284 : R = CH₂OH
285 : R = CH=NNH₂

Scheme 96

linear precursor *284* was converted to the diazo derivative *285*. Subsequent treatment with 2 equiv of CuI in THF afforded a mixture of products via intermediate *286*. *Casbene* (*287*) was obtained in only 14% yield from *284* after chromatography.

9 Acknowledgement

We thank the "Schweizerischer Nationalfonds zur Förderung wissenschaftlicher Forschung" for generous support.

10 List of Symbols and Abbreviations

Ac	acetyl
acac	acetylacetonate
AIBN	2,2′-azobis(2-methylpropionitrile)
Bn	benzyl
Boc	*t*-butoxycarbonyl
BOP-Cl	*N,N*-bis(2-oxo-3-oxazolidinyl)phosphordiamidic chloride
Bu	butyl
Bz	benzoyl
DABCO	1,4-diazabicyclo[2.2.2]octane
DBN	1,5-diazabicyclo[4.3.0]non-5-ene
DCC	dicyclohexylcarbodiimide
DEAD	diethyl azodicarboxylate
DEPC	diethyl phosphorocyanidate
DMAP	4-dimethylaminopyridine
DME	1,2-dimethoxyethane
DMF	*N,N*-dimethylformamide
DPDS	2,2′-dipyridyldisulfide
DPPA	diphenylphosphoryl azide
DPPB	1,4-bis(diphenylphosphino)butane
DPPP	1,3-bis(diphenylphosphino)propane
EE	ethoxyethyl
equiv	equivalent(s)
Et	ethyl
h	hour(s)
HOBt	hydroxybenztriazole
HOSu	*N*-hydroxysuccinimide
HMAP	hexamethylphosphoric triamide
Im	imidazole-1-yl
LDA	lithium diisopropylamide
Me	methyl
MEM	2-methoxyethoxymethyl
Mes	mesityl
min	minute(s)

mol	mole(s)
Ms	mesyl
MS	molecular sieve
MOM	methoxymethyl
MTM	methylthiomethyl
NCS	*N*-chlorosuccinimide
NMM	*N*-methylmorpholine
NMP	*N*-methylpyrrolidone
PDC	pyridinium dichromate
Ph	phenyl
Piv	pivaloyl
PMB	*p*-methoxybenzyl
4-PP	4-pyrrolidinopyridine
Pr	propyl
Py	pyridine
RT	room temperature
TBDMS	*t*-butyldimethylsilyl
TBDPS	*t*-butyldiphenylsilyl
TES	triethylsilyl
Tf	trifluoromethanesulfonyl
TFA	trifluoroacetic acid
TFAA	trifluoroacetic anhydride
THF	tetrahydrofuran
THP	tetrahydropyranyl
TIPS	triisopropylsilyl
TMS	trimethylsilyl
TPM	triphenylmethyl
Ts	*p*-toluenesulfonyl
TTN	thallium trinitrate
Z	benzyloxycarbonyl

11 References

1. Ruzicka L (1926) Helv Chim Acta 9: 230, 715, 1008
2. Woodward RB (1957) Angew Chem 69: 50
3. Trost BM, Verhoeven TR (1980) J Am Chem Soc 102: 4743
4. Omura S (1984) Macrolide antibiotics, academic, orlando
5. Guggisberg A, Hesse M (1983) In: Brossi A (ed) The Alkaloids, Academic, New York, vol 22, p 85
6. Krebs HC (1986) In: Herz W, Grisebach H, Kirby GW, Tamm C (eds) Progress in the chemistry of organic natural products, Springer, Vienna New York, vol 49, p 151
7. a) Nicolaou KC (1977) Tetrahedron 33: 683
   b) Back TG (1977) ibid 33: 3041
   c) Masamune S, Bates GS, Corcoran JW (1977) Angew Chem 89: 602; Int Engl ed 16: 585
   d) Paterson I, Mansuri MM (1985) Tetrahedron 41: 3569
   e) Boeckman RK, Goldstein SW (1988) in The total synthesis of natural products (ApSimon J ed), Wiley, New York, vol 7, p 1

8. Hesse M (1991) Ring enlargement in organic chemistry, Verlag Chemie, Weinheim
9. Meng Q, Hesse M (1990) Synlett 148
10. Ohtsuka Y, Niitsuma S, Tadokoro H, Hayashi T, Oishi T (1984) J Org Chem 49: 2326
11. Smith AB (1984) in Strategies and tactics in organic synthesis (Lindberg, T ed), Academic Press, Orlando, p 223
12. Rossa L, Vögtle F (1983) Top Curr Chem 113: 1
13. a) Illuminati G, Mandolini L (1981) Acc Chem Res 14: 95
    b) Galli C, Mandolini L (1982) J Chem Soc Chem Commun 251
14. Corey EJ, Nicolaou KC (1974) J Am Chem Soc 96: 5614
15. Corey EJ, Clark DA (1979) Tetrahedron Lett 2875
16. Plata DJ, Kallmerten J (1988) J Am Chem Soc 110: 4041
17. Gerlach H, Thalmann A (1974) Helv Chim Acta 57: 2661
18. Gerlach H, Oertle K, Thalmann A Servi S (1975) ibid 58: 2036
19. Corey EJ, Brunelle DJ (1976) Tetrahedron Lett 3409
20. Corey EJ, Kim S, Yoo S, Nicolaou KC, Melvin LS, Brunelle DJ, Falck JR, Trybulski EJ, Lett R, Sheldrake PW (1978) J Am Chem Soc 100: 4620
21. Mukaiyama T, Usui M, Saigo K (1976) Chem Lett 49
22. Ley, SV, Anthony NJ, Armstrong A, Brasca MG, Clarke T, Culshaw D, Greck C, Grice P, Jones AB, Lygo B, Madin A, Sheppard RN, Slawin AMZ, Williams DJ (1989) Tetrahedron 45: 7161
23. White JD, Bolton GL (1990) J Am Chem Soc 112: 1626
24. Narasaka K, Masui T, Mukaiyama T (1977) Chem Lett 763
25. Narasaka K, Maruyama K, Mukaiyama T (1978) ibid 885
26. Mukaiyama T, Narasaka K, Kikuchi K (1977) ibid 441
27. Narasaka K, Yamaguchi M, Mukaiyama T (1977) ibid 959
28. a) Masamune S, Kim CU, Wilson KE, Spessard GO, Georghiou PE, Bates GS (1975) J Am Chem Soc 97: 3512
    b) Masamune S, Yamamoto H, Kamata, S, Fukuzawa A (1975) ibid 97: 3513
    c) Masamune S, Kamata S, Schilling W (1975) ibid 97: 3515
29. Huang J, Meinwald J (1981) ibid 103: 861
30. Masamune S, Hayase Y, Schilling W, Chan WK, Bates GS (1977) ibid 99: 6756
31. Tatsuta K, Nakagawa A, Maniwa S, Kinoshita M (1980) Tetrahedron Lett 21: 1479
32. Tatsuta K, Amemiya Y, Maniwa S, Kinoshita M (1980) ibid 21: 2837
33. Roush WR, Blizzard TA (1983) J Org Chem 48: 758
34. Roush WR, Blizzard TA (1984) ibid 49: 4332
35. Jeker N, Tamm C (1988) Helv Chim Acta 71: 1904
36. Inanaga J, Hirata K, Saeki H, Katsuki T, Yamaguchi M (1979) Bull Chem Soc Jpn 52: 1989
37. a) Niwa H, Miyachi Y, Uosaki Y, Kuroda A, Ishiwata H, Yamada K (1986) Tetrahedron Lett 27: 4609
    b) Niwa H, Okamoto O, Yamada K (1988) ibid 29: 5139
38. Thijs L, Egenberger DM, Zwanenburg B (1989) ibid 30, 2153
39. Kaiho T, Masamune S, Toyoda T (1982) J Org Chem 47: 1612
40. Shishido K, Tanaka K, Fukumoto K, Kametani T (1985) Chem Pharm Bull 33: 532
41. Corey EJ, Hua DH, Pan BC, Seitz SP, (1982) J Am Chem Soc 104: 6818
42. Taub D, Girotra NN, Hoffsommer RD, Kuo CH, Slates HL, Weber S, Wendler NL (1968) Tetrahedron 24: 2443
43. White JD, Lodwig SN, Trammell GL, Fleming MP (1974) Tetrahedron Lett 3263
44. a) Mitsunobu O (1981) Synthesis page No. 1
    b) Kurihara T, Nakajima Y, Mitsunobu O (1976) Tetrahedron Lett 2455
45. Seebach D, Adam G, Zibuck R, Simon W, Rouilly M, Meyer WL, Hinton JF, Privett TA, Templeton GE, Heiny DK, Gisi U, Binder H (1989) Liebigs Ann Chem 1233
46. Zibuck R, Liverton NJ, Smith AB (1986) J Am Chem Soc 108: 2451
47. Vorbrüggen H, Krolikiewicz K (1977) Angew Chem 89: 914; Int Engl ed 16: 876

48. Asaoka M, Yanagida N, Takei H (1980) Tetrahedron Lett 21: 4611
49. Boden EP, Keck GE (1985) J Org Chem 50: 2394
50. Hanessian S, Ugolini A, Dubé D, Hodges PJ, André C (1986) J Am Chem Soc 108: 2776
51. a) Parmee ER, Steel PG, Thomas EJ (1989) J Chem Soc Chem Commun 1250
    b) White JD, Amedio JC (1989) J Org Chem 54: 736
52. Williams DR, Barner BA, Nishitani K, Philips JG (1982) J Am Chem Soc 104: 4708
53. Colvin EW, Purcell TA, Raphael RA (1976) J Chem Soc Perkin Trans I 1718
54. Gais HJ, Lied T (1984) Angew Chem 96: 143; Int Engl ed 23: 145
55. Narasaka K, Sakakura T, Uchimaru T, Guédin-Vuong D (1984) J Am Chem Soc 106: 2954
56. Steliou K, Szczygielska-Nowosielska A, Favre A, Poupart MA, Hanessian S (1980) ibid 102: 7578
57. a) Shanzer A, Mayer-Shochet N (1980) J Chem Soc. Chem Commun 176
    b) Shanzer A, Berman E (1980) ibid 259
58. Boeckman RK, Pruitt JR, (1989) J Am Chem Soc 111: 8286
59. Wasserman HH, Gambale RJ, Pulwer MJ (1981) Tetrahedron 37: 4059
60. Vedejs E, Ahmad S, Larsen SD, Westwood S (1987) J Org Chem 52: 3937
61. a) White JD, Amedio JC, Gut S, Jayasinghe L (1989) ibid 54: 4268
    b) Karim MR, Sampson P (1990) ibid 55: 598
62. Esmond R, Fraser-Reid B, Jarvis BB (1982) ibid 47: 3358
63. Deslongchamps P (1983) Stereoelectronic effects in organic chemistry, Pergamon, Oxford
64. Woodward RB, Logusch E, Nambiar KP, Sakan K, Ward DE, Au-Yeung B-W, Balaram P, Browne LJ, Card PJ, Chen CH, Chênevert RB, Fliri A, Frobel K, Gais H-J, Garrat DG, Hayakawa K, Heggie W, Hesson DP, Hoppe D, Hoppe I, Hyatt JA, Ikeda D, Jacobi PA, Kim KS, Kobuke Y, Kojima K, Krowicki K, Lee VJ, Leutert T, Malchenko S, Martens J, Matthews RS, Ong BS, Press JB, Rajan Babu TV, Rousseau G, Sauter HM, Suzuki M, Tatsuta K, Tolbert LM, Truesdale EA, Uchita I, Ueda Y, Uyehara T, Vasella AT, Vladuchick WC, Wade PA, Williams RM, Wong HN-C (1981) J Am Chem Soc 103: 3213
65. Stork G, Rychnovsky SD (1987) ibid 109: 1565
66. Hikota M, Tone H, Horita K, Yonemitsu O (1990) Tetrahedron 46: 4613
67. Smith AB, Schow SR, Bloom JD, Thompson AS, Winzenberg KN (1982) J Am Chem Soc 104: 4015
68. a) Gross E, Meinhofer J (1983) The peptides, Academic, New York, vol 1-5, 1979
    b) Bodanzsky M (1984) Principles of peptide synthesis, Springer, Berlin Heidelberg New York
    c) Bodanzsky M, Bodanzsky A (1982) The practice of peptide synthesis, Springer, Berlin Heidelberg New York
    d) Kopple KD, (1972) J Pharm Sci 61: 1345
69. a) Diago-Meseguer J, Palomo-Coll AL, Fernandez-Lizarbe JR, Zugaza-Bilbao A (1980) Synthesis 547
    b) Tung RD, Dhaon MK, Rich DH (1986) J Org Chem 51: 3350
    c) Nakata M, Akiyama N, Kojima K, Masuda H, Kinoshita M, Tatsuta K (1990) Tetrahedron Lett 31: 1585
    d) Nakata M, Akiyama N, Kamata J, Kojima K, Masuda H, Kinoshita M, Tatsuta K (1990) Tetrahedron 46: 4629
    e) Baker R, Castro JL (1990) J Chem Soc Chem Commun 378 (1989); J Chem Soc Perkin Trans I 47
70. a) Shioiri T, Ninomiya K, Yamada S (1972) J Am Chem Soc 94: 6203
    b) Yamada S, Kasai Y, Shioiri T (1973) Tetrahedron Lett 1595
    c) Takeuchi Y, Yamada S (1974) Chem Pharm Bull 22: 832, 841
    d) Shioiri T, Yamada S (1974) ibid 22: 849, 855, 859
    e) Yamada S, Ikota N, Shioiri T, Tachibana S (1975) J Am Chem Soc 97: 7174

f) Shioiri T, Yokoyama Y, Kasai Y, Yamada S (1976) Tetrahedron 32: 2211
g) Hamada Y, Rishi S, Shioiri T, Yamada S (1977) Chem Pharm Bull 25: 224
71. Kurokawa N, Ohfune Y (1986) J Am Chem Soc 108: 6043
72. a) de Laszlo SE, Ley SV, Porter RA (1986) J Chem Soc Chem Commun 344
    b) Hamada Y, Shibata M, Shioiri T (1985) Tetrahedron Lett 26: 5155
    c) Evans DA, Weber AE (1987) J Am Chem Soc 109: 7151
    d) Boger DL, Yohannes D (1988) J Org Chem 53: 487
73. Iio H, Nagaoka H, Kishi Y (1980) J Am Chem Soc 102: 7965
74. Durette PL, Baker F, Barker PL, Boger J, Bondy SS, Hammond ML, Lanza TJ, Pessolano AA, Caldwell CG (1990) Tetrahedron Lett 31: 1237
75. Corey EJ, Weigel LO, Chamberlin AR, Cho H, Hua DH (1980) J Am Chem Soc 102: 6613
76. Corey EJ, Clark DA (1980) Tetrahedron Lett 21: 2045
77. Endo Y, Shudo K, Itai A, Hasegawa M, Sakai S (1986) Tetrahedron 42: 5905
78. Nakatsuka S, Masuda T, Goto T (1987) Tetrahedron Lett 28: 3671
79. Nutt RF, Chen K-M, Joullié MM (1984) J Org Chem 49: 1013
80. a) Schmidt U, Lieberknecht A, Griesser H, Talbiersky J (1982) ibid 47: 3261
    b) Schmidt U (1986) Pure Appl. Chem 58: 295
81. Evans DA, Ellman JA (1989) J Am Chem Soc 111: 1063
82. a) Schmidt U, Lieberknecht A, Griesser H, Utz R, Beuttler T, Bartkowiak F (1986) Synthesis 361
    b) Schmidt U, Weller D (1986) Tetrahedron Lett 27: 3495
    c) Schmidt U, Griesser, H (1986) ibid 27: 163
    d) Schmidt U, Weller D, Holder A, Lieberknecht A (1988) ibid 29: 3227
    e) Schmidt U, Kroner M, Griesser H (1988) ibid 29: 4407
    f) Schmidt U, Kroner M, Griesser H (1988) ibid 29: 3057
    g) Heffner RJ, Joullié MM (1989) ibid 30: 7021
83. Wieland T, Vogeler K (1962) Angew Chem 74: 904
84. Wälchli-Schaer E, Eugster CH (1978) Helv Chim Acta 61: 928
85. Bald E, Saigo K, Mukaiyama T (1975) Chem Lett 1163
86. Jones TK, Reamer RA, Desmond R, Mills SG (1990) J Am Chem Soc 112: 2998
87. Nakatsuka M, Ragan JA, Sammakia T, Smith DB, Uehling DE, Schreiber SL (1990) ibid 112: 5583
88. Jones AB, Villalobos A, Linde RG, Danishefsky SJ (1990) J Org Chem 55: 2786
89. a) Boeckman RK, Weidner CH, Perni RB, Napier JJ (1989) J Am Chem Soc 111: 8036
    b) Paquette LA, Macdonald D, Anderson LG, Wright J (1989) ibid 111: 8037
90. Yamamoto H, Maruoka K (1981) ibid 103: 6133
91. Collum DB, Chen S-C, Ganem B (1978) J Org Chem 43: 4393
92. Ogawa M, Kuriya N, Natsume M (1984) Tetrahedron Lett 25: 969
93. Magnus P, Annoura H, Harling J (1990) J Org Chem 55: 1709
94. Ju J, Reddy BR, Khan M, Nicholas KM (1989) ibid 54: 5426
95. Smith AB, Guaciaro MA, Schow SR, Wovkulich PM, Toder BH, Hall TW (1981) J Am Chem Soc 103: 219
96. a) Mukaiyama T, Hayashi M (1974) Chem Lett 15
    b) Mukaiyama T, Ishida A (1975) ibid 319
    c) Banno K, Mukaiyama T (1975) ibid 741
97. Smith AB, Lupo AT, Ohba M, Chen K (1989) J Am Chem Soc 111: 6648
98. Cockerill GS, Kocienski P, Treadgold R (1985) J Chem Soc, Perkin Trans I 2093, 2101
99. Reeves RL (1966) in The chemistry of the carbonyl group (Patai, S, ed), Wiley, London, p 593
100. Takeda K, Yano S, Yoshii E (1988) Tetrahedron Lett 29: 6951
101. Takai K, Kimura K, Kuroda T, Hiyama T, Nozaki H (1983) ibid 24: 5281
102. Rowley M, Kishi Y (1988) ibid 29: 4909
103. Schreiber SL, Meyers HV (1988) J Am Chem Soc 110: 5198

104. a) Okude Y, Hirano S, Hiyama T, Nozaki H (1977) ibid 99: 3179
    b) Hiyama T, Okude Y, Kimura K, Nozaki H (1982) Bull Chem Soc Jpn 55: 561
    c) Buse CT, Heathcock CH (1978) Tetrahedron Lett 1685
105. Shibuya H, Ohashi K, Kawashima K, Hori K, Murakami N, Kitagawa I (1986) Chem Lett 85
106. Still WC, Mobilio D (1983) J Org Chem 48: 4785
107. a) Marshall JA, DeHoff BS, Crooks SL (1987) Tetrahedron Lett 28: 527
    b) Marshall JA, Crooks SL, DeHoff BS (1988) J Org Chem 53: 1616
108. Danishefsky SJ, Mantlo NB, Yamashita DS (1988) J Am Chem Soc 110: 6890
109. Moriya T, Handa Y, Inanaga J, Yamaguchi M (1988) Tetrahedron Lett 29: 6947
110. a) Maruoka K, Hashimoto S, Kitagawa Y, Yamamoto H, Nozaki H (1977) J Am Chem Soc 99: 7705
    b) Tsuji J, Mandai T (1978) Tetrahedron Lett 1817
111. a) Overman LE, Blumenkopf TA, Castaneda A, Thompson AS (1986) J Am Chem Soc 108: 3516
    b) Overman LE, Thompson AS (1988) ibid 110: 2248
    c) Blumenkopf TA, Bratz M, Castaneda A, Look GC, Overman LE, Rodriguez D, Thompson AS (1990) ibid 112: 4386
112. Trost BM, Lee DC (1988) ibid 110: 6556
113. Takahashi T, Nagashima T, Tsuji J (1981) Tetrahedron Lett 22: 1359
114. a) Takahashi T, Ikeda H, Tsuji J (1981) ibid 22: 1363
    b) Takahashi T, Nemoto H, Tsuji J (1983) ibid 24: 2005
    c) Takahashi T, Nemoto H, Kanda Y, Tsuji J (1987) Heterocycles 25: 139
    d) Takahashi T, Shimizu K, Doi T, Tsuji J, Fukazawa Y (1988) J Am Chem Soc 110: 2674
115. Marshall JA, Cleary DG (1986) J Org Chem 51: 858
116. a) Trost BM, Gowland FW (1979) ibid 44: 3448
    b) For a recent review, see: Trost BM (1989) Angew Chem 101: 1199; Int Engl ed 28: 1173
117. Trost BM, Ohmori M, Boyd SA, Okawara H, Brickner SJ (1989) J Am Chem Soc 111: 8281
118. Trost BM, Brickner SJ (1983) ibid 105: 568
119. a) Marshall JA, Andrews RC (1986) Tetrahedron Lett 27: 5197
    b) Marshall JA, Andrews RC, Lebioda L (1987) J Org Chem 52: 2378
120. Trost BM, Warner RW (1982) J Am Chem Soc 104: 6112
121. Kitagawa Y, Itoh A, Hashimoto S, Yamamoto H, Nozaki H (1977) ibid 99: 3864
122. Kodama M, Matsuki Y, Ito S (1975) Tetrahedron Lett 3065
123. Takahashi T, Kasuga K, Tsuji J (1978) ibid 4917
124. a) Shimada K, Kodama M, Ito S (1981) ibid 22: 4275
    b) Kodama M, Takahashi T, Kojima T, Ito S (1988) Tetrahedron 44: 7055
    c) Schwabe R, Farkas I, Pfander H (1988) Helv Chim Acta 71: 292
125. a) Nicholas KM (1987) Acc Chem Res 20: 207
    b) Schreiber SL, Klimas MT, Sammakia T (1987) J Am Chem Soc 109: 5749
126. Magnus P, Lewis RT, Huffman JC (1988) ibid 110: 6921
127. For a review, see: McMurry JE (1983) Acc Chem Res 16: 405
128. McMurry JE, Rico JG, Shih Y (1989) Tetrahedron Lett 30: 1173
129. McMurry JE, Dushin RG (1989) J Am Chem Soc 111: 8928
130. McMurry JE, Miller DD (1983) ibid 105: 1660
131. McMurry JE, Miller DD (1983) Tetrahedron Lett 24: 1885
132. a) Porter NA, Chang, VH-T (1987) J Am Chem Soc 109: 4976
    b) Porter NA, Chang, VH-T, Magnin DR, Wright BT (1988) ibid 110: 3554
    c) Porter NA, Lacher B, Chang VH-T, Magnin DR (1989) ibid 111: 8309
133. a) Cox NJG, Pattenden G, Mills SD (1989) Tetrahedron Lett 30: 621
    b) Hitchcock SA, Pattenden G (1990) ibid 31: 3641

134. Boger DL, Mathvink RJ (1990) J Am Chem Soc 112: 4008
135. a) Corey EJ, Hamanaka E (1964) ibid 86: 1641
     b) Corey EJ, Kirst HA (1972) ibid 94: 667
136. a) Corey EJ, Hamanaka E (1967) ibid 89: 2758
     b) Dauben WG, Beasley GH, Broadhurst MD, Muller B, Peppard DJ, Pesnelle P, Suter C (1974) ibid 96: 4724
     c) Crombie L, Kneen G, Pattenden G, Whybrow D (1980) J Chem Soc Perkin Trans I 1711
137. Inoue S, Uchida S, Kobayashi M, Sato S, Miyamoto O, Sato K (1985) Nippon Kagaku Kaishi 425; CA 103: 215039a
138. For a review, see: Sainsbury M (1980) Tetrahedron 36: 3327
139. a) Semmelhack MF, Ryono LS (1975) J Am Chem Soc 97: 3873
     b) Semmelhack MF, Helquist P, Jones LD, Keller L, Mendelson L, Ryono SL, Gorzynski Smith J, Stauffer RD (1981) ibid 103: 6460
140. a) Hart DJ, Hong WP (1985) J Org Chem 50: 3670
     b) Hart DJ, Hong WP, Hsu L-Y (1987) ibid 52: 4665
141. Brandt S, Marfat A, Helquist P (1979) Tetrahedron Lett 2193
142. Meyers AI, Reider PJ, Campbell AL (1980) J Am Chem Soc 102: 6597
143. a) Kumazawa S, Nakano Y, Kato T, Kitahara Y (1974) Tetrahedron Lett 1757
     b) Kato T, Kumazawa S, Kabuto C, Honda T, Kitahara Y (1975) ibid 2319
     c) Kato T, Kobayashi T, Kitahara Y (1975) ibid 3299
144. Kato T, Susuki M, Nakazima Y, Shimizu K, Kitahara Y (1977) Chem Lett 705
145. Utimoto K, Tanaka M, Kitai M, Nozaki H (1978) Tetrahedron Lett 2301
146. a) Gerlach H (1977) Helv Chim Acta 60: 3039
     b) Baker PM, Bycroft BW, Roberts JC (1967) J Chem Soc (C) 1913
147. Hurd RN, Shah DH (1973) J Org Chem 38: 390
148. a) Woodward RB (1973) Pure App Chem 33: 145
     b) Eschenmoser A (1988) Angew Chem 100: 5; Int Engl ed 27: 5
149. a) Fässler A, Pfaltz A, Kräutler B, Eschenmoser A (1984) J Chem Soc Chem Commun 1365
     b) Block MH, Zimmerman SC, Henderson GB, Turner SPD, Westwood SW, Leeper FJ, Battersby AR (1985) ibid 1061
150. a) Meyers AI, Comins DL, Roland DM, Henning R, Shimizu K (1979) J Am Chem Soc 101: 7104
     b) Smith AB, Dorsey BD, Visnick M, Maeda T, Malamas MS (1986) ibid 108: 3110
     c) Tius MA, Fauq A (1986) ibid 108: 6389
     d) Nicolaou KC, Daines RA, Chakraborty TK, Ogawa Y (1988) ibid 110: 4685
     e) Smith AB, Rano TA, Chida N, Sulikowski GA (1990) J Org Chem 55: 1136
     f) Williams DR, McGill JM (1990) ibid 55: 3457
151. Akita H, Yamada H, Matsukura H, Nakata T, Oishi T (1990) Tetrahedron Lett 31: 1735
152. McMurry JE, Kees KL (1977) J Org Chem 42: 2655
153. Dauben WG, Wang T, Stephens RW (1990) Tetrahedron Lett 31: 2393
154. McMurry JE, Bosch GK (1987) J Org Chem 52: 4885
155. Marshall JA, Jenson TM, DeHoff BS (1986) ibid 51: 4316
156. a) Marshall JA, Nelson DJ, (1988) Tetrahedron Lett 29: 741
     b) Krebs A, Wehlage T, Kramer C-P (1990) ibid 31: 3533
157. a) Nicolaou KC, Duggan ME, Hwang C-K (1986) J Am Chem Soc 108: 2468
     b) Nicolaou KC, Prasad CVC, Hwang C-K, Duggan ME, Veale CA (1989) ibid 111: 5321
158. a) Noda H, Niwa M, Yamamura S (1981) Tetrahedron Lett 22: 3247
     b) Nishiyama S, Yamamura S (1982) ibid 23: 1281
     c) Nishiyama S, Suzuki T, Yamamura S (1982) Chem Lett 1851
     d) Nishiyama S, Yamamura S (1985) Bull Chem Soc Jpn 58: 3453
159. Nishiyama S, Nakamura K, Suzuki Y, Yamamura S (1986) Tetrahedron Lett 27: 4481

160. Inaba T, Umezawa I, Yuasa M, Inoue T, Mihashi S, Itokawa H, Ogura K (1987) J Org Chem 52: 2957
161. Evans DA, Ellman JA, DeVries KM (1989) J Am Chem Soc 111: 8912
162. Suzuki Y, Nishiyama S, Yamamura S (1989) Tetrahedron Lett 30: 6043
163. Nishiyama S, Suzuki Y, Yamamura S (1989) ibid 29: 559 (1988), 30: 379
164. Kaseda T, Kikuchi T, Kibayashi C (1989) ibid 30: 4539
165. Trost BM, Cossy J (1982) J Am Chem Soc 104: 6881
166. Bailey TR, Garigipati RS, Morton JA, Weinreb SM (1984) ibid 106: 3240
167. Thomas EJ, Whitehead JWF (1989) J Chem Soc Perkin Trans I 499
168. a) Stork G, Nakamura E (1983) J Am Chem Soc 105: 5510
     b) Schreiber SL, Kiessling LL (1988) ibid 110: 631
169. a) Ko SS, Confalone PN (1985) Tetrahedron 41: 3511
     b) Confalone PN (1990) J Heterocyclic Chem 27: 31
170. Toma K, Miyazaki E, Murae T, Takahashi T (1982) Chem Lett 863

Expanded Porphyrins

Jonathan L. Sessler and Anthony K. Burrell

Department of Chemistry and Biochemistry, University of Texas at Austin, Austin, Texas 78712, USA

Table of Contents

Topics in Current Chemistry, Vol. 161
© Springer-Verlag Berlin Heidelberg 1991

Although considerable effort has been devoted to the synthesis and study of porphyrins and other tetrapyrrolic macrocycles, larger aromatic pyrrole-containing systems, the so-called "expanded porphyrins", have received considerably less attention. Such systems, by virtue of containing a greater number of π electrons, a greater number of donating (e.g. pyrrolic) groups, or a larger central binding core, however, might have properties which differ substantially from their far better studied porphyrin analogues. In this review, the synthesis and properties of various large, pyrrole-, furan-, and thiopene-containing macrocycles will be discussed and the aromaticity properties of several of the better known expanded porphyrins, namely the sapphyrins, superphthalocyanines, pentaphyrins, texaphyrins, and vinylogous porphyrin-like annulenes, highlighted. In addition, to the extent they are known, the metal binding properties of the various "expanded" pyrrolic systems will be discussed and reviewed in terms of such variables as core size, donor number, and ligand geometry. Potential applications of the expanded porphyrins will also be presented with the use of certain systems as possible gadolinium (III) chelating agents for use in magnetic resonance imaging enhancement and as far-red absorbing photosensitizers for use in photodynamic therapy detailed. Opportunities for further work, both in regards to these potential applications and in terms of general expanded porphyrin chemical development, will also be presented.

1 Introduction and Scope

Although considerable effort has been devoted to the synthesis and study of porphyrins and other tetrapyrrolic macrocycles [1], larger aromatic pyrrole-containing systems, the so-called "expanded porphyrins", have received considerably less attention. Such systems, by virtue of containing a greater number of π-electrons, a greater number of donating (e.g. pyrrolic) groups, or a larger central binding core might have properties which differ substantially from their far better studied porphyrin analogues. In this review, the syntheses and properties of several large, pyrrole, furan, and thiophene containing macrocycles will be discussed.

Prior to this review, only two surveys of "expanded" porphyrin-like macrocycles had been published [2, 3]. The previous reports were confined to macrocycles which were capable of coordinating metals in a pentadentate ligand environment, and to a preliminary review of the so-called sapphyrin class of expanded porphyrins. Here, we have enlarged the scope of the survey to encompass the synthesis and general chemical properties of a much wider range of expanded porphyrin-like macrocycles. In addition, to the extent they are known, the metal coordination capabilities of these systems will be discussed and reviewed in terms of such variables as core size, donor atom numbers, and ligand geometry. Potential applications of the expanded porphyrins will also be presented with the use of some systems as either possible gadolinium (III) chelating agents for use in magnetic resonance imaging (MRI) enhancement or as far-red absorbing photosensitizers for the use in photodynamic therapy (PDT) highlighted, with the latter application receiving particular emphasis. Opportunities for further work, both in regards to these potential applications and in terms of general "expanded porphyrin" chemical development, will also be presented. The present review was undertaken due to the excitement currently being generated by these potential applications and the perceived opportunities for further work they are presently engendering.

The porphyrins are arguably the best ligands in existence, forming coordination complexes with elements of almost the entire periodic table. The relative stability of the porphyrin macrocycle has also enabled an astoundingly diverse organic chemistry. This core and its unique properties have also been examined by almost every physical technique known. With the amount of effort that has been directed towards the study of the porphyrins and their properties, it is not surprising that other porphyrin-like macrocycles have begun to attract increasing attention. In the last few years, in particular, considerable effort has been devoted to exploring such systems. Much of this effort has been concerned with the chemistry of larger porphyrin-like systems, the so-called expanded porphyrins, and it is these systems which are the subject of the present review.

To limit the size of this report, a restriction was placed on the compounds that were eligible for inclusion. Firstly, to be included, the compound had to contain at least one five-membered heterocycle in its structure. Secondly, the number of atoms in the internal ring pathway of the macrocycle was required to contain at least 17 atoms (one more than the number of atoms present in the inner core of porphine; Fig. 1). Thus this definition, which is in two parts, eliminates conjugated macrocycles containing other small heterocycles such as pyridines. It also rules

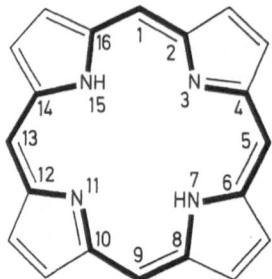

Fig. 1

out inclusion of various non-expanded porphyrin analogues such as, e.g. Vogel's recently reported porphycene [4]. It does, however, include macrocycles that are not aromatic. The justification for this latter decision derives from the potential benefit which might result from making comparisons between the spectral and coordination properties of a variety of closely related aromatic and non-aromatic systems.

This review is divided into sections based on type; each section deals with separate groups of related expanded porphyrins. Although there is a certain flexibility in these assignments, a clear attempt has been made to place similar macrocycles in the same section. Herein we describe the expanded porphyrin systems that have been reported up to January 1991. Every attempt has been made to ensure the thoroughness of this review, but due to the potential breadth of this field omissions may have been made. The authors would like to apologize in advance for any incompleteness or inaccuracy which might, therefore, result.

2 Homoporphyrins

The porphyrins are near-planar aromatic macrocycles. The insertion of a single carbon between the *meso* and α-pyrrolic carbons disrupts the aromaticity and forms a new group of expanded porphyrins referred to as "homoporphyrins".

The first reported formation of a homoporphyrin came from the group of Callot [5]. Treatment of N-ethoxycarbonylmethyl-*meso*-tetraphenylporphyrin 1 with Ni(acac)$_2$ led to the new expanded porphyrins 2 and 3 (Scheme 1) [5, 6]. The single crystal X-ray structure of 3 was determined [7, 8] (Fig. 2) and was found to reveal an essentially square planar nickel coordination involving the four nitrogens of the pyrrole bases. The pyrrole rings in 3 are planar but rotated and inclined with respect to the four-nitrogen least-squares plane. The striking distortions in the planarity of the macrocycle most likely result from the insertion of the saturated carbon into what was originally a methine bridge in the starting porphyrin. The atoms of the interior ring of the homoporphyrin skeleton form a delocalized (but non-aromatic) π-electron system. As a result, although the carbon atoms do not lie in a single plane, the bonding about each core carbon atom is in every case planar (the sum of the angles equals 360°).

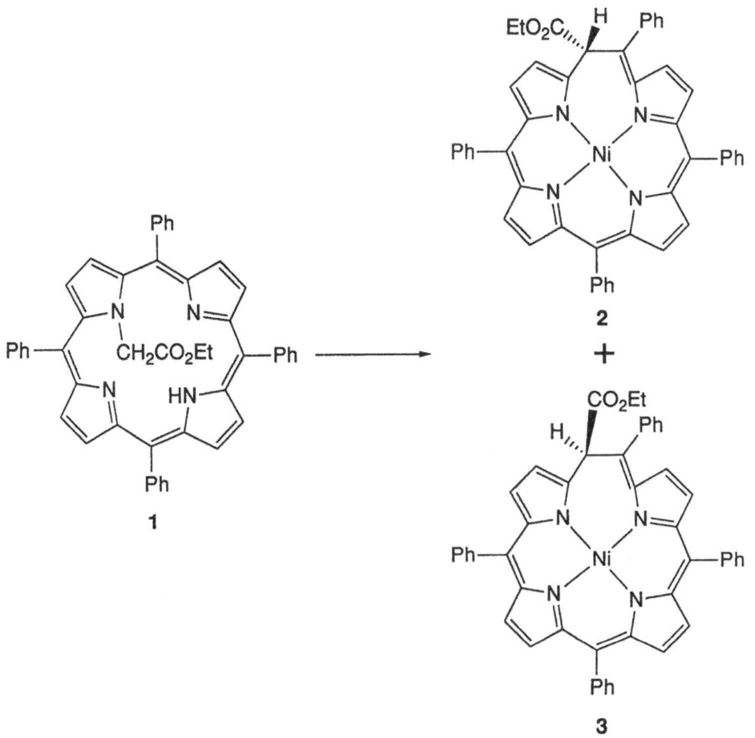

Scheme 1

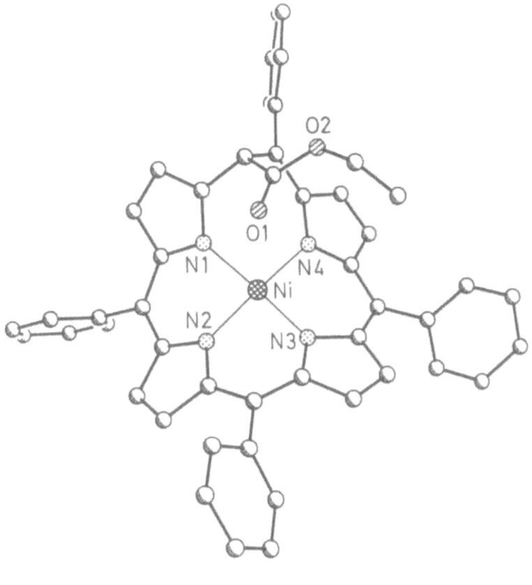

Fig. 2

181

At higher temperatures an equilibrium between **2** and **3** (60/40) is established which can be explained by ring inversion. Kinetic measurements are consistent with inversion barriers of $\Delta G^{*} = 29.8 \pm 0.2$ kcal/mol and 30.0 ± 0.2 kcal/mol for **2** → **3** and **3** → **2** interconversion, respectively [5, 6]. In addition to the inversion reaction, at higher temperatures these complexes also exhibit hydrogen migration followed, eventually, by cyclopropane migration. For instance, when either **2** or **3** is heated to temperatures greater than 160 °C, an equilibrium is attained between the four products **2**, **3**, **4**, and **5** (Scheme 2) [9]. The same mixture can be obtained by treating **2** or **3** with acid at room temperature [10]. Heating the equilibrium mixture of compounds **2–5** further (to temperatures higher than 200 °C) produces a mixture of four cyclopropyl annulated chlorins **6–9** [5, 6]. The visible spectra of all of these homoporphyrins (**2–5**) are similar to one another. They all display an intense absorption at approximately 450 nm ($\varepsilon = 8 \times 10^4$ cm · mol^{-1}) and two other bands in the 580–760 nm spectral region.

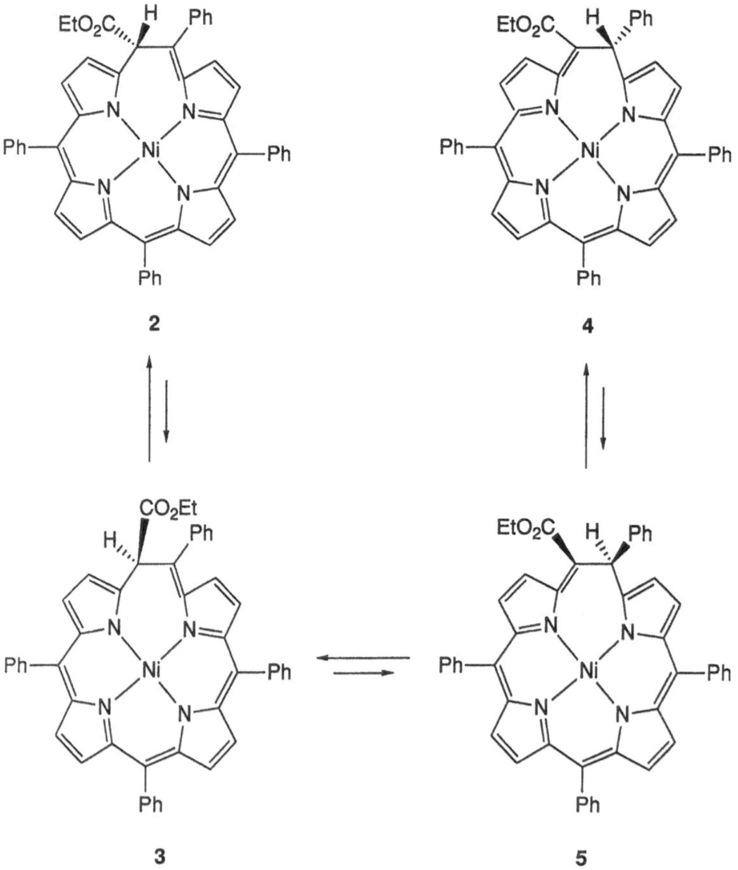

Scheme 2

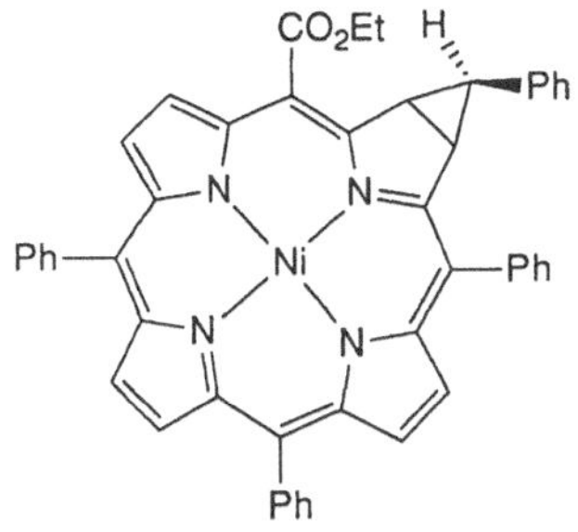

6

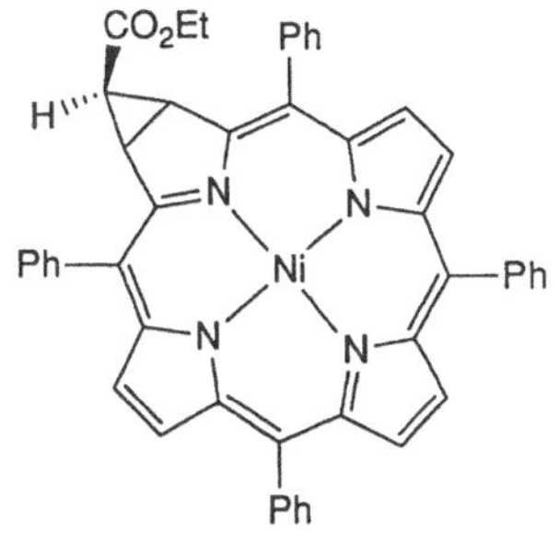

8

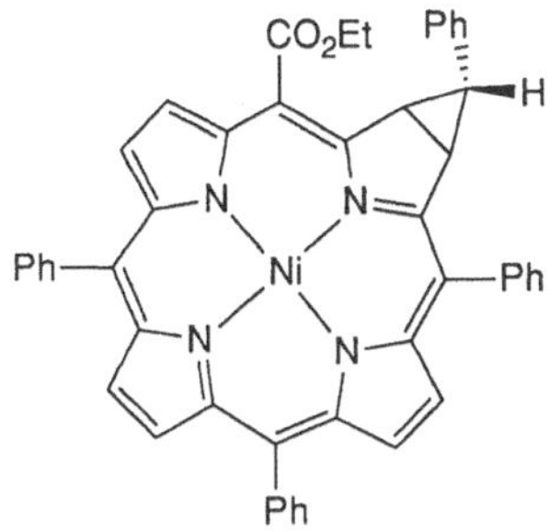

7

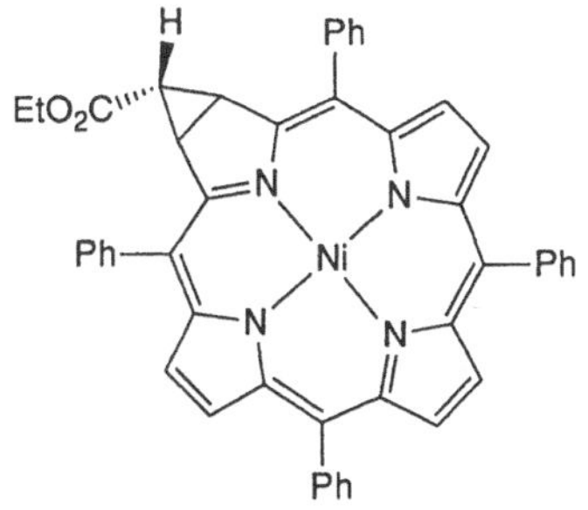

9

Subjecting **2** or **3** to strongly acidic conditions (e.g. 1 M CF_3CO_2H or concentrated HCl) leads to rapid demetalation and the production of two isomeric free-base macrocycles **10** and **11** [11, 12]. The structure of **10** was confirmed by its spectral properties and its remetalation to form the nickel complex **5**. On the other hand, the structure of **11** was determined solely by its spectral properties. For instance, the ^{13}C NMR spectrum of **11** demonstrated the presence of a fully unsaturated system. Typical also was the presence of pyrrolic proton signals in the 6–7 ppm range of the ^1H NMR spectrum [12]. Compound **11** was very unstable

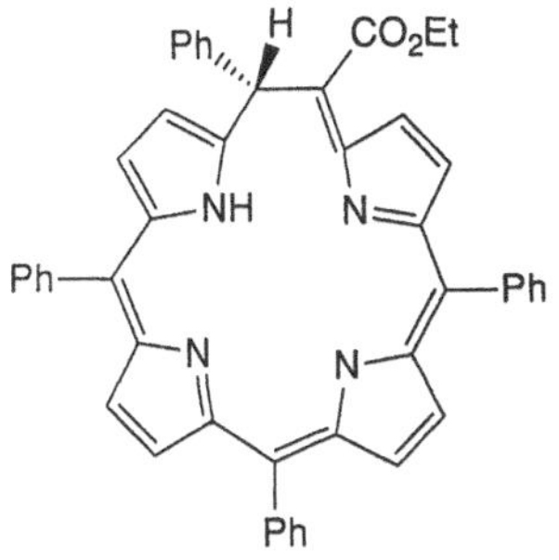

10

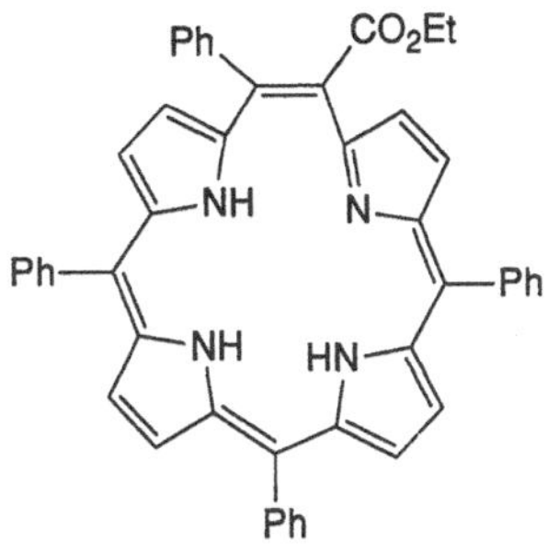

11

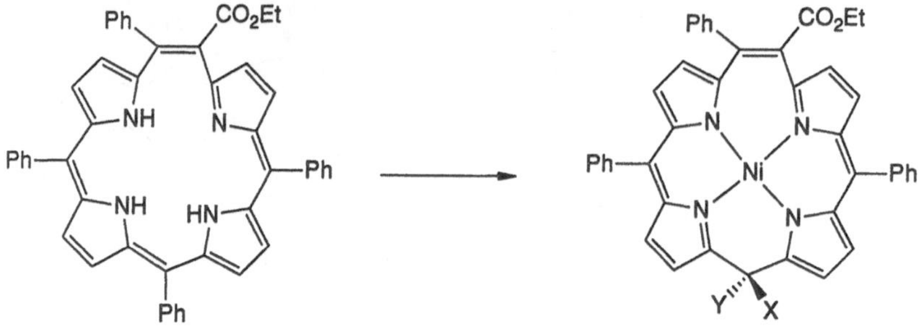

Scheme 3

12 X = OMe, Y = Ph
13 X = OH, Y = Ph
14 X = Ph, Y= OMe
15 X = Ph, Y = OH

and was even found T₀ undergo decomposition in the solid state at 0 °C to yield a multitude of polar products. However, metalation of **11** with Ni(II) under an inert atmosphere gave **2** in 15% yield. When the analogous metalation reaction was carried out in nucleophilic solvents in the presence of oxygen, the compounds **12–15** were obtained (Scheme 3). The structure of the alcohol **12** was determined by single crystal X-ray analysis [13]. As was the case for **2**, the nickel was found to be coordinated in a square planar environment. The homoporphyrin skeleton in this structure, however, is far from planar. In fact, the overall effect of the distortions of the skeleton is to generate a saddle-shaped surface (Fig. 3). Furthermore, the ethylene bridge was found to lie in a "chimney-like" position, and, as a consequence, it precludes any conjugation between the neighboring pyrrole rings. Not surprisingly, macrocycle **12** was found to display little, if any aromatic character.

Further treatment of **12–15** with acids enabled isolation of the intermediate cationic macrocycle **16** (Scheme 4) [12]. Acidic solutions of **16** were stable. Pure

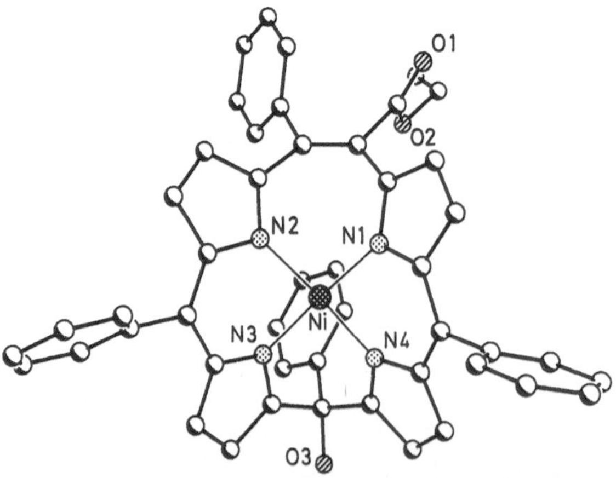

Fig. 3

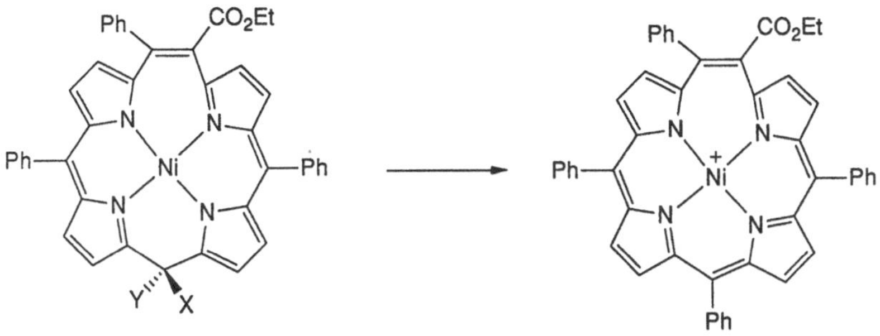

12 X = OMe, Y = Ph
13 X = OH, Y = Ph
14 X = Ph, Y= OMe
15 X = Ph, Y = OH

16

Scheme 4

crystals of **16**, however, could not be obtained free of the hydrolysis products **13** and **15**. Methanolysis of **16**, gave isomer **12** exclusively. On the other hand, zinc-acetic acid reduction of **16**, yielded small amounts of **3** and **4**, and a new homoporphyrin **17**, which on prolonged heating at reflux in o-dichlorobenzene gave isomer **18** (Scheme 5). Here, it is interesting to note that in the presence of base, compound **17** is converted to **2**, while under the same reaction conditions isomer **18** is stable and remains unchanged.

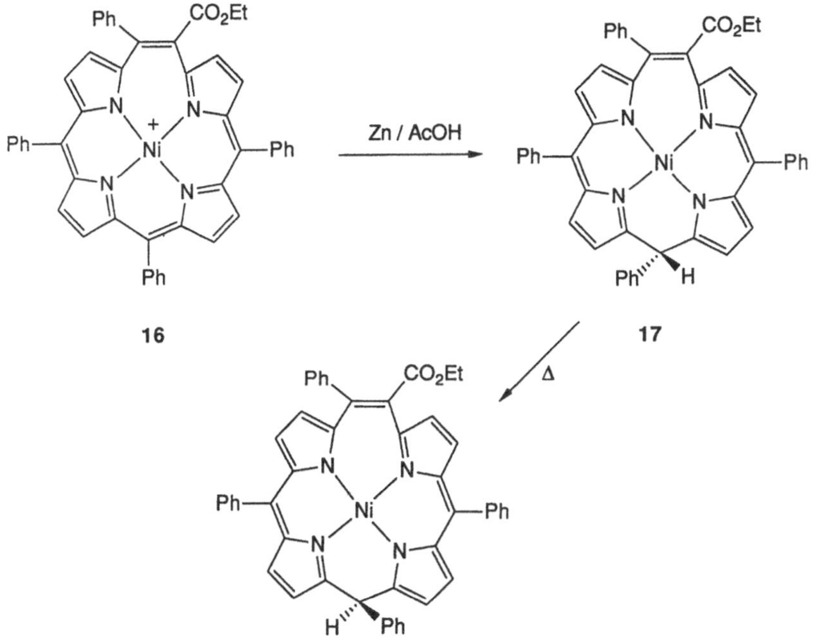

16

Zn / AcOH

17

Δ

18

Scheme 5

185

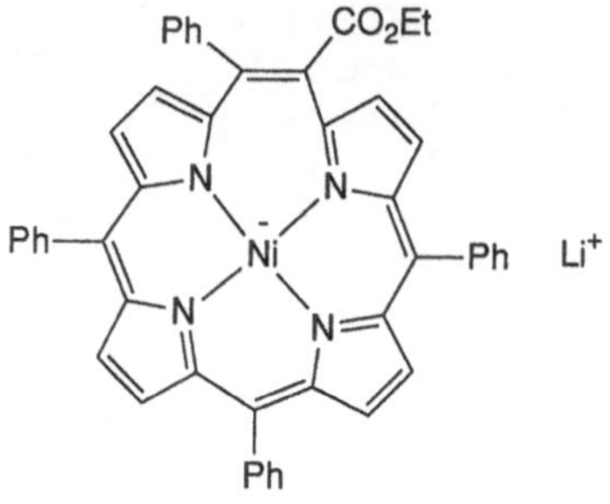

19

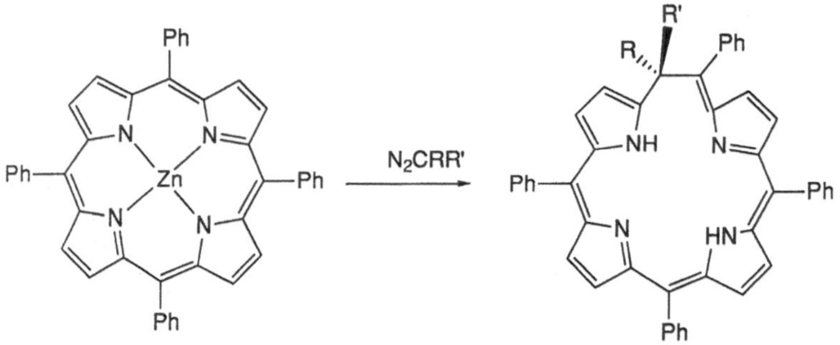

20

N_2CRR'

21 R = CO$_2$Me, R' = Me
22 R = CO$_2$Et, R' = Me
23 R = CO$_2$Me, R' = CH$_2$COOMe
24 R = CO$_2$Me, R' = CH$_2$Ph
25 R = P(O)(OCH$_3$)$_2$, R' = Me

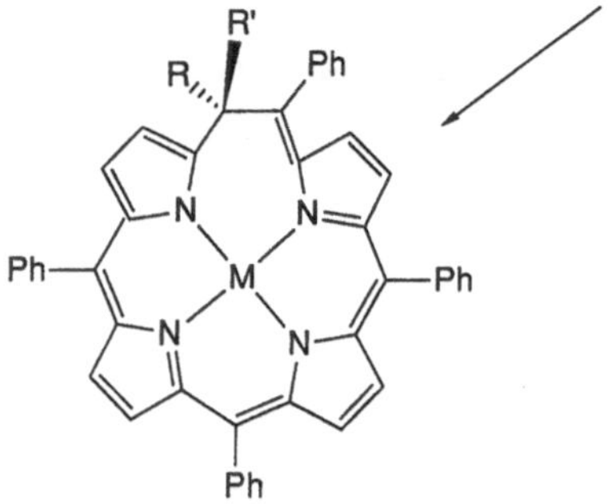

26 R = CO$_2$Me, R' = Me, M = Ni
27 R = CO$_2$Et, R' = Me, M = Ni
28 R = CO$_2$Me, R' = CH$_2$CO$_2$Me, M = Ni
29 R = CO$_2$Me, R' = CH$_2$Ph, M = Ni
30 R = P(O)(OCH$_3$)$_2$, R' = Me, M = Ni
31 R = CO$_2$Me, R' = Me, M = Cu
32 R = CO$_2$Me, R' = Me, M = Zn
33 R = CO$_2$Me, R' = Me, M = Co
34 R = CO$_2$Me, R' = Me, M = FeCl

Scheme 6

A new anionic homoporphyrin **19** is produced when either **2** or **3** is treated with a strong base such as $i\text{-}Pr_2NLi$. Protonation of **19** gives a mixture of the macrocycles **2**, **3** and **16**.

Four years after their initial report, the first alternative synthesis of homoporphyrins was reported by Callot and coworkers [14]. In this work it was found that the reaction of zinc tetraphenylporphyrin **20** with disubstituted diazoalkanes (Scheme 6) gives, after workup, the corresponding free-base homoporphyrins, namely **21–25**. Treatment of these expanded porphyrins with Ni(II) gives the corresponding nickel complexes, namely **26–30**. Reactions of **21** with other metals also gives the expected metal complexes **31–34** [15–17].

Ring contraction of the metallohomoporphyrins to the respective metallotetra-phenylporphyrin complexes **36–41** occurs when compounds **26–34** are reduced electrochemically (Scheme 7) [15]. The only homoporphyrin examined that did not give the corresponding tetraphenylporphyrin was the ethyl ester functionalized macrocycle **27**. This compound gave a triphenylporphyrin with the ester group substituted at the final *meso* position **40**. The electrochemical reduction of the macrocycles **12–15** was also examined [18]. Although the first reduction potential for species **12–15** is very close to that of **2** and **3**, the observced reversibility of this first reduction wave rules out the possibility that further follow-up ring contractions take place after the initial electrochemical reduction [15].

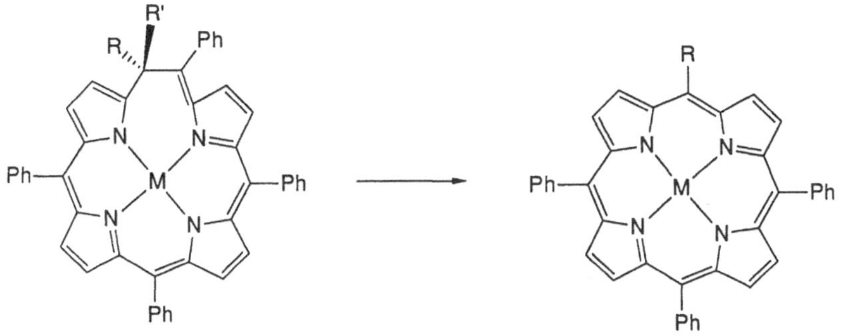

21 R = CO_2Me, R' = Me, M = H_2	**35** R = Ph, M = H_2
26 R = CO_2Me, R' = Me, M = Ni	**36** R = Ph, M = Ni
27 R = CO_2Et, R' = Me, M = Ni	**37** R = Ph, M = Cu
31 R = CO_2Me, R' = Me, M = Cu	**38** R = Ph, M = Zn
32 R = CO_2Me, R' = Me, M = Zn	**39** R = Ph, M = Co
33 R = CO_2Me, R' = Me, M = Co	**40** R = CO_2Et, M = Ni
34 R = CO_2Me, R' = Me, M = FeCl	**41** R = Ph, M = Fe

Scheme 7

Recently a new synthesis of homoporphyrins was discovered serendipitously by Smith and coworkers [19]. These investigators found that treating the unsymmetrically substituted a,c-biladiene salt **42** with copper(II) afforded the copper containing macrocycle **43**. Attempts to remove copper with sulfuric acid and trifluoroacetic acids resulted in ring expansion to the free-base homoporphyrin **44**

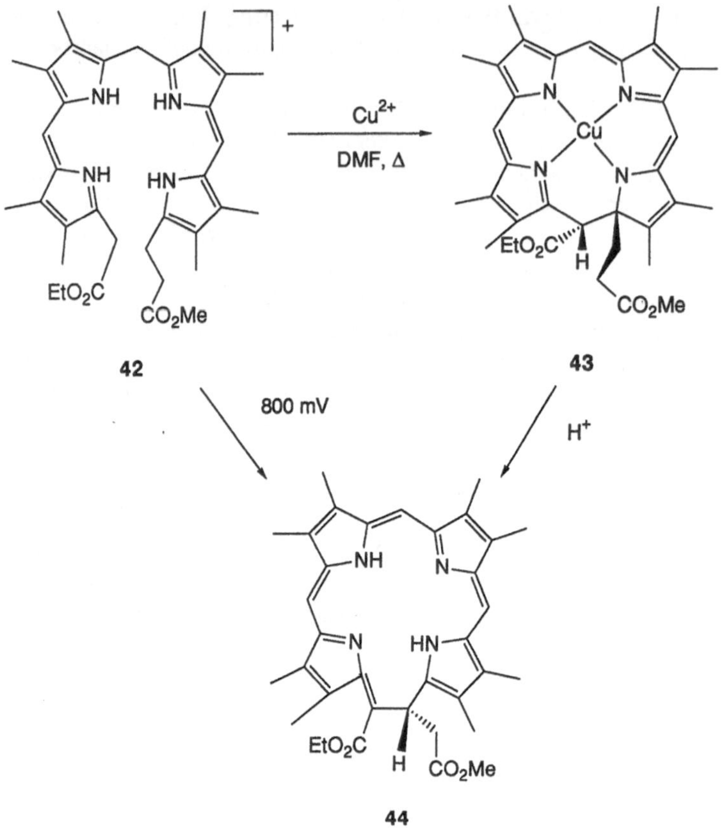

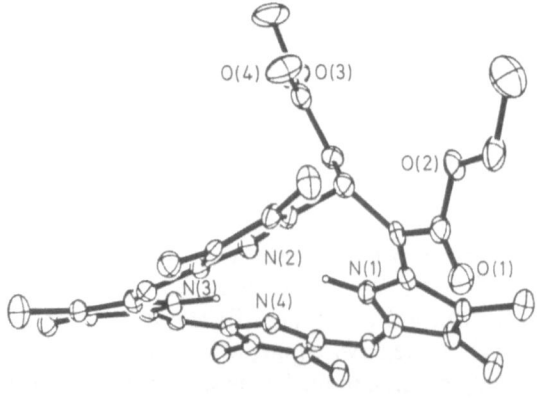

Scheme 8

(Scheme 8). The structure of **44** was confirmed by a single crystal X-ray analysis (Fig. 4). The reaction is thought to proceed *via* an initial ring opening, followed by closure in a different sense. This ring opening if favoured by the presence of

Fig. 4

an enolizable proton on the bridging methylene. An alternative synthesis of **44** has been achieved *via* the anodic oxidation of **42** at 0.8 V (vs. Ag/AgCl) [20,21].

The homoporphyrins have the most diverse organic chemistry of any of the expanded porphyrins. This is largely due to their inherent instability, which allows for rearrangements and ring contractions under both thermal and electrochemical conditions. Interestingly, however, this reactivity is not so great as to preclude the isolation and characterization of new materials. Taken together, these two factors, make the homoporphyrins a particularly interesting class of expanded porphyrins.

3 Heterohomoporphyrins

There is only one fully characterized example of this class of expanded porphyrin [22, 23]. While other macrocycles of this type are thought to exist as intermediates

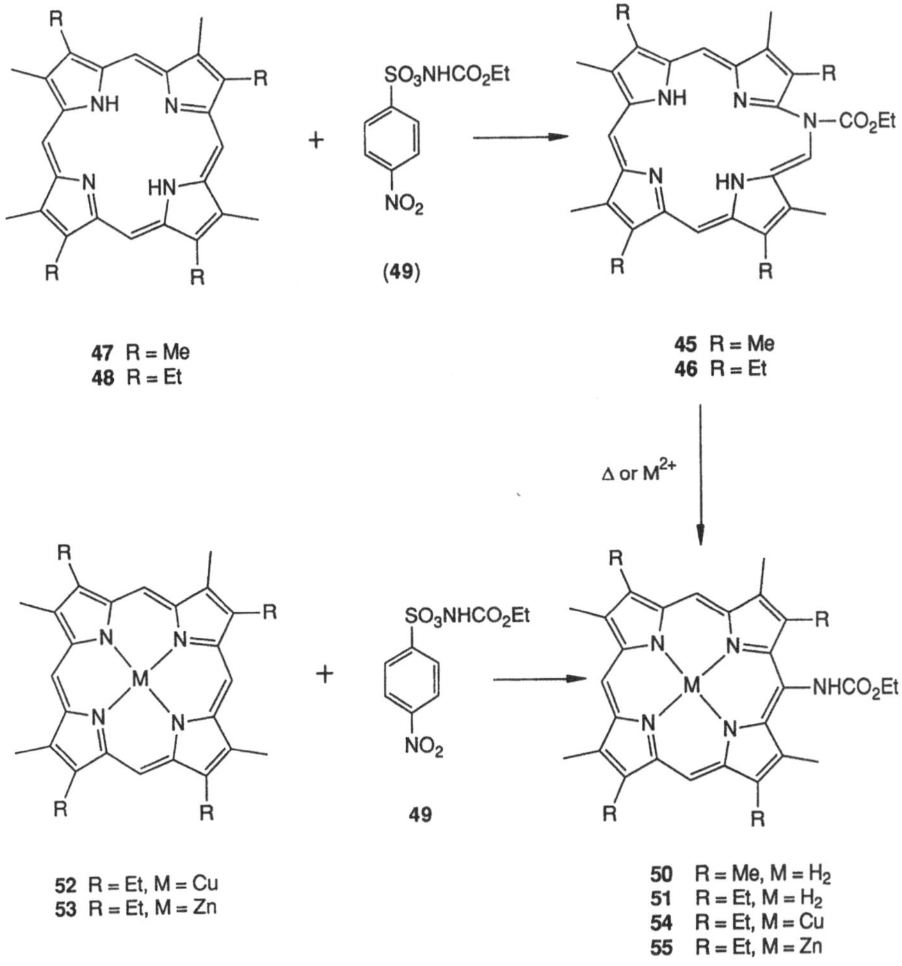
R
R
NH
N
N
HN
R
R
SO_3NHCO_2Et
NO_2
(49)
R
R
NH
N
N—CO_2Et
N
HN
R
R
47 R = Me
48 R = Et
45 R = Me
46 R = Et
Δ or M^{2+}
M
SO_3NHCO_2Et
NO_2
49
$NHCO_2Et$
52 R = Et, M = Cu
53 R = Et, M = Zn
50 R = Me, M = H_2
51 R = Et, M = H_2
54 R = Et, M = Cu
55 R = Et, M = Zn

Scheme 9

in a number of reactions used to form macrocycles, notably the sapphyrins their existence is still speculative [24–27]. Thus for the purposes of this review, discussion will be limited to the one well-characterized example, homoazaporphyrin **45** and **46** (see Scheme 9).

As was the case with the initial synthesis of the homoporphyrins [5–7], the only well characterized heterohomoporphyrin was discovered serendipitously [22, 23]. Here the initial work involved investigating the reaction of ethoxycarbonylnitrene with porphyrins and metalloporphyrins, in the interest of probing the reactivity of porphyrins towards electrophiles [22, 23]. It was found that when the octaalkylporphyrins **47** and **48** were treated with **49** a rapid reaction occurred. Careful isolation of the products gave the *meso*-homoazaporphyrins **45** and **46**, respectively (Scheme 9). When the expanded porphyrins **45** or **46** were heated in chloroform, a ring contraction occurred and the *meso*-ethoxycarbonylaminopor-phyrins **50** or **51** were obtained, respectively. The reactions of metallo-octaalkylpor-phyrins Cu(II) (**52**), Zn(II) (**53**), Ni(II) and Co(III) with ethoxycarbonylnitrene were also examined. The copper (**52**) and zinc (**53**) porphyrins gave metal complexes of the corresponding non-expanded *meso*-substituted derivatives **54** and **55**. No reaction was noted with the nickel porphyrin, and the cobalt(III) containing porphyrin was found only to undergo anion exchange, with the *p*-nitrophenyl sulfonate replacing the original counter anion associated with the Co(III) center.

In addition to the above reactions, the preparation of metal complexes directly from **46** was also studied. Upon treatment of **46** with either Zn(II) or Cu(II) an immediate ring contraction occurred and the *meso*-substituted porphyrin deriv-atives **54** and **55** were again obtained.

While these expanded porphyrins represent a unique and potentially interesting class of expanded porphyrins, their stability towards ring contraction severely limits their utility, except perhaps as intermediates in the production of *meso*-substituted metalloporphyrin systems.

4 Schiff Base Expanded Porphyrins

Condensation reactions between carbonyl compounds and primary amines have played a central role in the synthesis of new macrocyclic ligands [28–34]. Usually, though not in all cases, such reactions are conducted in the presence of metal ions which can serve to direct the condensation preferentially to cyclic rather than oligomeric/polymeric products and to stabilize the macrocycle once formed. The relative atomic radius of the templating ion has a considerable effect on the size of the macrocycle formed. For instance, in what is now classic work, cations such as Mg(II) (r = 0.72 Å) were found to stabilize the formation of macrocycles such as **60** from "1 + 1" condensations [35], while larger cations such as Sr(II) (r = 1.16 Å), Ba(II) (r = 1.36 Å), Ag(I) (r = 1.15 Å), and Pb(II) (r = 1.18 Å) were found to produce macrocycles, such as **61**, that are the result of "2 + 2" condensations [31, 36, 37]. Even larger macrocycles have been stabilized by using metal clusters as templates [38, 39]. In many cases these Schiff base macrocycles are not stable in the absense of a coordinating metal. In an attempt to distinguish

60

61

between the macrocycles that exist only as metal stabilized ligands and those that can be isolated as the free macrocycles, two separate labeling systems are used in this section. Schiff base macrocycles that are not stable in the absence of a coordinating metal are labeled using an L^n designation; other stable macrocycles are labeled by normal conventions.

While the majority of macrocycles formed by this type of Schiff base condensation reaction are derived from pyridine containing fragments, considerable attention has also been devoted to the use of other heterocycles, including five membered ones, as the primary macrocyclic precursors. Although these latter ligands are for the most part not completely conjugated, they form an important group of expanded porphyrin-type macrocycles. It is for this reason that they are included in the present review.

4.1 Furan-Containing Schiff Base Macrocycles

The condensation reaction between 2,5-diformylfuran and 1,3-diaminopropane in MeOH using $Ba(ClO_4)_2$ as a template gave the complex $[Ba(L^1)_2(H_2O)][ClO_4]_2$ (**62**), in >70% yield [40]. The macrocycle (L^1) is formed from a "2 + 2" condensation. A single crystal X-ray investigation of **62** is shown in Figs. 5 and 6 [41]. The two independent macrocycles which are coordinated to the barium have different conformations. While one of the macrocycles is coordinated *via* all six donor atoms (the four nitrogen atoms and both of the furan oxygen atoms), the other ligand is bound by only three of the donor atoms (one of the furan oxygen atoms and two of the nitrogen atoms). There is also a water molecule occupying one of the coordination sites on the barium. The result is an eleven coordinate Ba(II) cation.

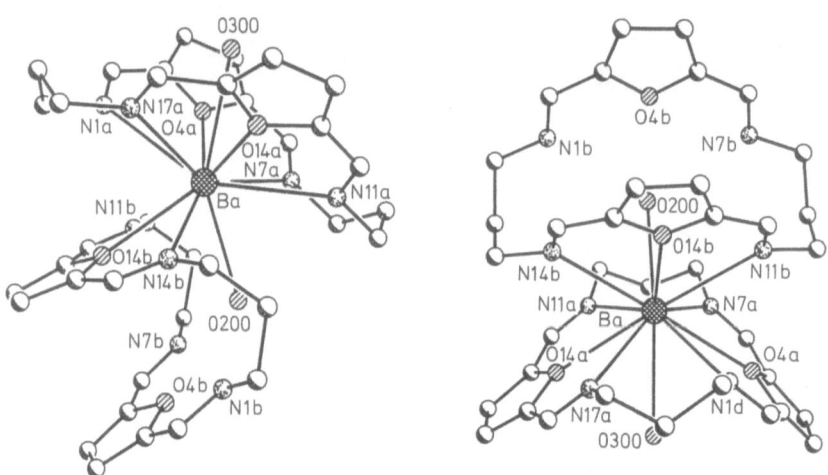

Fig. 5   **Fig. 6**

Treatment of **62** with Cu(II) gave the trans-metalated, di-μ-hydroxo-di-Cu(II) complex, $Cu_2(L^1)(OH)_2(ClO_4)_2 \cdot H_2O$ (**63**), in good yield. The structure of **63** was inferred from the antiferromagnetic behavior ($\mu_{eff}/Cu = 1.37\,\mu_B$ at 293 K and 0.70 μ_B at 93 K) of the complex [41]. The obvious binucleating character of this ligand prompted further investigations into these and other related bimetallic complexes as models for the Type 3 copper proteins. To date, a variety of Cu(II) complexes of (L^1) have been isolated $[Cu_2(L^1)(OH)_2][ClO_4]_2 \cdot H_2O$ (**63**), $[Cu_2(L^1)(OR)_2(MeCN)_2][BPh_4]_2$ (**64**), and $[Cu_2(L^1)(OR)_2(NCS)_2]$ (**65**), (R = Me, Et, n-Pr) [42]. The structure of **65** (R = Et) is shown in Fig. 7. Each Cu(II) ion is bonded to two imino nitrogens of the macrocycle, the nitrogen of one (terminally bound) thiocyanate ion, and to two bridging ethoxide groups in an approximate trigonal-bipyramidal geometry. All of these dicopper(II) complexes undergo reduction upon heating in MeCN [42]. For the complex **65**, the product is

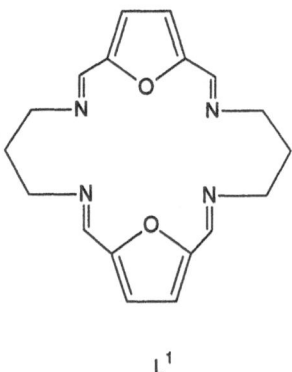

L^1

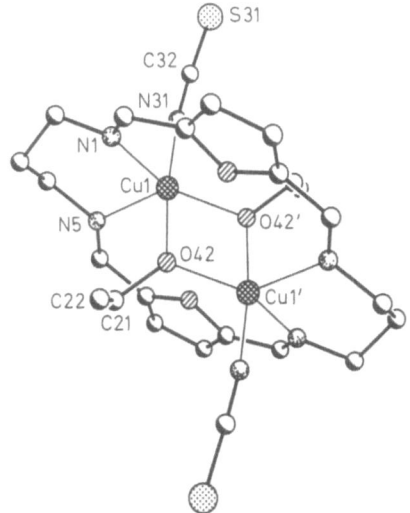

Fig. 7

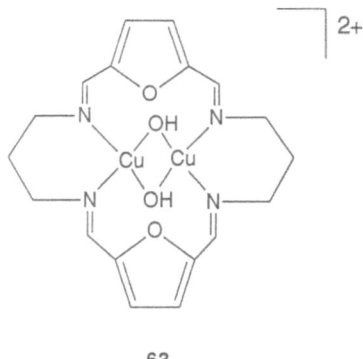

63

64 R = Me, Et, n-Pr; L = MeCN
65 R = Me, Et, n-Pr; L = NCS

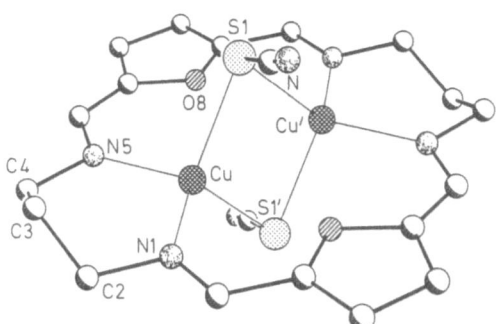

Fig. 8

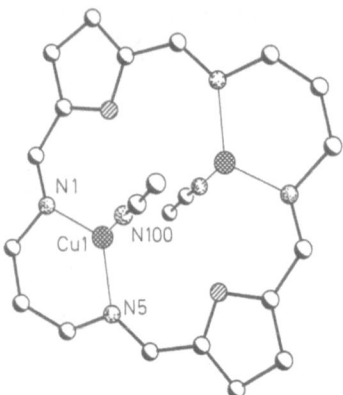

Fig. 9

$[Cu_2(L^1)(NCS)_2]$ **(66)**, which has been shown (c.f. Fig. 8) [43] to contain two tetrahedrally coordinated Cu(I) ions held 2.796(8) Å apart and linked intermolecularly *via* the sulfur atoms of the thiocyanate ions. For the complexes **63** and **64**, the reduction product is the diamagnetic complex $[Cu_2(L^1)(MeCN)_2](Y)_2$ **(67)** (Y = ClO_4 or BPh_4) in which each three coordinate Cu(I) ion is bonded to two of the four macrocyclic nitrogen atoms and to the nitrogen of one of the two MeCN molecules (Fig, 9) [43]. In the presence of certain substrates the reduction of **63** or **64** is accompanied by substrate oxidation. For example, PhSH, PhC≡CH, hydrazobenzene, catechols, hydroquinone, and ascorbic acid afford PhSSPh, PhC≡CC≡Ph, azobenzene, *o*-quinones, *p*-quinone, and dehydroascorbic acid, respectively, together with the reduced species **67** and/or other copper complexes [42]. When carried out in a dimethylformamide solution in the presence of O_2, several of these substrate oxidations proved to be catalytic in **63** or **64**. The reaction

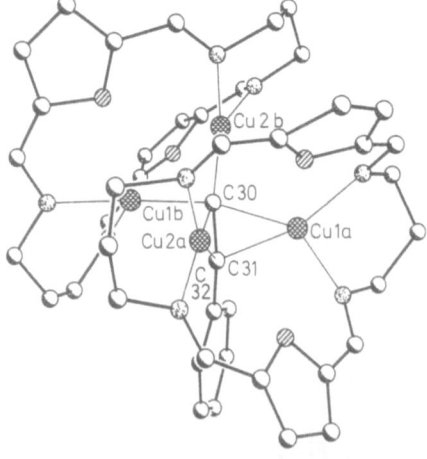

Fig. 10

of **63** with phenyl acetylene gave orange crystals of $[Cu(I)_4(L^1)_2(C\equiv CPh)]$ · $[ClO_4]_3$ · $0.5(dpda)$ (**68**), (dpda = diphenyldiacetylene) and white crystals of dpda. The structure of **68** was determined by X-ray analysis [40]. It is depicted in Fig. 10. In this structure each macrocycle is bonded to a pair of copper atoms *via* the four imino-nitrogen atoms. The conformation of each $'Cu_2N_4'$ moiety is such that the two metal atoms sit outside the approximate $'N_4'$ plane on the surface of the saddle shaped macrocycle (Fig. 10). An interesting feature of this structure is that the terminal carbon of the phenyl acetylide is five coordinate, being bonded to all four copper atoms as well as being linked to the other carbon atom of the acetylide. The complex **67** and the related complex **69** may also be formed by treatment of the appropriate Ba(II) complex with a three-fold excess of $[Cu(MeCN)_4][ClO_4]$ in MeCN-EtOH at 60 °C under anaerobic conditions [44].

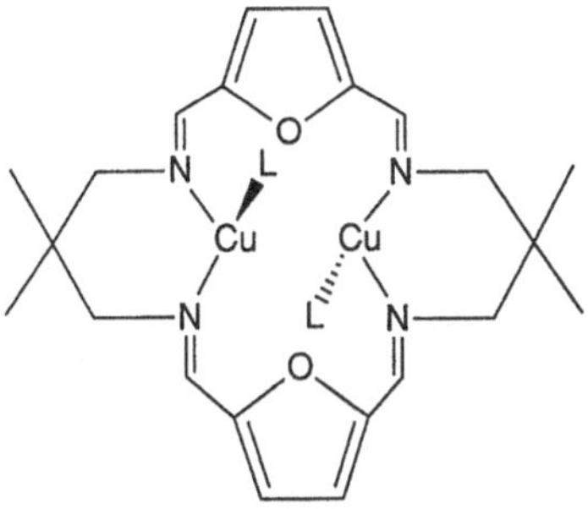

69 L = MeCN

However, when the transmetalation was carried out at reflux temperature in "wet" solvent with access to air, the pentanuclear complexes $[Cu_5(L^1)_2(dmt)_2][ClO_4]_3$ (**70**), and $[Cu_5(L^2)_2(dmt)_2][ClO_4]_3$ (**71**), (dmt = 3,5-dimethyl-1,2,4-triazolate anion) were obtained. The structures of these complexes were confirmed by an X-ray diffraction analysis of complex **71** (Fig. 11) [44]. Separate experiments established

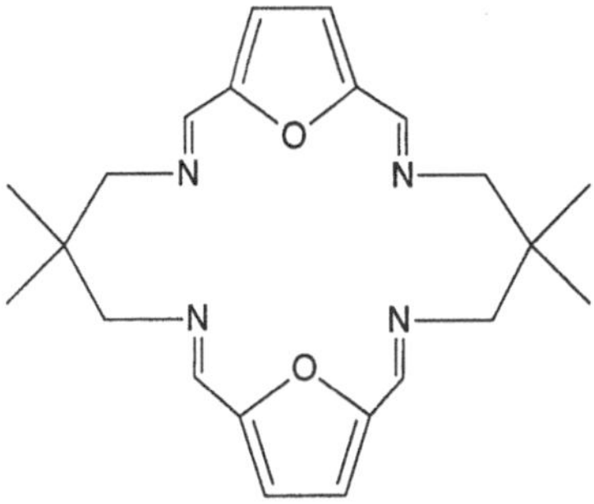

L^2

the necessity of both H_2O and O_2 in the formation of the triazolate ring. Moreover, no triazolate was formed when $[Cu(MeCN)_4][ClO_4]$ was used in place of the dinuclear complexes **67** or **69**. The use of EtCN in the reaction mixture afforded the corresponding pentametallic complex derived from a 3,5-diethyl-1,2,4-triazolate anion. Until this report [44], the only reported formation of a 1,2,4-triazole *via* N ··· N coupling involved reacting the sodium salt of an amine with CuCl [45]. Since the aggregation of simple Cu(I) salts is well known in non-coordinating solvents [46], it would seem that the stabilization of copper clusters provided by ligands such as L^1 and L^2 supplies a facilitating environment for this type of reaction.

The most practical use for these expanded porphyrin macrocycles is in complexation of larger cations. Preliminary investigations into lanthanide complexes of L^1 indicate that most of the lanthanides form complexes with L^1 *via* template reactions [47]. However, to date this chemistry has not been explored extensively nor exploited in terms of any practical applications.

A series of molecular mechanical investigations involving the furan- and thiophene-containing analogues of macrocycle L^1 [48] were used to investigate the relationship between this type of 2 + 2 Schiff base ligand and the corresponding binuclear copper complexes, especially with regards to their relative stability. In all cases, the steric stability of the complexes was confirmed by the calculations. Based on this, the viability of this method for determining *a priori* which macrocycles would be most suitable for modeling Type 3 copper proteins was then suggested [48].

While the furan containing complexes of ligands of type L^1 have attracted the most attention, other ostensibly related furan-derived ligand systems have also been considered as being likely to provide complexes of interest [49]. For instance, macrocycles which contain more than one type of heterocycle have been reported

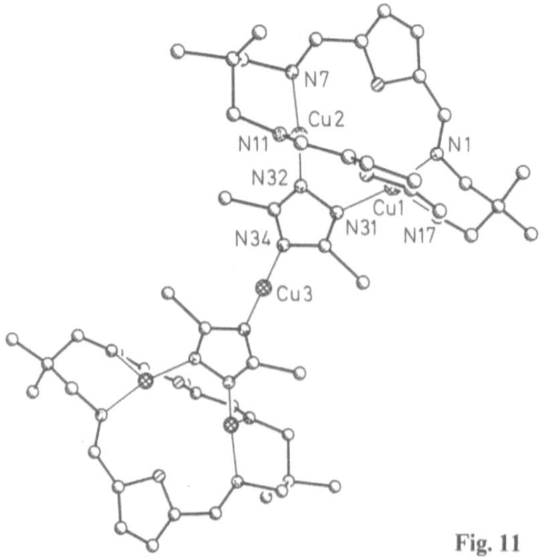

Fig. 11

[49] from the condensation of 2,5-diformylfuran and 2,6-bis(2-aminophenoxyme-thyl)pyridine in the presence of $Pb(ClO_4)_2$. This gives the complex $[Pb(73)(ClO_4)_2]$ (72). Treatment of complex 72 with sodium tetrahydroborate afforded the parent metal-free macrocycle 73 [49].

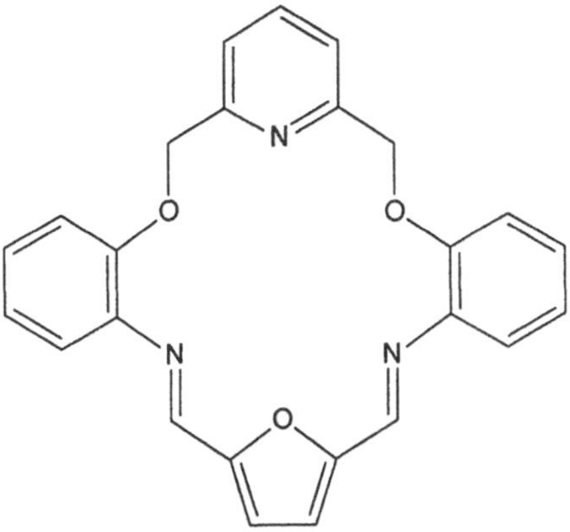

73

Macrocyclic ligands that provide endogenous bridges were expected to be even more versatile ligands than their simple alkyl linked counterparts. The incorporation of hydroxyl groups into the macrocycle framework, for instance, was thought to provide a source of bridging alkoxide groups, derived from the ligand itself [50, 51]. It was envisioned that this would provide a more sophisticated series of bimetallic complexes. To date, some tests of this general idea have been carried out. For instance, the reactions of 1,5-diamino-3-hydroxypentane with

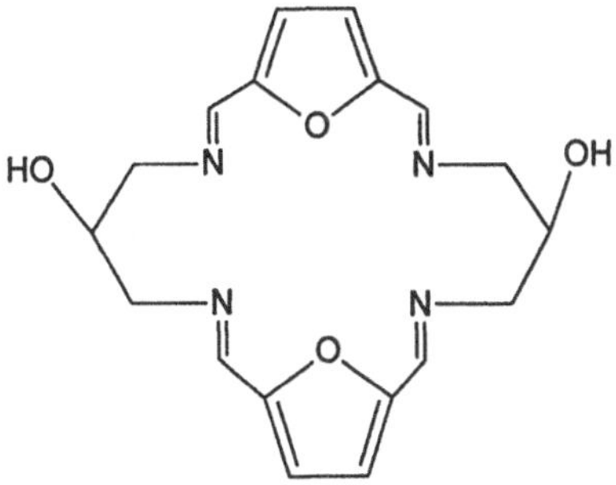

L^4

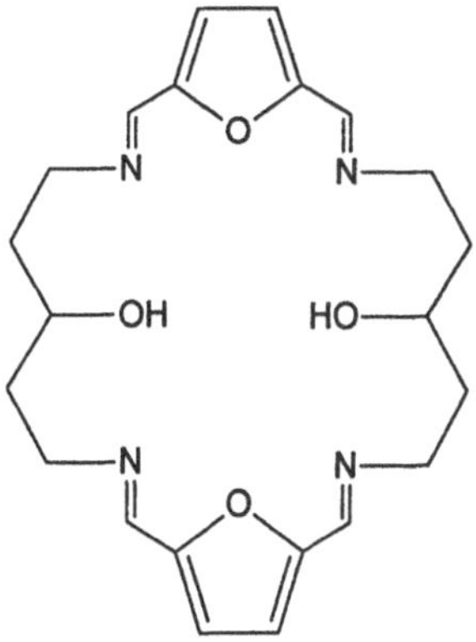

L^5

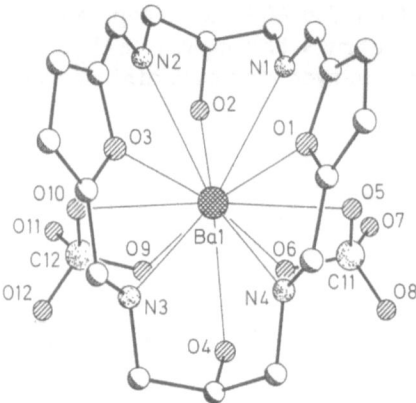

Fig. 12

2,5-diformylfuran in the presence of $Ba(ClO_4)_2$ as a template give colorless crystals of $Ba(L^4)(ClO_4)_2$ (74) [50, 51]. The structure of 74 was determined by a single crystal X-ray investigation (Fig. 12) [51]. Using a longer amine chain in the condensation reaction gives, as expected, $Ba(L^5)(ClO_4)_2 \cdot EtOH$, (75). This too was characterized by X-ray crystallography (Fig. 13) [51]. The major difference between 74 and 75 is the number of metal ligating centers which derive from the macrocycle. With the smaller macrocycle 74 all of the potential donor atoms are coordinated to the Ba(II) center. In contrast, the larger system 75 has only five of its potentially available eight donor atoms actually involved in coordination to the metal center. This difference reflects a point of prime importance, namely that due caution must be exercised when attempting to form a particular type of complex from a particular type of ligand and chosen metal center: The macrocycle and the ligand must be suitably matched.

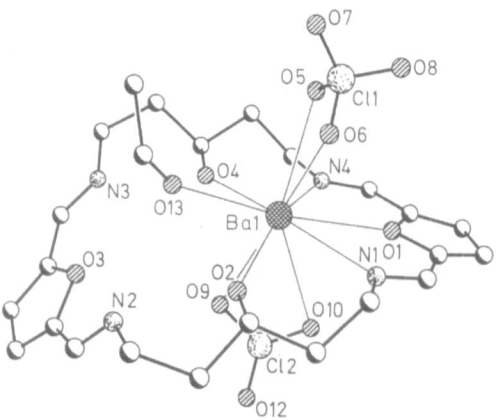

Fig. 13

4.2 Thiophene-Containing Schiff Base Macrocycles

Thiophene-containing Schiff base macrocycles have also been investigated as potential binucleating ligands. For instance, Fenton and coworkers found that reaction of equimolar amounts of 2,5-diformylthiophene with various α,ω-amino-ethers gives a facile synthesis of the corresponding metal-free tetraimine macrocycles **76** and **77** [52, 53]. An X-ray investigation of the free ligand **76** is depicted in Fig. 14 [52]. The molecule **76** adopts a folded conformation in which two planar thiophene rings lie aligned and approximately parallel (although not fully eclipsed). Treatment of **76** or **77** with Ag(ClO$_4$) gives the di-silver complexes Ag$_2$(**76**)[ClO$_4$]$_2$ (**78**), and Ag$_2$(**77**)[ClO$_4$]$_2$ (**79**), respectively. The X-ray structure of **79** (Fig. 15) confirmed that the sulfur atoms of the thiophene moiety are not involved in coordination to the silver(I) centers [52]. In fact, the structure of **79** given in Fig. 15 reveals that the skeleton of the macrocycle has a twisted-loop conformation such that the two silver atoms are bonded only to the two imine nitrogen atoms and also, rather remotely, to a water molecule but not to a thiophene sulfur site. Other thiophene containing macrocycles prepared *via* similar condensations have also been isolated as the free ligands [53]; c.f. structures **80–87**. Here, it is of interest

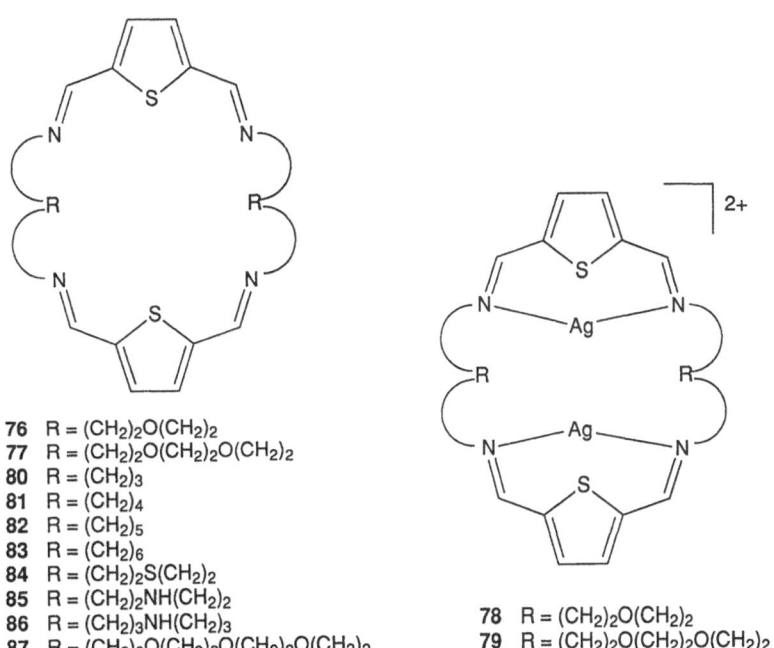

76   R = (CH$_2$)$_2$O(CH$_2$)$_2$
77   R = (CH$_2$)$_2$O(CH$_2$)$_2$O(CH$_2$)$_2$
80   R = (CH$_2$)$_3$
81   R = (CH$_2$)$_4$
82   R = (CH$_2$)$_5$
83   R = (CH$_2$)$_6$
84   R = (CH$_2$)$_2$S(CH$_2$)$_2$
85   R = (CH$_2$)$_2$NH(CH$_2$)$_2$
86   R = (CH$_2$)$_3$NH(CH$_2$)$_3$
87   R = (CH$_2$)$_2$O(CH$_2$)$_2$O(CH$_2$)$_2$O(CH$_2$)$_2$

78   R = (CH$_2$)$_2$O(CH$_2$)$_2$
79   R = (CH$_2$)$_2$O(CH$_2$)$_2$O(CH$_2$)$_2$

to note that in at least one case the condensation reaction is complicated by a side reaction which leads to ring contraction and formation of a macrocycle containing imidazolidine rings (**88**) [53]. While many of the macrocycles **80–87**

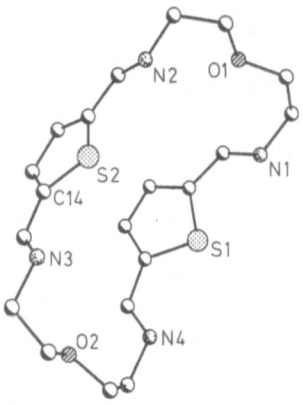

Fig. 14

Fig. 15

88

89   **90**

appear to form bimetallic complexes with Ba(II), none of the complexes has been characterized structurally. Thus, it is difficult at present, to appreciate whether thiophene ligation is playing an important role in this chemistry. In the formation of 87 another macrocycle 89 was also formed. The structure of this 1:1 condensation product was confirmed by reduction with $NaBH_4$; this gave the reduced macrocycle 90 [53]. In any case, the thiophene-derived, silver-containing macrocycle 78 appears to behave in a manner similar to that of furan ligands L^1 and L^2: When treated with $[Cu(MeCN)_4][ClO_4]$ in "wet" MeCN-EtOH in the presence of air, 78 gave the pentanuclear complex $[Cu_5(L^6)_2(dmt)_2][ClO_4]_3$ (91), which is reported [44] to have a structure similar to its furan analog 71 which, in turn, is depicted in Fig. 11.

4.3 Pyrrole-Containing Schiff Base Macrocycles

The use of pyrrole as the source of a donor atom in Schiff base ligand macrocycles is relatively rare. This probably just reflects the long and tedious synthesis of the required starting 2,5-diformylpyrrole (92), rather than any kind of general philosophic opposition to using this heterocycle. Consistent with this supposition is the realization that the publication [54] of a relatively facile route to 2,5-diformylpyrrole (92) in 1981 was reflected soon thereafter in terms of the synthesis by Fenton and coworkers of a series of macrocycles containing a 2,5-substituted pyrrole moiety 93–98 [50, 55, 56].

93 R = $(CH_2)_2$
94 R = $CH_2CH(CH_3)CH_2$
95 R = $(CH_2)_3$
96 R = $(CH_2)_4$
97 R = $(CH_2)_5$
98 R = $(CH_2)_6$

The preparation of the Cu(II) complexes of these macrocycles differs from that of the other Schiff base expanded porphyrins described above in that first a copper complex is formed from 2,5-diformylpyrrole and copper(II) acetate in methanol or ethanol and only then are the resulting complexes (e.g. **99** and **100**) treated with the appropriate diamine to give the desired product macrocycles (Scheme 10). As before, the size of the macrocycle formed influences the type of metal complex obtained. 1,2-Diaminoethane and 1,3-diaminopropane result in the monometallic copper complexes **101** and **102**, respectively. Using longer interamine bridging chain lengths gives the bimetallic complexes **103–105** (Scheme 10). The X-ray structure of one of the mononuclear complexes **102** was determined. The structure (Fig. 16) demonstrated that the Cu(II) ion is coordinated at one end of the macrocycle and coordinated to both of the pyrrole nitrogen atoms. It is, however, bound to just one pair of the four imine nitrogen atoms [56].

The last class of Schiff base macrocycles to be discussed in this subsection are the only ones which can truly be called expanded porphyrins. This is because in all cases reliance is not made on simple pyrroles but on linked polypyrroles. The first

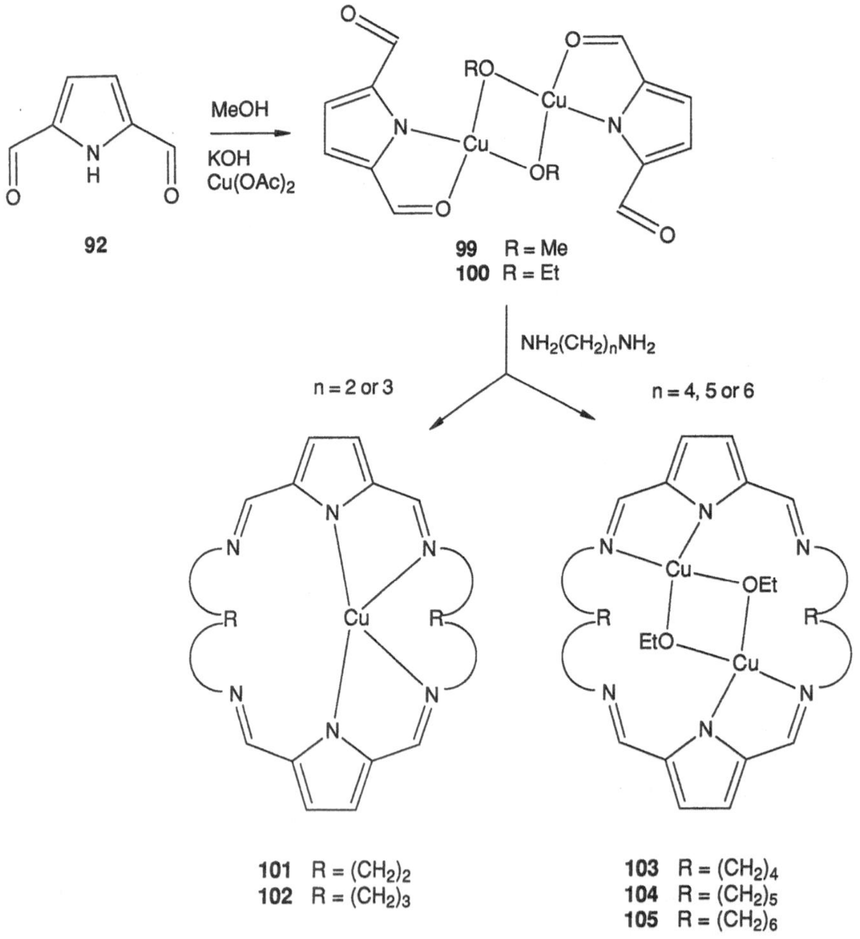

Scheme 10

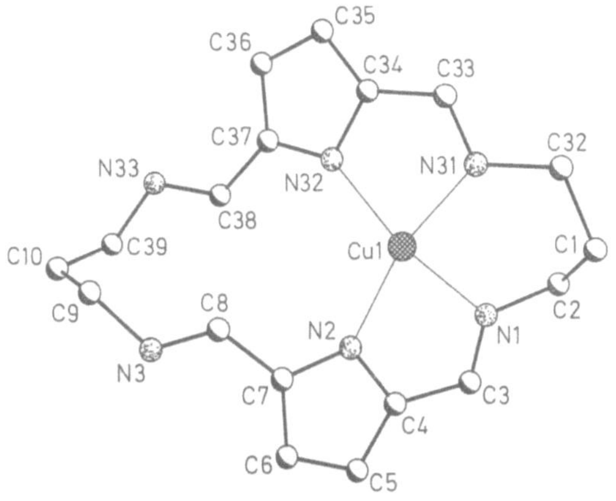

Fig. 16

of this class of compounds to be prepared is the so-called "accordion" porphyrin of Mertes and coworkers (Scheme 11) [57]. Here, the reaction of 5,5′-diformyldipyrromethane **106** with 1,2-diaminoethane or 1,3-diaminopropane in the presence

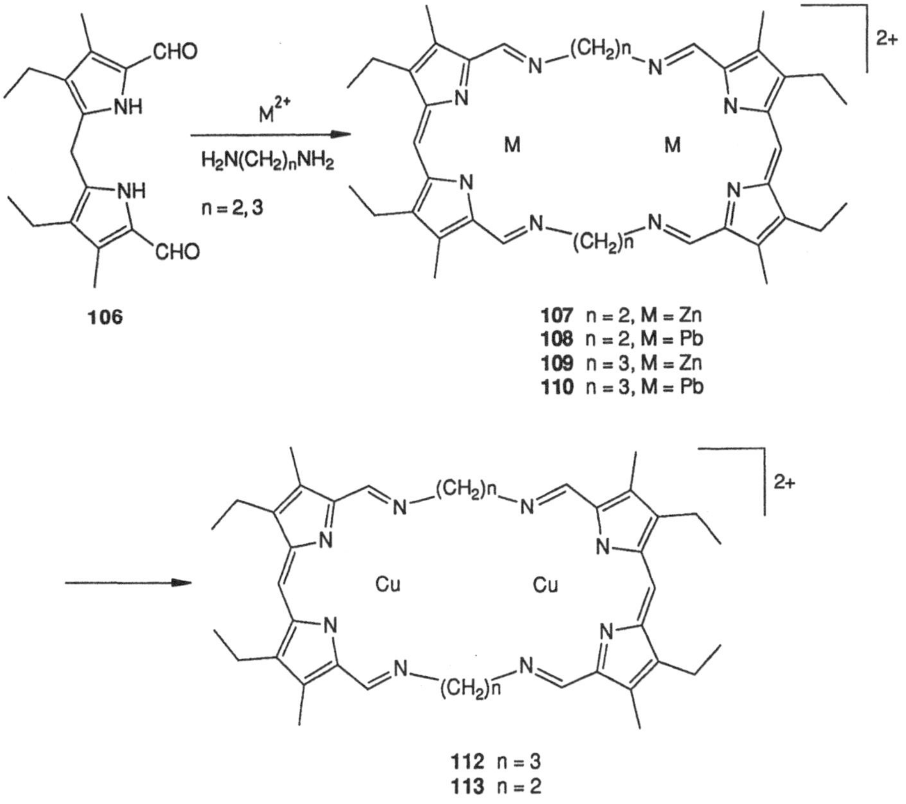

106

107 n = 2, M = Zn
108 n = 2, M = Pb
109 n = 3, M = Zn
110 n = 3, M = Pb

112 n = 3
113 n = 2

Scheme 11

203

of either zinc(II) or lead(II) was used to obtain bimetallic complexes of the general form **107–110** [58]. The free ligand **111** was obtained when Ba(II) was used as the metal template in the condensation between 1,3-diaminopropane and the di-formyldipyrromethane **106** (Scheme 12) [58]. The copper(II) complexes of these

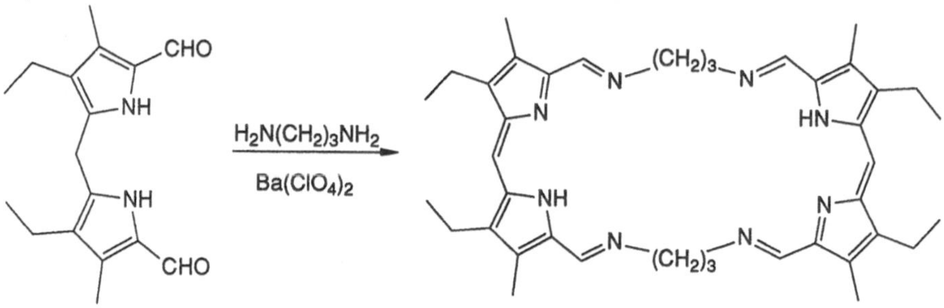

106 $H_2N(CH_2)_3NH_2$ $Ba(ClO_4)_2$ **111**

Scheme 12

macrocyclic systems could be obtained either from the direct reaction of the free ligand **111**, to give **112**, or *via* transmetalation of the complexes **107–110**, which gave the respective bis-Cu(II) complex **112** or **113**. All of the copper complexes obtained appear to be binuclear in nature. In fact, a single crystal X-ray structural analysis of the bis-azide adduct of complex **112** confirmed the presence of two coppers. The actual structure (Fig. 17 and 18) revealed a somewhat unusual geometry [58]. The coordination geometry about each copper(II) ion is a distorted square pyramid made up of the nitrogen atoms of the macrocycle and a nitrogen from the azide anion. The ligand is in an "in-out" geometry with respect to the imine conformation. Moreover, rather than being in a flattened, near-planar

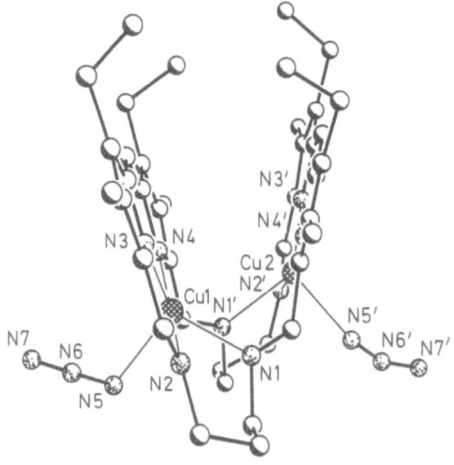

Fig. 17

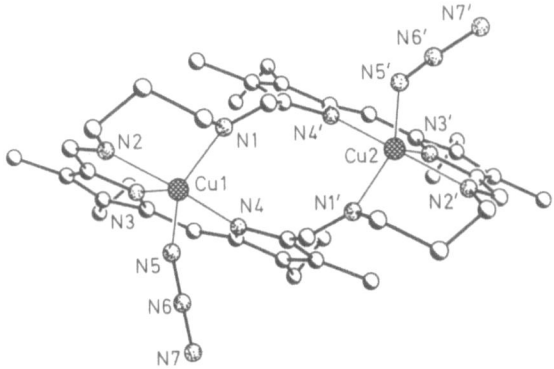

Fig. 18

conformation, the ligand has twisted and folded so that the two dipyrromethenes are almost facing each other and ligation to the imines takes place "across" the macrocycle. That said, it is important to appreciate that the dipyrromethene fragments each show extensive electron delocalization and, including the imines,

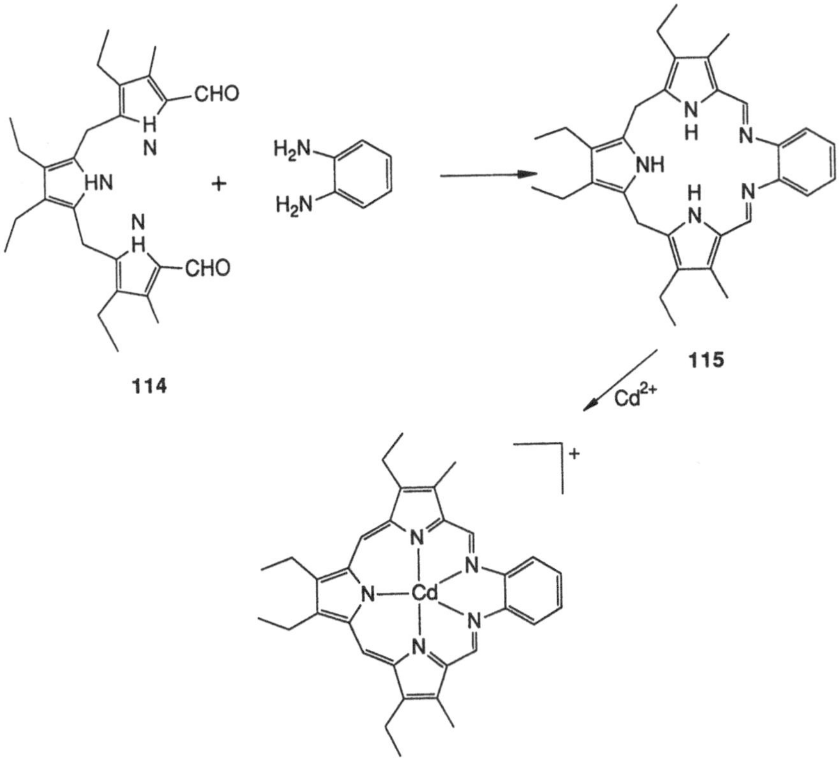

Scheme 13

205

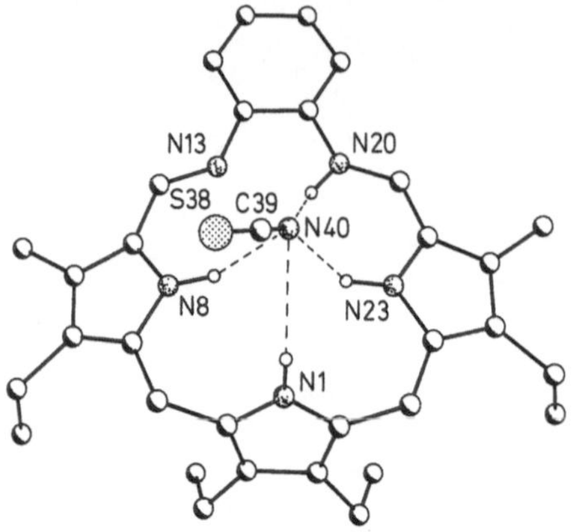

Fig. 19

are planar. Moreover, within each dipyrromethene fragment the bond lengths resemble those in porphyrin structures.

In a manner similar to that used to prepare **112**, the reaction of the diformyl-tripyrrane **114** with o-phenylenediamine was found by Sessler and coworkers to result in the synthesis of a pentaazamacrocycle **115** (Scheme 13) [59]. An X-ray structure of a derivative of **115** is shown in Fig. 19. Unfortunately, no structurally characterized metal complexes of **115** have been reported to date. However, oxidation of **115** in the presence of cadmium(II) was found to give the aromatic pentaaza-macrocycle metal complex **116**, which has been characterized by X-ray diffraction [60]. The properties and chemistry of these tripyrroledimethine-derived "texaphyrins" is reported in the next section.

Other large macrocycles **117** and **118** have also been reported from similar condensations between the polypyrrole dialdehydes **119** and **120** and substituted phenylenediamines (Schemes 14 and 15) [61]. However, at present, little published information is available for these systems.

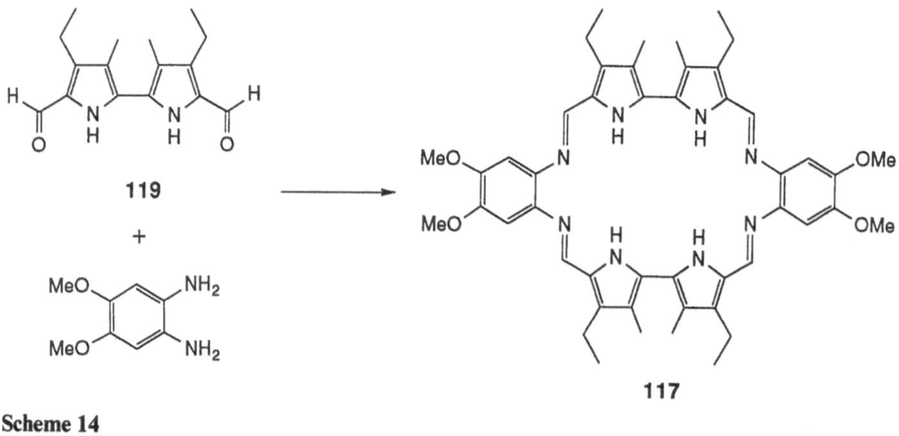

Scheme 14

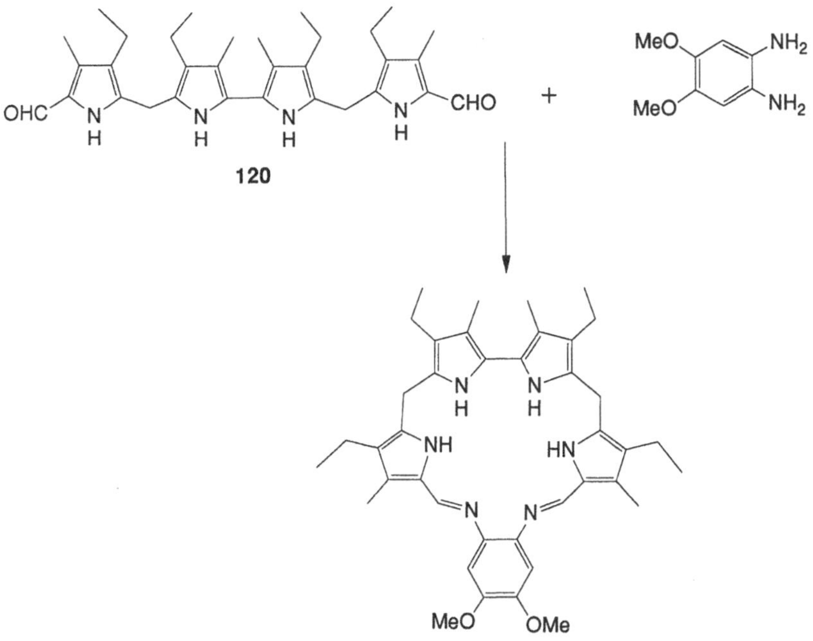

Scheme 15

The Schiff base complexes provide the widest range of expanded porphyrins available from the same general reaction sequence. This ease of synthesis coupled with the large potential variability in ligand size and structure, enables extensive conceivable modification of the macrocycle. As a result it is possible to envision the design of Schiff base expanded porphyrins to fit, presumably, any given metal that one may wish to complex. Thus, it is expected that this class of expanded porphyrins will continue to be among the most extensively investigated in the years to come.

5 Texaphyrins

The synthesis of a tripyrrane containing porphyrinogen-like macrocycle was reported by Sessler *et al.* in 1987 [59]. As discussed in the previous section (4.3), the first representative of this new class of expanded porphyrins (e.g. **115**) was formed by the Schiff base condensation between a diformyltripyrrane **114** and *o*-phenylenediamine. Subsequent to this initial report, this approach has been generalized to afford a wide range of new macrocycles **115, 121–134**. Initial attempts to form coordination complexes between a variety of metals and these methylene-linked ligands, however, proved to be unsuccessful [62]. In fact, to this date no well-characterized metal complexes of the reduced macrocycles **115, 121–134** have been reported. However, it was expected that oxidation of these compounds (by four electrons), and the accompanying aromatization, would improve the potential utility of these systems as ligands [59, 62].

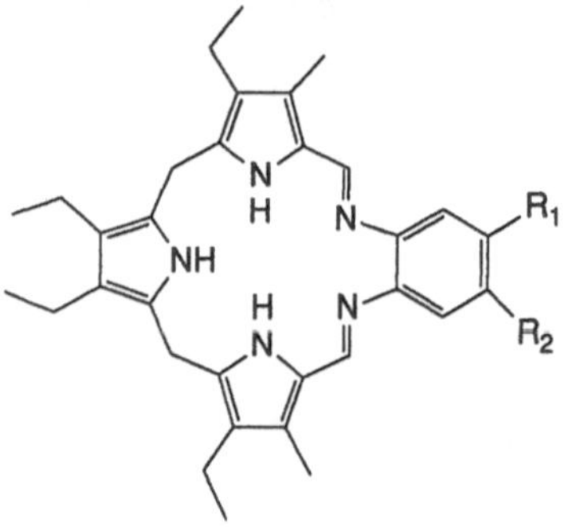

115 $R_1 = R_2 = H$
121 $R_1 = R_2 = Me$
122 $R_1 = H, R_2 = Me$
123 $R_1 = H, R_2 = OMe$
124 $R_1 = H, R_2 = Cl$
125 $R_1 = H, R_2 = CO_2H$
126 $R_1 = H, R_2 = NO_2$
127 $R_1 = R_2 = OCH_2CH_2CH_2OH$

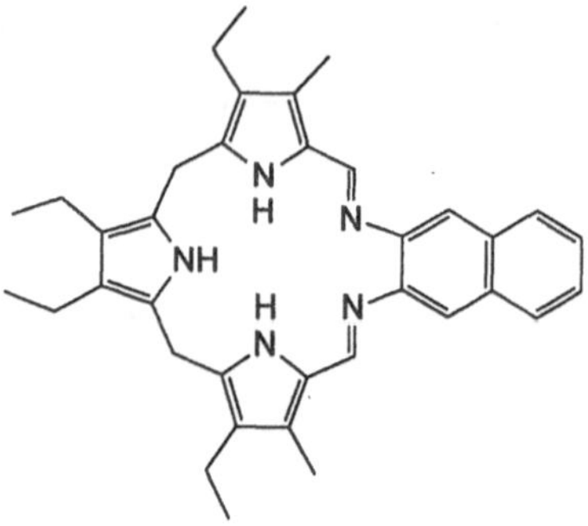

128

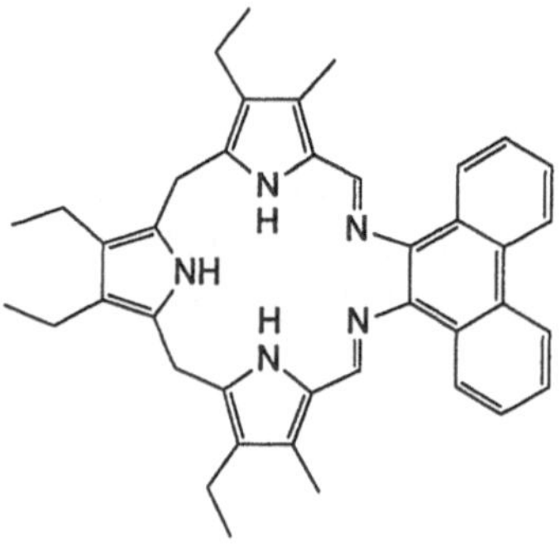

129

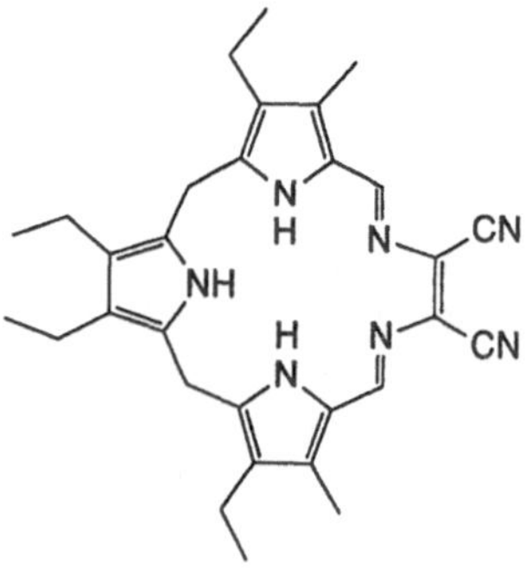

130

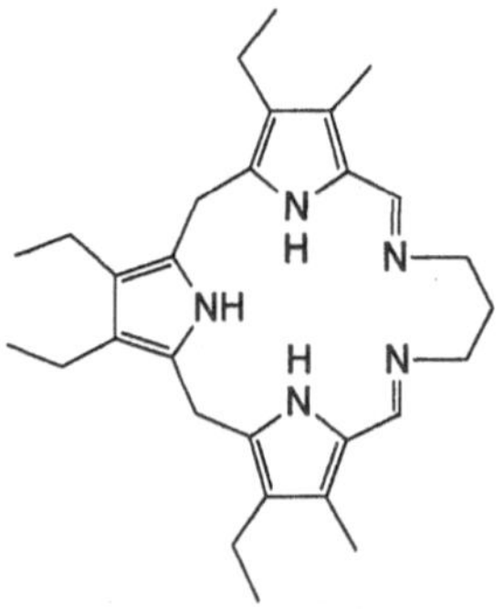

131

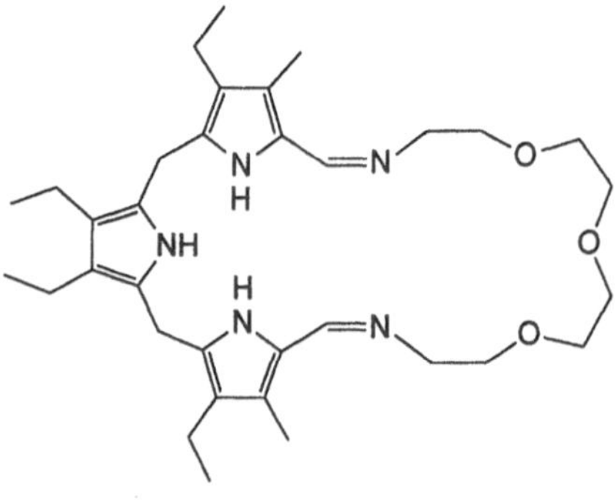

132

133   134

Unfortunately, however, considerable effort was required before conditions could be found which would enable this key 4-electron oxidative transformation to be effected. Standard organic oxidations of macrocycle **115** gave no sign of producing the desired oxidized product **135**. Under a range of conditions and in the presence of a variety of oxidants, including *o*-chloranil, DDQ, PbO$_2$, and PtO$_2$, only the starting porphyrinogen-like species **115**, or decomposition products were obtained [60]. Finally suitable oxidation conditions were found [59, 60]. Exposing **115** to oxygen in the presence of a non-nucleophilic base was found to provide the aromatic macrocycle, albeit in low yield. Specifically, stirring the reduced macrocycle **115** in air-saturated chloroform-methanol containing N,N,N',N'-tetra-methyl-1,8-diaminonaphthalene [60] gave the product **135** as a green solid in ca. 10% yield. Although the yield for the oxidation is low, the aromatic macrocycle once formed appears to be quite stable, decomposing at a slower rate than its precursor [62].

This enhanced stability (of **135**) is attributed to the aromatic stabilization present in the oxidized form [6]. This new aromatic expanded porphyrin **135** can be considered as a 22 π-electron benzannulene with an 18 π-electron delocalization pathway. Due to its large core size, this aromatic expanded porphyrin was assigned the trivial name "texaphyrin", (for Texas-sized porphyrin) [63]. Further confirmation of the aromatic nature of the texaphyrin macrocycle **135** derived from the chemical shift of the NH proton (at δ 0.9 ppm) which was shifted upfield by over 10 ppm as compared to the pyrrolic protons present in the reduced macrocycle **115**. This shift parallels that seen when the *sp*^{3}-linked macrocycle, octaethyl-porphyrinogen (δ(NH) = 6.9 ppm), is oxidized to the corresponding octaethyl-porphyrin (OEP) (δ(NH) = −3.74 ppm) [64]. This suggests that the diamagnetic ring current present in **135** is similar in strength to that present in the porphyrins [60].

If the oxidation of the reduced macrocycle **115** is carried out in the presence of an appropriate metal salt the result is not only oxidation of the macrocycle, but also coordination of the metal to form a metallotexaphyrin (Schemes 16–19)

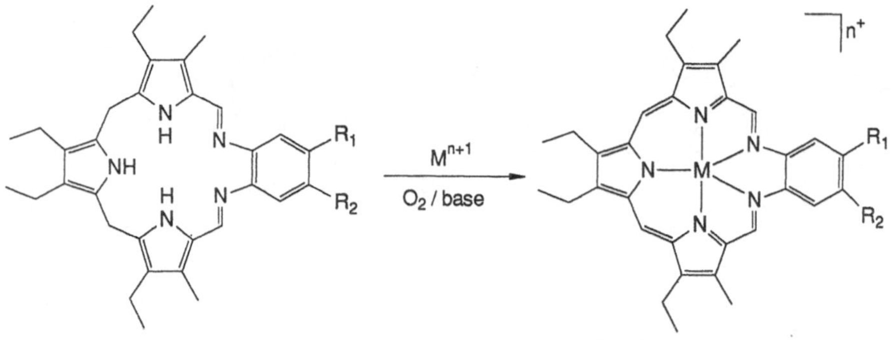

115 $R_1 = R_2 = H$
121 $R_1 = R_2 = Me$
122 $R_1 = H, R_2 = Me$
123 $R_1 = H, R_2 = OMe$
124 $R_1 = H, R_2 = Cl$
125 $R_1 = H, R_2 = CO_2H$
126 $R_1 = H, R_2 = NO_2$

116 $M = Cd, R_1 = R_2 = H, n = 1$
135 $M = H, R_1 = R_2 = H, n = 0$
136 $M = Zn, R_1 = R_2 = H, n = 1$
137 $M = Mn, R_1 = R_2 = H, n = 1$
138 $M = Nd, R_1 = R_2 = H, n = 2$
139 $M = Sm, R_1 = R_2 = H, n = 2$
140 $M = Eu, R_1 = R_2 = H, n = 2$
141 $M = Sm, R_1 = R_2 = H, n = 2$
142 $M = H, R_1 = R_2 = Me, n = 0$
143 $M = Cd, R_1 = R_2 = Me, n = 1$
144 $M = Zn, R_1 = R_2 = Me, n = 1$
145 $M = Sm, R_1 = R_2 = Me, n = 2$
146 $M = Eu, R_1 = R_2 = Me, n = 2$
147 $M = Gd, R_1 = R_2 = Me, n = 2$
148 $M = Cd, R_1 = H, R_2 = OMe, n = 1$
149 $M = Cd, R_1 = H, R_2 = Cl, n = 1$
150 $M = Cd, R_1 = H, R_2 = CO_2H; n = 1$
151 $M = Cd, R_1 = H, R_2 = NO_2, n = 1$
152 $M = Cd, R_1 = H, R_2 = Me, n = 1$

Scheme 16

[60, 65]. For example, the oxidation of **115** in the presence of cadmium chloride yields a dark green powder which formulates as **116** · Cl [60]. The optical spectrum of cation **116** bears some resemblance to those of other aromatic pyrrole-containing macrocycles [24–27, 66]. The dominant transition, in chloroform, is a Soret-like

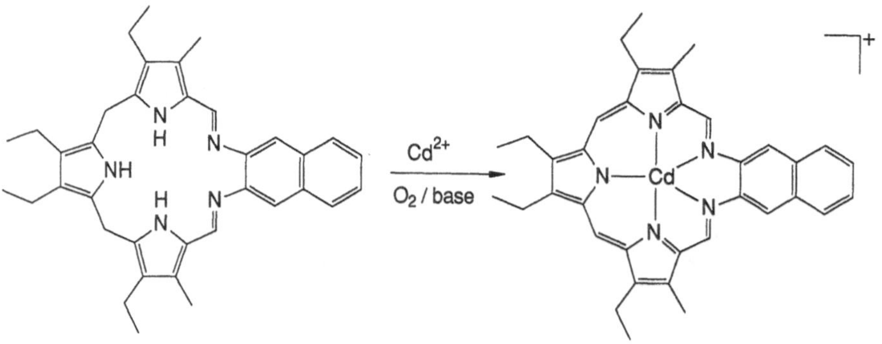

128

153

Scheme 17

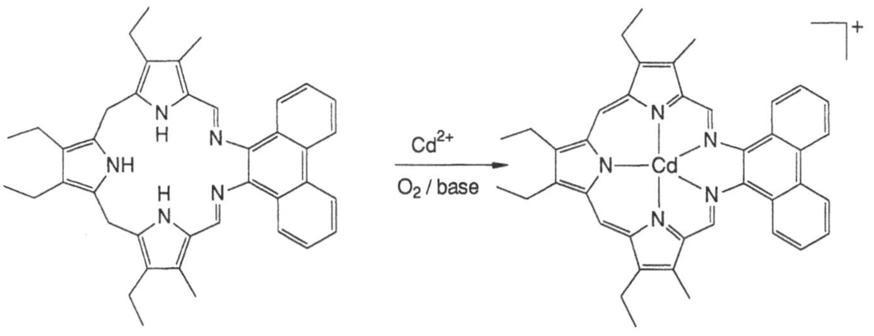

129 → Cd^{2+}, O$_2$ / base → **154**

Scheme 18

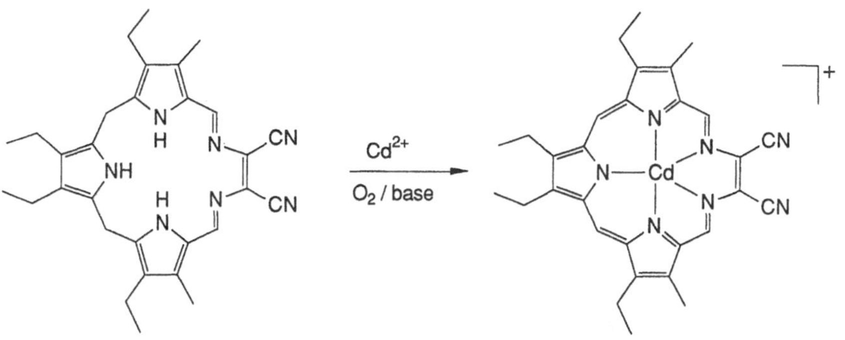

130 → Cd^{2+}, O$_2$ / base → **155**

Scheme 19

band at 427 nm ($\varepsilon = 72{,}700$ cm · mol^{-1}) which is considerably less intense than that seen for Cd(OEP)(Py) (Py = pyridine) [67]. This absorption band is flanked by exceptionally strong N- and Q-like features at higher and lower energies, respectively. As would be expected for a larger π-system, both the lowest energy Q-like absorption ($\lambda_{max} = 767.5$ nm, $\varepsilon = 41{,}200$ cm · mol^{-1}) and emission ($\lambda_{max} = 792$) bands of **116** · Cl are substantially red-shifted (by ca. 200 nm) as compared to those of typical cadmium porphyrins [67, 68].

Using a different metal salt, Cd(NO$_3$)$_2$, instead of CdCl$_2$, the oxidation reaction results in a slightly higher yield of the cadmium texaphyrin complex [60, 65]. However, upon purification, the product is obtained as a mixture of crystalline and non-crystalline solids. A single crystal X-ray diffraction study of the crystalline portion of the sample gave an unexpected result. The structure obtained (Fig. 20) revealed a six-coordinate pentagonal pyramidal cadmium(II) complex **156** (c.f. Scheme 20) where one of the two possible axial ligation sites is occupied by a benzimidazole [65]. The five donor atoms of the pentadentate texaphyrin macrocycle complete the coordination sphere about the cadmium with the cadmium

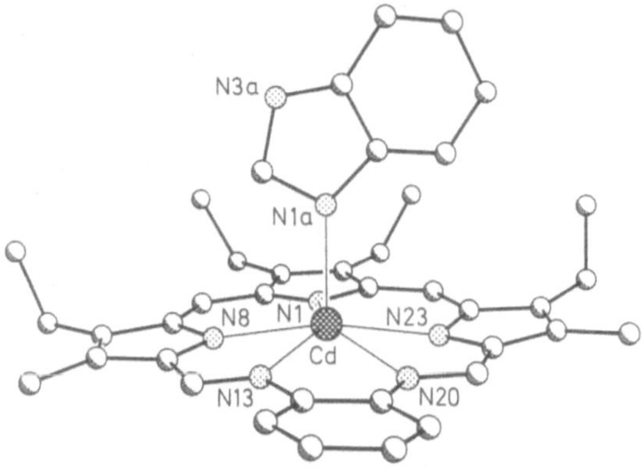

Fig. 20

atom displaced, towards the benzimidazole by 0.338(4) Å from the plane of the five texaphyrin nitrogens [65]. The benzimidazole is thought to result from electrophilic aromatic deacylation of a tripyrrane α-carbon and subsequent condensation with *o*-phenylenediamine [65].

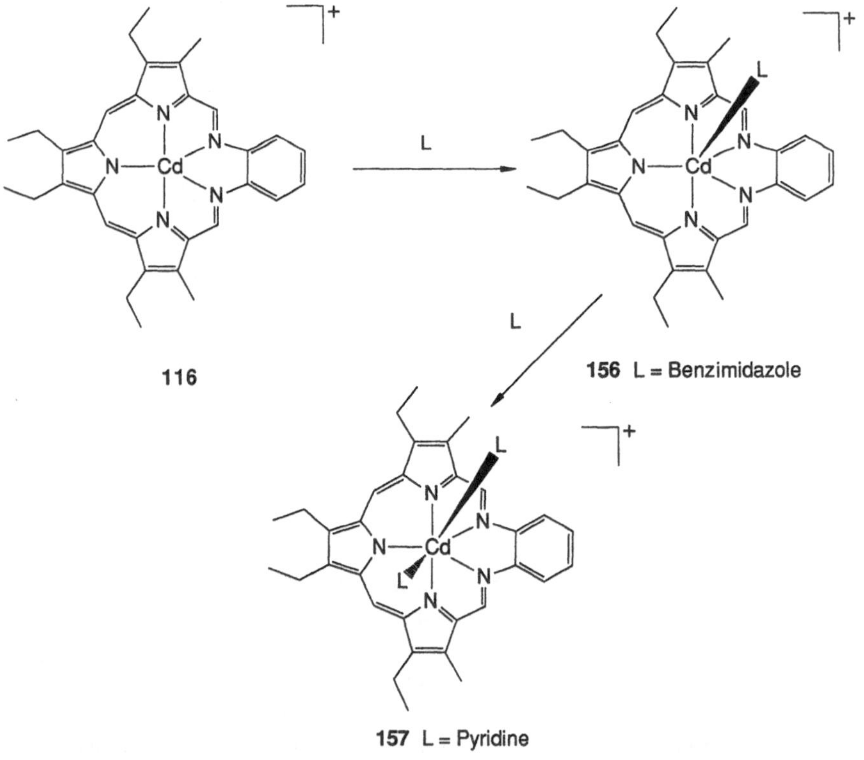

116

156 L = Benzimidazole

157 L = Pyridine

Scheme 20

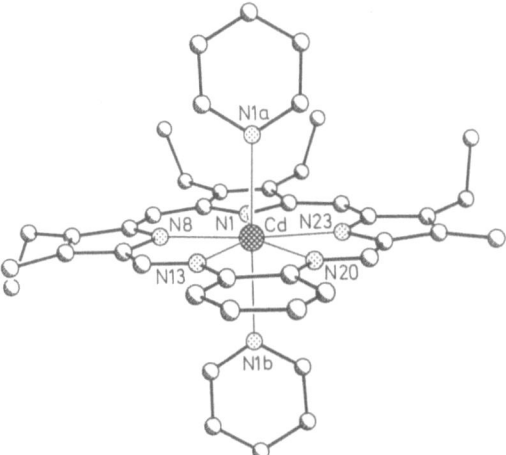

Fig. 21

Treatment of the inhomogeneous material, formed from the oxidation of **115** in the presence of $Cd(NO_3)_2$, with pyridine gives only a single crystalline aromatic product **157**. A single crystal X-ray investigation (Fig. 21) of **157** confirmed the coordination of pyridine to the cadmium [60, 65]. In this complex the cadmium atom is held essentially within the plane of the macrocycle, the two pyridine ligands coordinated to the axial binding sites of the seven-coordinate, pentagonal bipyramidal cadmium center. These two structures (Figs. 20 and 21) confirm the core of the texaphyrin macrocycle as being roughly 20% larger (center-to-nitrogen radius) than that found in typical metalloporphyrins [69].

A quantitative ^1H NMR investigation into the binding of pyridine and benzimidazole to the cadmium texaphyrin complex **116** was carried out [65]. It revealed that the five-coordinate cadmium texaphyrin complex **116**, with no axial ligands, can exist in pure form and that it, along with varying concentrations of six and seven coordinate species (e.g. **156** and **157**) may be produced under certain conditions (Scheme 20). This study thus helps account for the production of inhomogeneous, mixed ligation, materials during the oxidation of the reduced macrocycle **115** in the presence of $Cd(NO_3)_2$ [65].

Support for the above conclusions was also obtained from an analysis of the ^{113}Cd NMR and solid state MAS spectra of the five, six, and seven coordinate cadmium texaphyrin complexes **116, 156,** and **157,** respectively [70]. From the solid-state MAS results, a single tensor was observed for the five-coordinate complex **116** with a corresponding isotropic chemical shift of 194 ppm. The MAS spectrum of a complex prepared in the presence of pyridine, however, revealed two tensors, presumed to be due to a mixture of six and seven coordinate species. Based upon the isotropic chemical shifts and the symmetry of the tensors the six coordinate species was assigned to the isotropic shift at 188 ppm and the seven coordinate species assigned to the isotropic shift at 221 ppm. The MAS spectrum of the benzimidazole complex of cadmium texaphyrin consists of only a single tensor, at an isotropic shift of 188 ppm, assumed to represent the six coordinate species. This work, thus, helped not only to define the coordination properties of cadmium(II) texaphyrins but also to demonstrate further the versatility of ^{113}Cd NMR as a useful probe for studying cadmium ligation in different coordination environments.

In addition to the above coordination-based studies, detailed investigations of the photoexcited tripled state of the free-base 135 and cadmium complex 116 have been carried out using laser-excitation-time-resolved EPR spectroscopy [71]. Measurements were carried out at low temperatures, using frozen glass matrices or uniaxial liquid crystals. The free-base 135 exhibits in-plane intersystem crossing (ISC) rates, independent of the matrix, while the ISC selectivity of 116 depends strongly on the matrix, i.e. out-of-plane in the frozen glass, and in-plane in the liquid crystal matrices. This was attributed to an enhancement in the optical transition moment of the absorption for the in-plane (long X-axis) absorption. Here, the direction of the transition moment in the molecular frame was determined from the anisotropy of the EPR signal intensities in the liquid crystal [71]. The strong dependence of the EPR triplet characteristics on the media in which the texaphyrins 135 and 116 are embedded is quite unusual and, at present, the actual extent to which it is unique to texaphyrins is still being investigated. These two texaphyrins have also proved useful in the development of a new method for detecting high-temperature EPR in fluid liquid crystals (i.e. above their melting points) [72]. This has general implications for the gathering of data on guest chromophores, as well as guest-guest and guest-host dynamics which depend upon temperature and magnetic field.

In other physical chemical analyses, the perimeter model [73] has been employed to analyze the spectral intensities and MCD signals for a series of porphyrinoid macrocycles derived from the $C_{20}H_{20}^{2+}$ perimeter, including the parent, benz-free analogue of texaphyrin 158 [74]. These calculations were then compared with the MCD spectra of a number of substituted cadmium texaphyrins (e.g. 116, 143, and 148–152, c.f. Scheme 16) [75]. The results confirmed that the perimeter model accounts in a simple way for the signs of the MCD B terms associated with the low-lying electronic transitions of these metallotexaphyrins.

158

Considerable effort has also been devoted to exploring the ground and excited state optical properties of metallotexaphyrins using more conventional means. Here, much of the original interest derived from the observation that the texaphyrins absorb strongly in the 720–780 nm spectral region and the accompanying realization that these systems could be of possible use in photodynamic therapy. As detailed in greater depth in Sect. 12, photodynamic methods are among the more promising of the approaches currently being considered for the treatment of localized neoplasia [76–80] and for the eradication of viral contaminants in blood [81]. To date, porphyrins and their derivatives, e.g. phthalocyanines and naphthalocyanines have been among the most widely studied compounds considered in the context of developing an effective photochemotherapeutic agent [82–88]. Unfortunately, all of these dyes suffer from disadvantages. While the various porphyrins have high triplet yields and long triplet lifetimes their absorption in the Q-band region often parallels that of heme-containing tissues which reduces their efficacy. Phthalocyanines and naphthalocyanines absorp in a more convenient spectra region but have significantly lower singlet oxygen quantum yields [89]. Because the texaphyrins absorb in the spectral region where living tissues are relatively transparent (i.e. 700–1,000 nm), they could possibly represent a viable new alternative [90].

Investigations into the potential of the texaphyrins as photosensitizers for photodynamic therapy began with an examination of their ability to generate singlet oxygen [91, 92]. The photophysical properties of a number of metallotexaphyrins, namely **116, 136, 137, 144–146** and **149**, were found to parallel those of the corresponding metalloporphyrins. The diamagnetic metallotexaphyrin complexes investigated demonstrated three important and near-unique optical properties: They were found to 1) absorb strongly in a physiologically important region, 2) form long-lived triplet states in high yield, and 3) act as efficient photosensitizers for the formation of singlet oxygen [91].

Further investigations with a variety of functionalized cadmium texaphyrins (**143,** and **148–155**) [92, 93] served to show that the absorption maxima and redox potential of the texaphyrin complexes could be influenced strongly by the nature of the substitutents on the phenyl ring as well as the extent of π-electron conjugation in the texaphyrin macrocycle. In fact, a linear relationship between the energies of the Q-type band maximum and the difference in the first oxidation and reduction potentials was observed for the substituted cadmium(II) texaphyrin complexes **143,** and **148–155**. The Q-type band absorption maxima of the texaphyrins reported in this particular study [93] could be varied from 629 to 864 nm without causing a substantial reduction in the corresponding singlet oxygen quantum yield [94]. This, coupled with their high chemical stability and appreciable solubility in polar media, led to the suggestion that these cation complexes could serve as viable photosensitizers for photodynamic therapy [93, 94].

In fact, the above-described results were encouraging enough to warrant a preliminary investigation of the *in vitro* photodynamic anti-viral and anti-bacterial activity of the cadmium texaphyrins [94]. In addition, they also provided the impetus for several *in vitro* cell localization experiments [94]. The anti-viral studies were carried out using herpes simplex virus (HSV-1) in 50% human serum containing varying concentrations of complexes **155** or **116**. The results indicated that these cadmium texaphyrins were moderately effective as photosensitizers for the inactivation of HSV-1. However, they were found to be considerably less efficient (by several orders of magnitude) than other available porphyrin-type dyes. This lack of higher activity was rationalized in terms of the high charge density of the cadmium texaphyrins which, it was thought, would preclude binding to the relatively hydrophobic membrane of HSV-1 [94].

It was precisely this lack of activity towards non-polar substrates that provided an important motivation for the antibacterial and cell localization studies [94]. The cadmium texaphyrin **116** was investigated with regard to photo-activity against a strain of an antibiotic-resistant bacteria (*S. aureus*). The texaphyrin complex **116** proved to be an effective photosensitizer for the photoinactivation of *S. aureus* cells, being comparable but somewhat less active than hematoporphyrin at any given concentration tested.

The first cell localization study, involving mononuclear cells, was carried out as a complement to the anti-viral photodynamic work. Here, the basic motivation for the study derived from the realization that HIV-1 replicates in human T-4 lymphocytes and that, as such, selective light-derived inactivation of such cells would be of benefit in possible photodynamic blood purification processes (see Sect. 12). Texaphyrin proved to be quite effective in the photodestruction of mononuclear cells. In fact it was found that the cadmium texaphyrin complex **116** is as effective on a *per mole* basis as the dihematoporphyrin ether (DHE) and slightly more so on a *per photon* basis.

A second set of cell localization studies was carried out with epithelial cells. Strong evidence was seen for a photodynamic inactivation effect [94]. This effect was manifest in observable morphological damage to the cells. Importantly, in this and other photodynamic studies, evidence was obtained to suggest that the macrocycle remained unaffected by the irradiation process. This, it was suggested, augers well for the eventual use of the texaphyrins in a number of photodynamic applications [93, 94].

The final *in vitro* photodynamic studies involved the use of human cancer cells. Here, K562 leukemic cells of myelocutic origin were used. The texaphyrin complex **116** was very effective in the photo-eradication of K562 leukemic cells, being considerably more effective than hematoporphyrin under similar conditions. This suggests that the texaphyrin expanded porphyrins, as a class of photosensitizers, could have further application in the photo-killing of these and other cancer cells [94].

In work along different lines, efforts have been made recently to explore further the metal chelation properties of the texaphyrin series of ligands. As expected, it was found that these large macrocycles provide a very stable coordination environment for large cations such as those of the lanthanide series [95]. As detailed in Sect. 12, this is of interest in terms of magnetic resonance imaging (MRI) applications. At present gadolinium (III) complexes derived from strongly binding anionic ligands, such as diethylenetriaminepentaacetic acid (DTPA) [96–98], 1,4,7,10-tetrazacyclododecane-*N,N',N'',N'''*-tetraacetic acid (DOTA) [96, 99, 100], and 1,10-diaza-4,7,13,16-tetraoxacyclooctadecane-*N,N'*-diacetic acid (dacda) [96,

101], are among the most widely studied of the paramagnetic contrast agents currently being considered for use in MRI [96]. The complex $[Gd(DTPA)]^{2-}$ is now being used clinically in the United States in certain enhanced tumor detection protocols. However, all of these systems suffer from deficiencies (including, to varing extents, kinetic lability). As such, the syntheses of other Gd(III) complexes including those of the texaphyrin series, which might have greater kinetic stability, superior relaxivity, and/or better biodistribution properties, would be of interest. This is especially important since such systems cannot be prepared for porphyrins; the large Gd(III) cation does not fit inside the core of the porphyrin and is easily displaced both *in vitro* and *in vivo* [102, 103].

The formation of a number of lanthanide texaphyrin complexes has been reported [95]. In all cases, metal insertion and oxidation proceeds smoothly (Scheme 16) [95]. The complexes demonstrate fair water solubility and good stability towards hydrolysis. Detailed kinetic studies of complex 147, for instance indicated that the half-life for decomplexation and/or decomposition of this complex is 37 days in a 1:1 mixture of $MeOH:H_2O$ (pH7). Thus, it appears that gadolinium(III) complexes of texaphyrins could provide the basis for a new approach to paramagnetic MRI contrast reagent development [95].

Paramagnetic properties of certain lanthanide complexes are also being studied in the context of ^{23}Na NMR. As is true for more classic MRI, ^{23}Na-based magnetic methods are being exploited as a noninvasive means for the study of cells in various physiological states [104]. Here again, paramagnetic shift reagents have been found to be of use. They are particularly attractive in terms of distinguishing intracellular Na^+ concentrations from extracellular ones. In this regard, Dy(III) complexes, which generate large hyperfine shifts, have been shown to be particularly effective as contrast agents [105–109]. Unfortunately, the metal chelates examined to date have all suffered from problems associated with metal dissociation and dysprosium toxicity *in vivo* [105]. With the known *in vitro* stability of the lanthanide complexes of texaphyrin [95], the study of an elaborated texaphyrin dysprosium complex 159 bearing a crown ether binding functionality was undertaken [110]. The synthesis of the reduced porphyrinogen-like precursor 133 was achieved from the condensation of 4′,5′-diamino{benzo-15-crown-5} with the diformyltripyrrane 114. Treatment of 133 with dysprosium nitrate in the presence of oxygen and base then gave the dysprosium (III) containing texaphyrin 159 (Scheme 21). The hyperfine ^{23}Na NMR shift induced by 159, of 0.86 ppm, is nearly identical to that induced by a simpler texaphyrin-free Dy(III) tetraazatetraoxo macrocyclic complex which binds sodium in the crown ether portion of the macrocycle [111]. The shift direction observed for the Na^+ caused by 159 is also consistent with the Na^+ cation being bound perpendicular to the principle magnetic axis.

Although quite new on the scientific scene, the lanthanide complexes of the texaphyrins are considered to be of particular interest. This derives in larger measure from their documented stability in aqueous solution and the relative ease with which the basic core structure can be subjected to variations in substituents and functionality. With the increasing use of lanthanide-based shift reagents in medical diagnosis, the continued study of these new systems could prove to be of considerable interest.

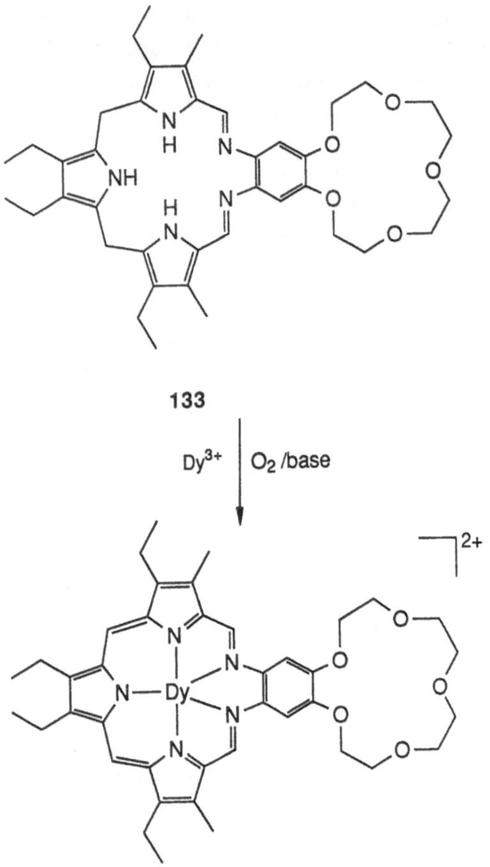

133

Dy^{3+} | O$_2$/base

159   **Scheme 21**

6 Uranyl Superphthalocyanines

The first structurally characterized expanded porphyrin system to be reported in the literature was the so-called "superphthalocyanine" ligand [112]. This compound, which represents the first example of a well-characterized pentaligated complex prepared from *any* aromatic pentadentate macrocycle ligand, was obtained as an outgrowth of early efforts to prepare uranyl phthalocyanine and not as the product of a directed step-by-step synthesis. As such, the early literature associated with this species remains somewhat clouded and incomplete.

Reports of initial work indicated that uranyl phthalocyanine complexes could be obtained from the reaction of bis(dimethylformamide)uranyl acetate with lithium phthalocyanine [113], or the reaction of uranyl acetate with phthalocyanine [114, 115]. However, in 1964, Bloor *et al.* [116] reported that the reaction of uranyl dichloride and *o*-dicyanobenzene in dimethylformamide at 180 °C gave rise to a presumed-to-be uranyl phthalocyanine complex with infrared and visible properties that differed substantially from those originally reported [113]. On the basis of their spectral evidence they concluded that the uranyl phthalocyanine obtained by the previous workers was essentially a mixture of metal-free phthalocyanine and inorganic uranium salts. Interestingly, mass spectral data of this new uranyl phthalocyanine complex [117] suggested that five dicyanobenzene subunits might be coordinated to the uranyl ion (Scheme 22). In addition, it was found that the reaction of anhydrous uranyl chloride with *o*-dicyanobenzene in dry dimethylformamide yielded, after extractive work-up, a blue-black crystalline material, which analyzed for (dicyanobenzene)$_5$UO$_2$ (**160**) [112, 118, 119]. While phthalocyanine complexes of stoichiometry (phthalocyaninato)M(*o*-dicyanobenzene) were then known [120, 121], in these cases, the extra nitrile is coordinated as an independent, displaceable ligand readily identifiable by the $v(C \equiv N)$ band (at 2200–2300 cm^{-1}), a band not seen in the (dicyanobenzene)$_5$UO$_2$ complex.

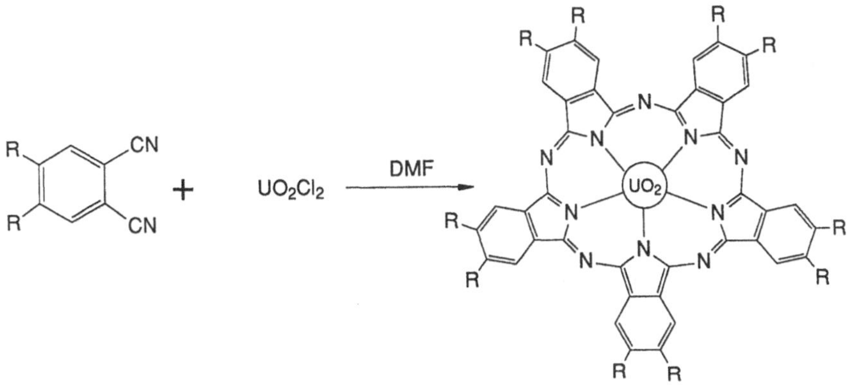

160 R = H
161 R = CH$_3$
162 R = *n*-C$_4$H$_9$

Scheme 22

The apparent contradiction between the empirical stoichiometry and the spectral characteristics of these new uranyl complexes was finally resolved by X-ray crystallography. Specifically, a single crystal X-ray structural analysis of the blue-black material formed from the reaction of the anhydrous uranyl chloride and *o*-dicyanobenzene [112] (Figures 22 and 23) revealed that the complex obtained was in fact an expanded five-subunit superphthalocyanine macrocycle in which a pentagonal bipyramidal coordination geometry pertains about the centrally-bound uranium atom.

As is apparent from Figs. 22 and 23 the superphthalocyanine ligands forms an hexagonal girdle around the uranium atom that is essentially perpendicular to

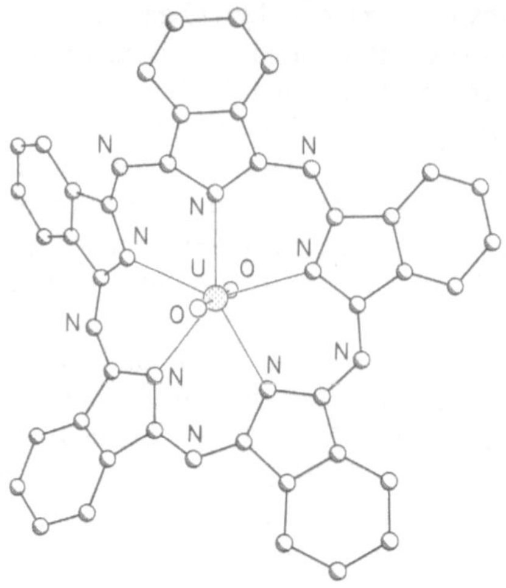

Fig. 22

the UO$_2$ axis. Unlike the four-subunit phthalocyanine macrocycle, which is essentially planar as the free acid [122, 123] and even in many of its metal complexes [124–126], the superphthalocyanine ligand is decidedly nonplanar (Fig. 23). The wave-like nature of the ligand is probably required to minimize steric strains imposed upon the "inner ring" of 20 atoms, surrounding the uranyl group. The radius of the central core in uranyl superphthalocyanine at 2.55 Å is ideally suited for UO$_2^{2+}$ coordination [118].

A variety of substituted uranyl superphthalocyanine complexes, such as the more soluble methyl **161** and butyl **162** substituted systems [118, 17] can be obtained from the general condensation reaction (Scheme 22). However, when the condensation reaction was carried out using 1,2-dicyanobenzenes with electron withdrawing substituents, or those which impose a greater steric congestion, no five subunit-containing macrocyclic products could be detected.

In terms of spectroscopic properties, the uranyl superphthalocyanine complexes **160–162** display features which, although reminescent of, differ substantially from those of the phthalocyanine. The IR spectrum exhibits a strong v(OUO) stretching transition at 925 cm^{-1} (KBr pellet) [112, 118, 119] or 933 cm^{-1} (evaporated film)

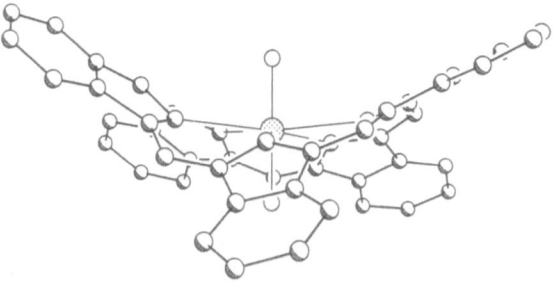

Fig. 23

220

[128]. In addition, the electronic spectrum of uranyl superphthalocyanine **160** is significantly different from those of known metal phthalocyanine complexes [129]. It exhibits an intense absorption at 914 nm ($\varepsilon = 6.67 \times 10^4$ cm $\cdot$ mol^{-1}) with a pronounced shoulder at 810 nm, and a strong absorption at 424 nm ($\varepsilon = 5.02 \times 10^4$ cm $\cdot$ mol^{-1}). Finally, the ^{1}H NMR spectrum of uranyl superphthalocyanine **160** exhibits an AA'BB' pattern for both of the benzo protons at δ 9.06 and 7.68 ppm [127]. However, these benzo protons are shielded in relation to those of the corresponding diamagnetic phthalocyanine complexes. Interestingly, an analysis of the nuclear magnetic shielding in unsubstituted planar phthalocyanines and superphthalocyanines in terms of diamagnetic ring-current displacements led to a prediction of an opposite trend [118]. Thus, the smaller downfield ring-current shift observed for the benzo protons of uranyl superphthalocyanine macrocycle **160** (with respect to those of the phthalocyanines) was attributed to the severe distortion from planarity inherent in the superphthalocyanine ligand [127].

Many of the other properties of the uranyl superphthalocyanine complex **160** may also be explained in terms of the severe strain within the macrocycle. The reaction of **160** with acids, for instance, under conditions which readily demetalates many phthalocyanine and porphyrin complexes [130, 131], results in an unprecedented ring contraction giving free-base phthalocyanine as the product (Scheme 23) [132]. A similar ring contraction also occurs when the uranyl superphthalo-

160 → acid → $+ UO_2X_2 +$ (o-dicyanobenzene, CN, CN)

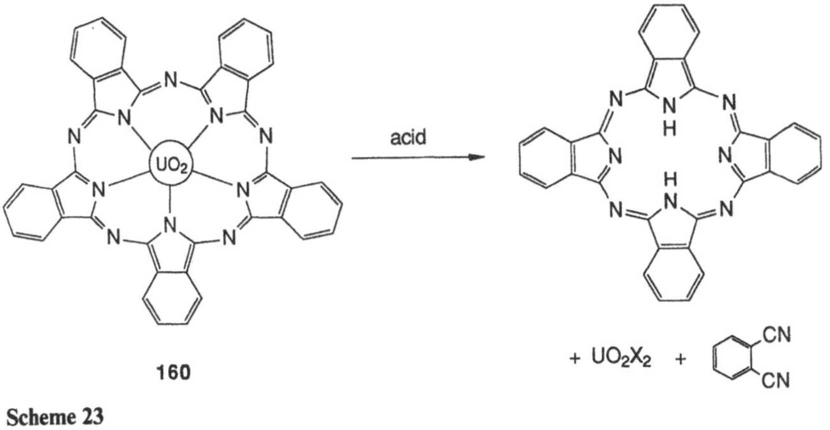

Scheme 23

cyanine complex **160** is treated with reducing reagents such as P(n-Bu)$_3$, thiophenol, and 2-mercaptoethanol [133]. The reaction with P(n-Bu)$_3$ appears to proceed through a purple intermediate thought to be a U(IV) superphthalocyanine species [133]. Reactions of the uranyl superphthalocyanine complex **160** with anhydrous metal salts (e.g. CoCl$_2$, NiCl$_2$, FeCl$_3$, CuCl$_2$, ZnCl$_2$, SnCl$_2$, and PbCl$_2$) also results in a ring contraction. In this case, the corresponding metal phthalocyanine complexes are formed (Scheme 24) [132, 133]. These contraction reactions indicate

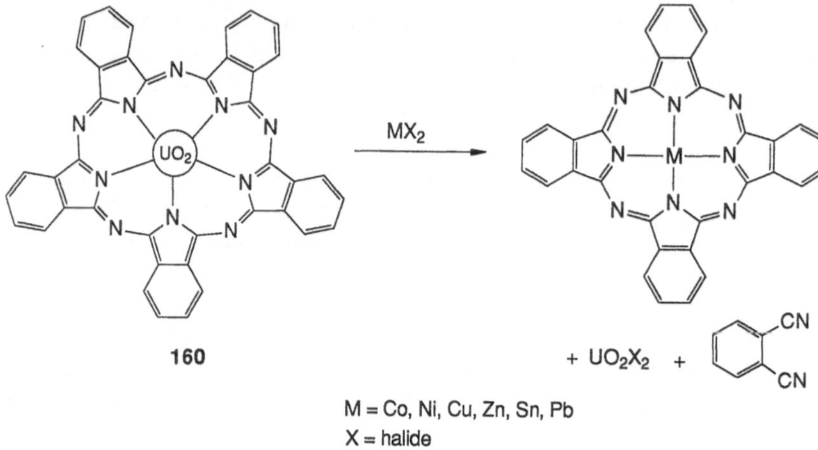

160

$+ UO_2X_2 +$

M = Co, Ni, Cu, Zn, Sn, Pb
X = halide

Scheme 24

that the uranyl ion plays a significant role in stabilizing the superphthalocyanine system. Even with large cations, such as Sn^{2+}, out of plane phthalocyanine complexes are formed. The actual mechanism(s) of the superphthalocyanine contraction is (are) still not known. However, the possibilities that have been considered [118] involve initial, rate determining, metal mediated displacement of the UO_2^{2+} from the macrocycle to form a new transient superphthalocyanine bimetallic complex which subsequently contracts (Scheme 25). Alternatively, the

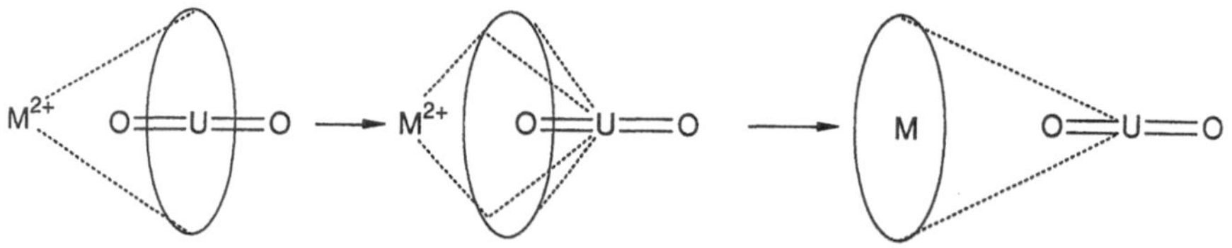

Scheme 25

rate determining step could involve coordination of the attacking metal at an imino nitrogen on the uranyl superphthalocyanine periphery (Fig. 24), followed by Lewis acid promoted ring opening. In any case, it is evident that the superphthalocyanine ligand is a chemically fragile entity and one that, apparently, can not exist without the stabilizing influence of a centrally coordinated uranyl cation. Thus, while uranyl superphthalocyanine stands as an excellent demonstration of the range and power of metal-template condensation reactions and a notable landmark in expanded porphyrin history, it appears that it is but an intriguing dead-end in terms of further chemical development.

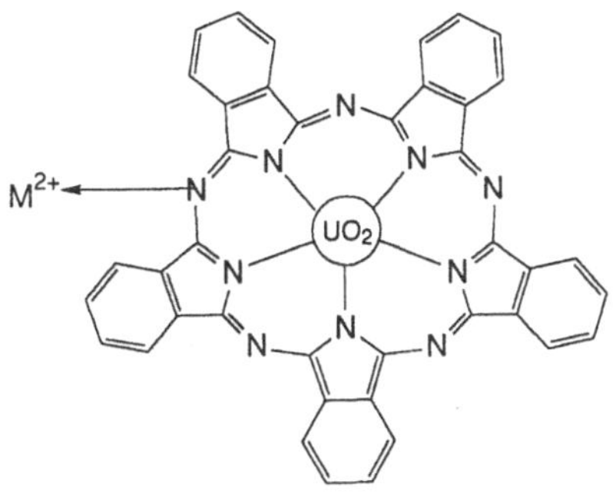

Fig. 24

7 Furan-Containing Annulenes and Annulenones as Expanded Porphyrins

During the late nineteen sixties and the nineteen seventies, questions involving the experimental and theoretical limits of aromaticity [134–136] and the relationship between the number of π-electrons and observable diamagnetic or paramagnetic ring currents were topics of considerable interest [137–141]. Theoretical calculations indicated that the Hückel $(4n + 2)$ rule should break down at higher values of n, with the onset of bond alternation [142], and it was predicted that the limit should lie between 22 π- and 26 π-electron containing ring compounds [143]. As a result, attention turned to the production of larger and larger annulene macrocycles. Unfortunately, one of the major problems associated with the formation of "aromatic" or "anti-aromatic" systems was that as the size of the annulene ring increased, there was observed to be a corresponding increase in the flexibility of the macrocycle which, in general, made the syntheses correspondingly more difficult. In fact, the inherent flexibility of the precursors and intermediates tended to encourage polymerization over cyclization. The practical solution to this problem was to form the large macrocycles from more rigid starting components. Here, the general approach was to use precursors containing one or more internal bridges. While this strategy often proved successful, in the production of a range of annulenes, only those annulenes that contain at least one five-membered-heterocyclic ring and which incorporate an inner macrocyclic core containing 17 or more atoms are of relevance to this review. In other words the "expanded porphyrins" component of this now-historic work is entirely limited to several furan-containing macrocyclic systems. Here, almost all the materials were obtained *via* the Wittig reaction and are included in an earlier (1975) review [144].

A typical example of a Wittig-based reaction, leading to a furan-containing macrocycle, is the base-induced self-condensation of 2-formyl-5-triphenylphosphoniomethylfuran chloride (**163**) [145]. This reaction, as expected, gives rise to several products, all in poor yields (Scheme 26). The $4n$ π-electron systems **165**

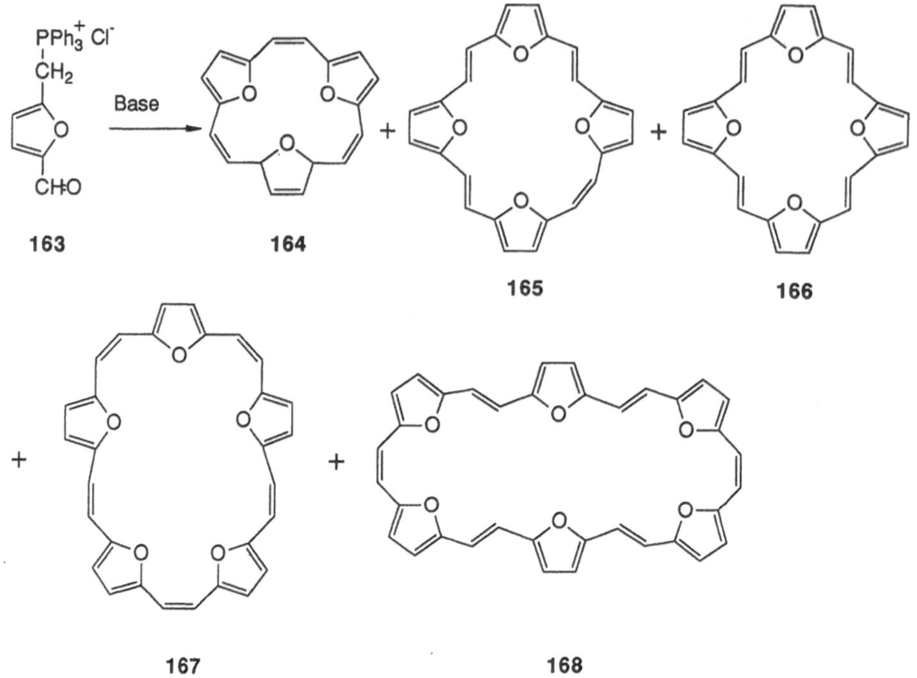
PPh3+ Cl-
CH2
CHO
Base
163
164
165
166
167
168

Scheme 26

and **166** sustain paramagnetic ring currents in an applied magnetic field. In addition to these products, two isomers of **167** as well as trace quantities of **168** were also isolated. However, their respective configurations remain unknown. The physical properties of the two isomers of **167** resemble those of similar unbridged annulenes, but exhibit ^{1}H NMR spectral features characteristic of nonaromatic compounds [145].

More rational syntheses of related furan-containing systems involve the condensation of bis-ylides with bis-aldehydes. The simplest such approach involved the condensation of 2,5-diformylfuran (**169**) with *o*-xylylene-bis[triphenylphosphonium bromide] (**170**) under basic conditions. The result was the production of three isomers of the formal "2 + 2" condensation product (Scheme 27) [146]. Irradiation of the three isomers **171–173** gave a single new isomer **174**. Compound **174** may also be generated by the reaction of the bis-ylide **170** with **175** (Scheme 28) [147].

A variety of other poly-furans have been employed in Wittig reactions as either the aldehyde or the ylide component. Condensation of 5,5′-thiodi-2-furaldehyde (**176**), for instance, with the 2,2′-bifuryl-5,5′-diylbis-(methylenetriphenyl-phosphonium bromide) (**177**) under basic conditions, gave the macrocycle **178** as orange crystals in 1.3% yield (Scheme 29) [148]. Two macrocyclic annulenones **182** and **184** were also obtained [149] from the base induced condensation of **181** with the dialdehydes **180** or **183**, respectively. Both products are obtained in moderate yields (12–15%) and are highly colored solids (Schemes 30 and 31, respectively).

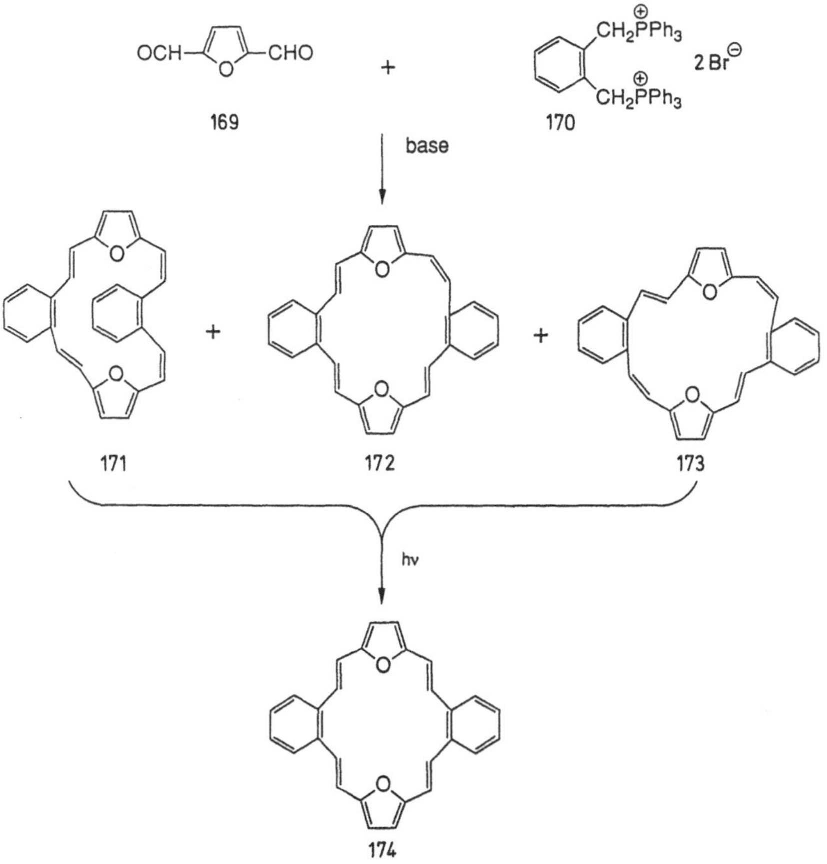

169 + 170

base

171 + 172 + 173

hv

174

Scheme 27

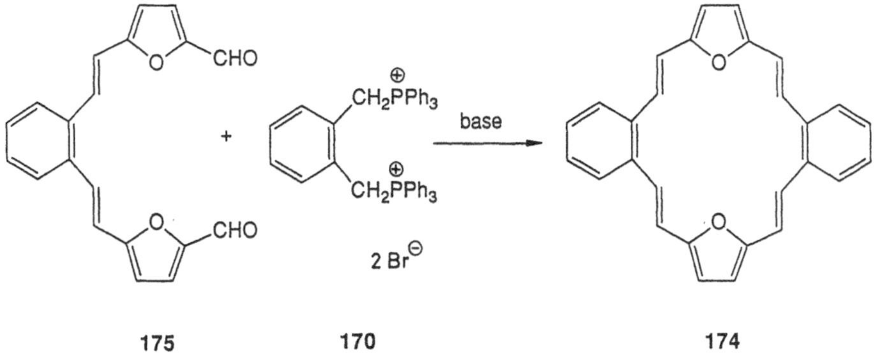

175 + 170 → 174

base

Scheme 28

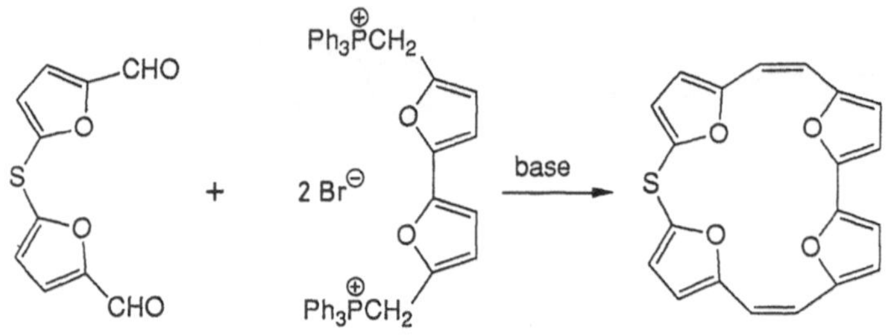

176   **177**   **178**

Scheme 29

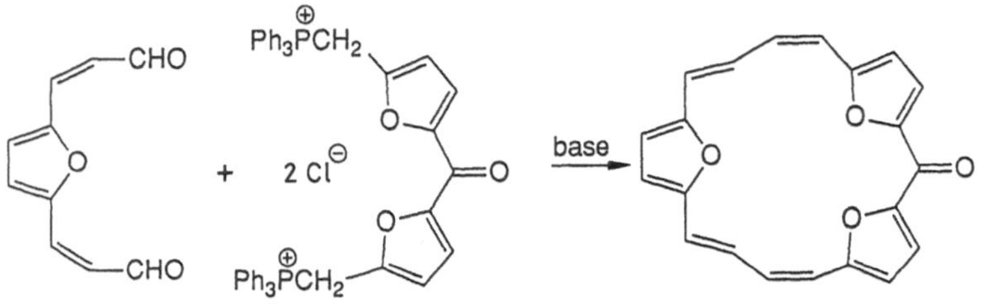

180   **181**   **182**

Scheme 30

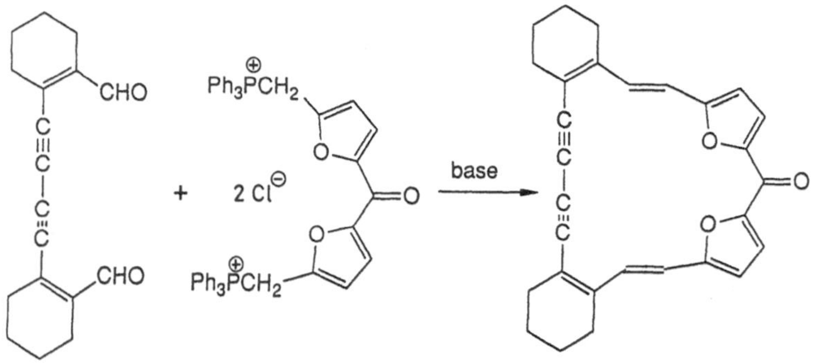

183   **181**   **184**

Scheme 31

In these condensation reactions, the relative size of macrocycles could be varied by simply changing the length of the linking chain. Specifically, the effect that the changes in the macrocycle size have upon the deshielding of the internal protons were investigated by the preparation of a series of non-conjugated ketones [150].

Here, for instance, the reaction of bis-5-(β-formylvinyl)-2-furyl ketone (**185**) with phosphonium ylides of differing chain lengths was used to give the ketones **186a–c** as shown in Scheme 32. Interestingly, in these reactions the yield of the corresponding macrocyclic product **186** increases as the chain length of the starting phosphonium salt becomes greater. All of the ketones (*viz.* **182**, **184**, and **186**) can be reduced by lithium aluminum hydride to give the respective alkanes, namely **187**, **189** and **188** (Schemes 33–35) [150]. As the ring size of the annulenones increased, the resonances of the internal protons were shifted to higher field (in the ^{1}H NMR) but the external protons continue to resonate at fairly constant field. The qualitative assessment given by the authors was that the low field resonances were due to steric compression rather than being due to a paramagnetic ring current [150].

To the best of our knowledge, there is only one other report of an annulated furan-derived macrocycle which conforms to our operative definition of an expanded porphyrin. This is the hexafuran compound **190**, which was reported by LeGoff [151] at the 196th A.S.C. National Meeting in 1988. At present, no details as to the synthesis and/or general chemical properties of this system are readily available.

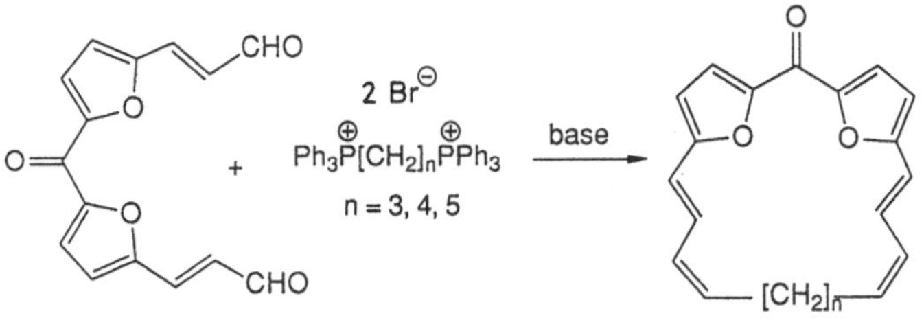

185

186a n=1
186b n=2
186c n=3

Scheme 32

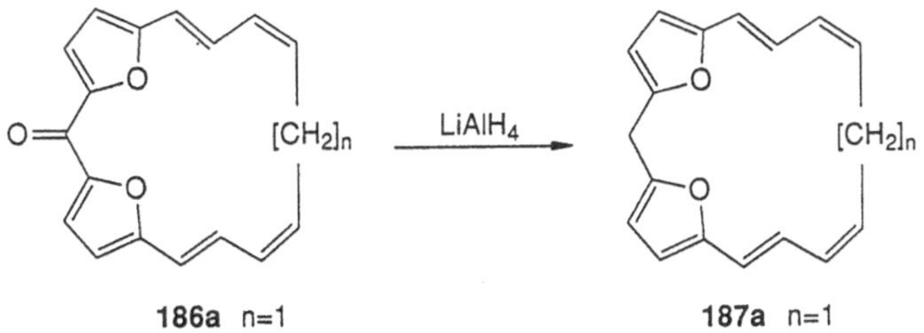

186a n=1
186b n=2
186c n=3

187a n=1
187b n=2
187c n=3

Scheme 33

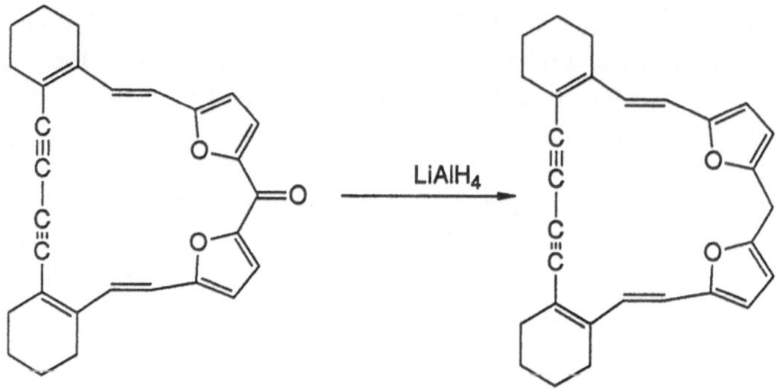

184   **189**

Scheme 34

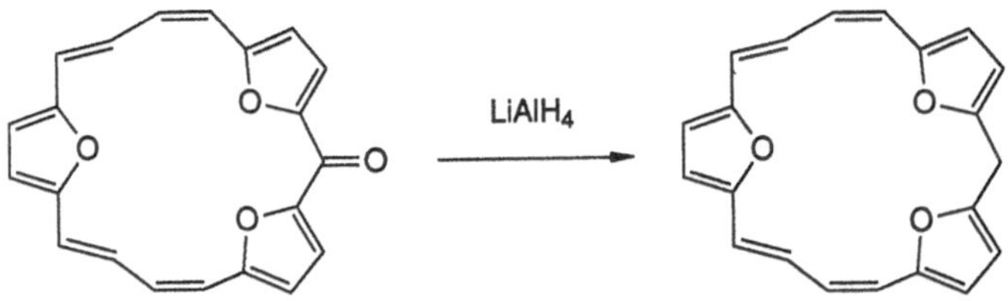

182   **188**

Scheme 35

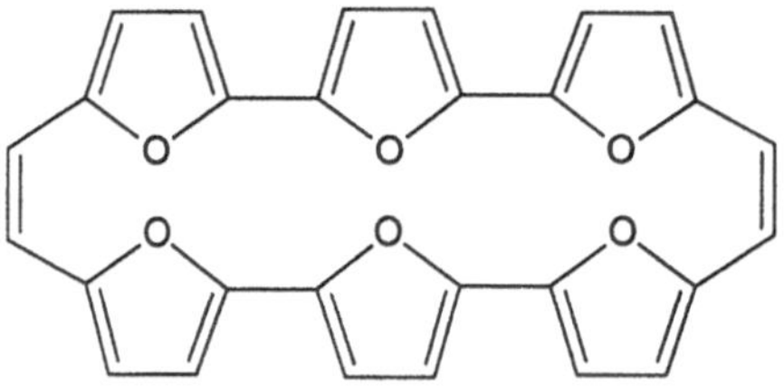

190

In closing this near historical overview, it is important to stress that the impetus for synthesizing these furan-containing systems was exclusively to examine their aromatic properties. While many of them may have uses as ligands, this aspect of their chemistry has yet to be explored. However, as will be detailed in the ensuing sections, pyrrole-containing congeners of several of these furan-based systems are now known and some of these display a rich chemistry indeed.

8 Sapphyrins and Heteroatom-Containing Analogues

With the exception of the texaphyrins of Sessler *et al.* [59, 60, 65, 95] (*vide supra*) the "sapphyrin" macrocycle and its heteroatom analogues (c.f. Scheme 36) are perhaps the best studied of the expanded porphyrins prepared to date. Sapphyrin was the first example of an expanded porphyrin to be reported, being discovered

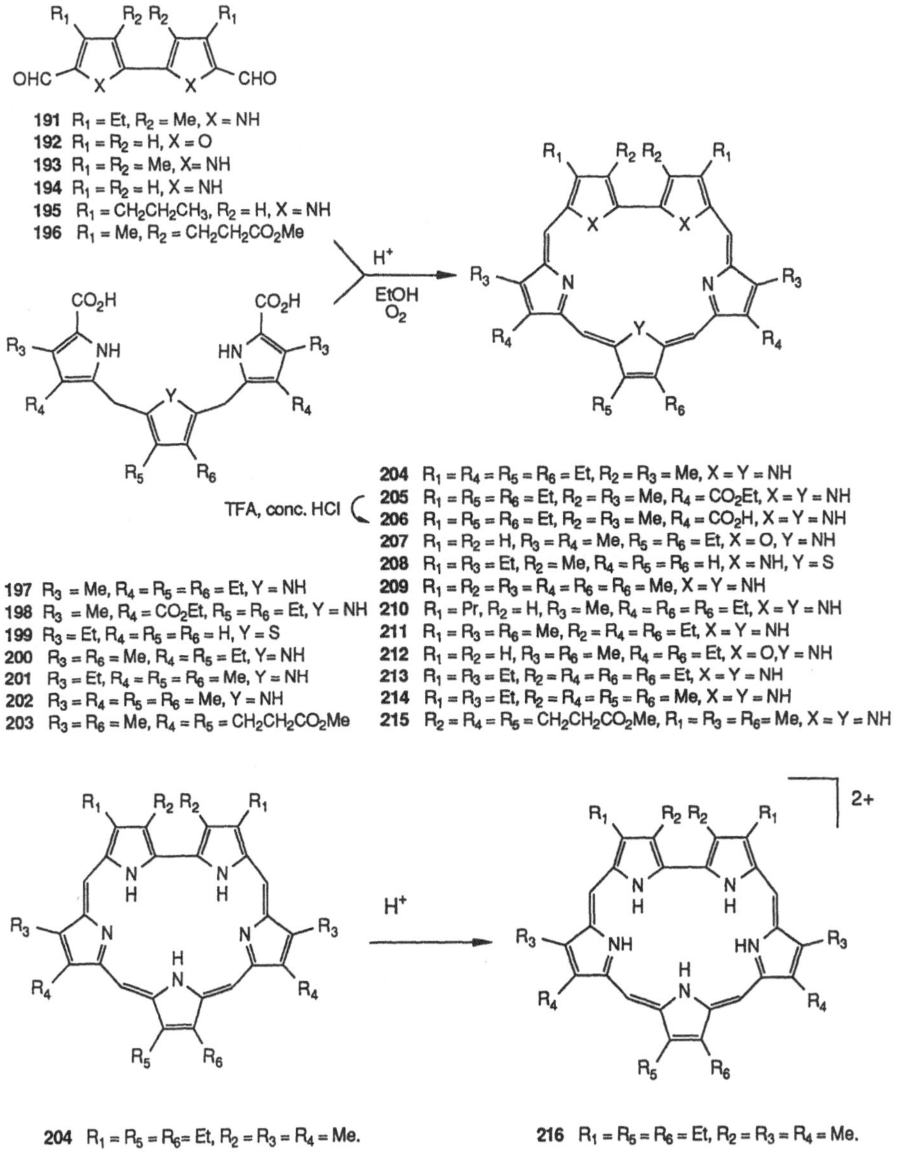

191 $R_1 = Et, R_2 = Me, X = NH$
192 $R_1 = R_2 = H, X = O$
193 $R_1 = R_2 = Me, X = NH$
194 $R_1 = R_2 = H, X = NH$
195 $R_1 = CH_2CH_2CH_3, R_2 = H, X = NH$
196 $R_1 = Me, R_2 = CH_2CH_2CO_2Me$

TFA, conc. HCl

197 $R_3 = Me, R_4 = R_5 = R_6 = Et, Y = NH$
198 $R_3 = Me, R_4 = CO_2Et, R_5 = R_6 = Et, Y = NH$
199 $R_3 = Et, R_4 = R_5 = R_6 = H, Y = S$
200 $R_3 = R_6 = Me, R_4 = R_5 = Et, Y = NH$
201 $R_3 = Et, R_4 = R_5 = R_6 = Me, Y = NH$
202 $R_3 = R_4 = R_5 = R_6 = Me, Y = NH$
203 $R_3 = R_6 = Me, R_4 = R_5 = CH_2CH_2CO_2Me$

204 $R_1 = R_4 = R_5 = R_6 = Et, R_2 = R_3 = Me, X = Y = NH$
205 $R_1 = R_5 = R_6 = Et, R_2 = R_3 = Me, R_4 = CO_2Et, X = Y = NH$
206 $R_1 = R_5 = R_6 = Et, R_2 = R_3 = Me, R_4 = CO_2H, X = Y = NH$
207 $R_1 = R_2 = H, R_3 = R_4 = Me, R_5 = R_6 = Et, X = O, Y = NH$
208 $R_1 = R_3 = Et, R_2 = Me, R_4 = R_5 = R_6 = H, X = NH, Y = S$
209 $R_1 = R_2 = R_3 = R_4 = R_6 = R_6 = Me, X = Y = NH$
210 $R_1 = Pr, R_2 = H, R_3 = Me, R_4 = R_6 = R_6 = Et, X = Y = NH$
211 $R_1 = R_3 = R_6 = Me, R_2 = R_4 = R_6 = Et, X = Y = NH$
212 $R_1 = R_2 = H, R_3 = R_6 = Me, R_4 = R_6 = Et, X = O, Y = NH$
213 $R_1 = R_3 = Et, R_2 = R_4 = R_6 = R_6 = Et, X = Y = NH$
214 $R_1 = R_3 = Et, R_2 = R_4 = R_5 = R_6 = Me, X = Y = NH$
215 $R_2 = R_4 = R_5 = CH_2CH_2CO_2Me, R_1 = R_3 = R_6 = Me, X = Y = NH$

204 $R_1 = R_5 = R_6 = Et, R_2 = R_3 = R_4 = Me.$

216 $R_1 = R_5 = R_6 = Et, R_2 = R_3 = R_4 = Me.$

Scheme 36

229

serendipitously by Woodward and co-workers [66, 152, 153] during the course of early efforts directed towards the synthesis of vitamin B_{12}. The system contains five pyrroles and possesses an overall aromatic 22 π-electron annulene framework. The sapphyrins are blue-green solids (hence the name). The free-base form of the sapphyrins (e.g. **204** in Scheme 36) have three normal protonated pyrroles and two sp^2 hybridized nitrogen atoms which are effectively pyridine-like. These two nitrogen atoms are relatively basic and are readily protonated by weak acids, such as silica gel. The basic nature of these nitrogen atoms is such that the free-base form of sapphyrin undergoes facile protonation even, over time, in the solid state. However, as yet the pK_b value for these nitrogen atoms have not been determined. In the free-base form the sapphyrins (e.g. **204**) display an intense Soret-like absorption at ca. 450 nm, in the visible spectrum. Other less intense Q-like bands are also observed in the 680–710 nm spectral region. Upon protonation the Soret-like absorbances shift slightly, but noticeably, and increase in intensity by roughly an order of magnitude. The Q-like absorbances, on the other hand, shift to the blue (to between 615–680 nm) and increase in intensity.

8.1 Synthesis and Spectroscopic Properties of Pyrrole-, Furan-, and Thiophene-Containing Sapphyrins

The original sapphyrin syntheses of Woodward [66, 152, 153], Johnson [26, 27, 154], along with later reports [3, 21] involved a MacDonald-type "3 + 2" condensation between a functionalized bipyrrole or analogue, such as **191–196**, and a dicarboxyl-substituted tripyrrane, such as **197–203**, as shown in Scheme 36. Although yields were good for the size of the macrocycle being formed, the syntheses of the requisite precursors were long and tedious. Recently, however, a simple, three-step, high yielding synthesis of the dicarboxyl-substituted tripyrrane **197** was reported [12] that involves as a critical step, the condensation of 3,4-diethylpyrrole and benzyl 5-(acetoxymethyl)-4-ethyl-3-methylpyrrole-2-carboxylate. This improvement in synthesis reduces by half the length of the previously published sequence [66]. It, combined with improvements in bipyrrole synthesis [155], has made this hitherto obscure class of molecules readily available on a laboratory scale.

Related systems which contain either furan (**207** or **212**) and thiophene **208** in place of one or more pyrroles, were also synthesized in accordance with the sequence shown in Scheme 36 [3, 24, 26, 27, 154]. Another synthesis of a dioxasapphyrin similar to **207** was achieved when bis(formylfuryl) sulfide (**217**) was condensed with the tripyrrane **200** (Scheme 37). The product resulting from this condensation is the dioxosapphyrin **212**. The mechanism proceeds in analogy to that invoked to rationalize the formation of corrole from *meso*-thiophlorin [25].

A particularly interesting feature of the above-described heteroatom "substitutions" is that such variations, which involve changing the number and type of central core heteroatoms, may provide a convenient means of modifying the

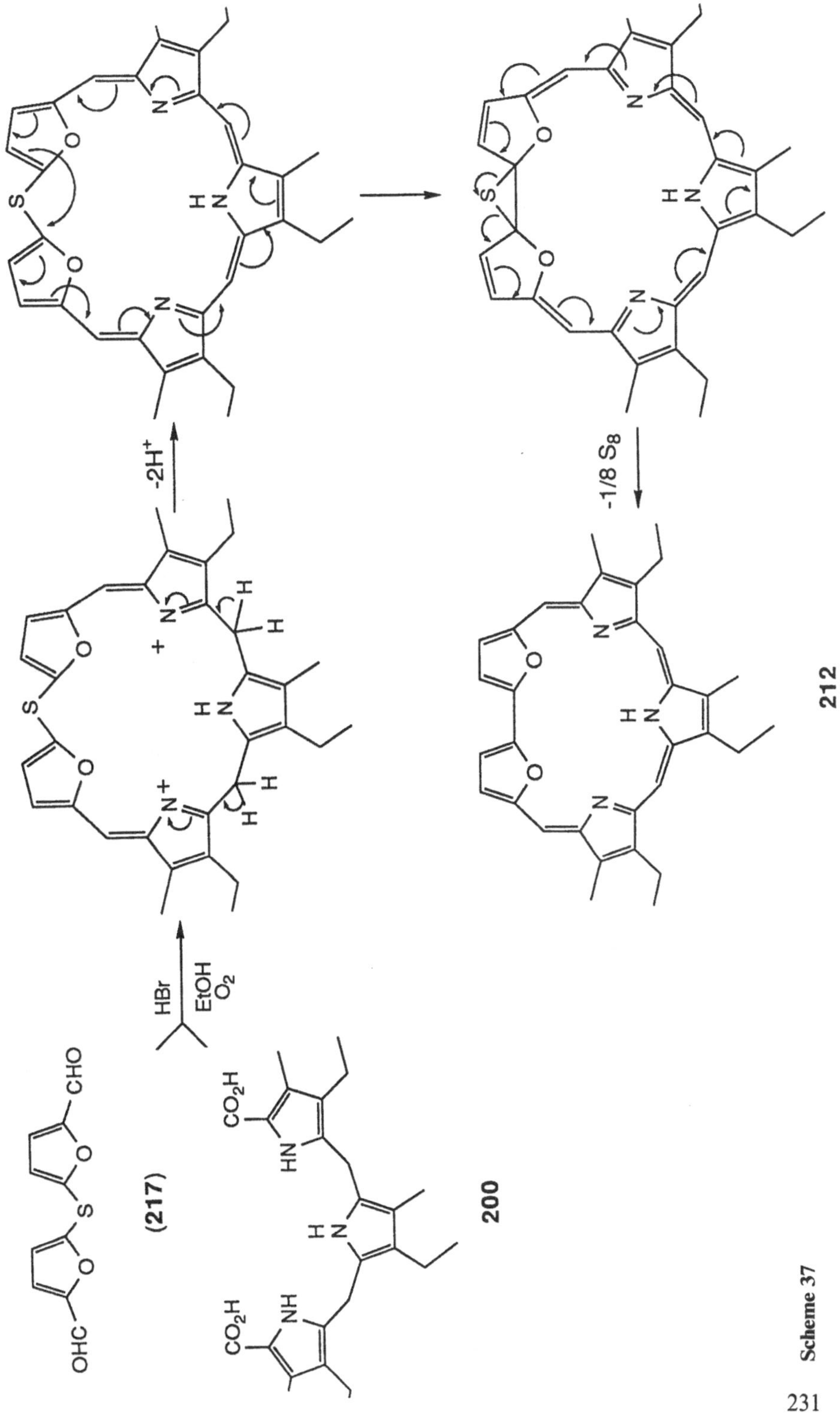

Scheme 37

231

electronic properties of the sapphyrin core. This could prove useful as the potential chemistry of these expanded porphyrins becomes better understood.

The spectral properties of sapphyrin and its heteroatom analogues, mentioned in the introductory paragraph given above, are consistent with their formulation as aromatic systems. The UV-visible spectra of free-base sapphyrins, such as **204**, display a dominant and very intense Soret-like absorption at approximately 456 nm. In addition four Q-like absorptions in the 620 to 720 nm spectral region are also observed. Also the protonated dicationic species **216** gives a simplified spectrum with only two Q-type bands, and a slightly shifted but greatly intensified Soret-like absorption 454 nm. The ^{1}H NMR spectrum of the dicationic sapphyrin **216** displays well resolved signals for the methine protons at δ 11.66 and 11.70 ppm. The internal pyrrole NH signals for the species at δ −4.31, −4.64, and −4.97 ppm, are upfield of TMS as would be expected for an aromatic system and in the 2:1:2 ratio expected for sapphyrins.

The hetero-substituted sapphyrins display spectral features similar to those of the parent systems. However, the dioxosapphyrin absorbs at much shorter wavelengths than the all-nitrogen analogue, with the Soret-like band being observed at ca. 435 nm. The ^{1}H NMR spectra for the dioxosapphyrin **207** is also consistent with its formulation as an aromatic system. The various *meso* protons resonate at δ 10.48, 10.45, 10.38 and 10.34 ppm, and the internal pyrrole NH protons appear as a single unresolved peak at δ −6.55 ppm [66]. The thiosapphyrin **208** on the other hand absorbs at longer wavelengths than the all-nitrogen macrocycle **204**, or the various dioxosapphyrin derivatives (e.g. **207** and **212**). In fact the Soret-like absorption of this macrocycle occurs at 461 nm. In the ^{1}H NMR, the methine signals of **208** appear at δ 10.12 ppm and the pyrrole NH resonances are observed together as a singlet at δ −2.2 ppm [66]. Collectively, these optical and magnetic resonance results indicate a low energy HOMO-LUMO gap and the presence of a large induced diamagnetic ring current. As such, they are completely consistent with the postulated aromatic formulations.

8.2 Metal Coordination Properties of Sapphyrins

To date very little is known about the coordination chemistry of the sapphyrin macrocycle, and no complexes have been reported for the dioxosapphyrin or the thiosapphyrin. In the free-base form, sapphyrin is a potential trianionic ligand and thus, on paper at least, seems perfectly suited for complexing the normally trivalent cations of the lanthanide series. Presumably, lanthanide(III) complexes of sapphyrin, which would be neutral (and potentially useful for magnetic resonance imaging applications; see Sect. 12.2), would be expected to form easily under typical porphyrin metalation conditions. However, in spite of the apparent correspondence in the sapphyrin core size and the ionic radii of the lanthanides, to date no lanthanide cation has been inserted into the core of the sapphyrin macrocycle using a variety of standard metal insertion techniques [156]. Nor, have any other pentaligated complexes of any other metal cations been reported to date. Nonetheless, some interesting results have emerged from various attempted metal insertion studies.

Bauer *et al.* [66] reported that metals such as Zn^{2+}, Co^{2+}, and Ni^{2+} formed tetraligated sapphyrin complexes, where four of the five pyrroles contribute to coordination. Such incomplete ligation by the sapphyrin is consistent with the more recent ^{1}H NMR data of Sessler *et al.* [157]. However, this latter work suggests that the static system initially proposed by Bauer *et al.* [66] may not necessarily be an accurate representation. For instance, Sessler *et al.* [157] reported that the diamagnetic zinc sapphyrin complex, Zn · HSap (Sap = 3,8,12,13,17,22-hexaethyl-2,7,18,23-tetramethylsapphyrinato trianion), showed a splitting of the *meso* proton signals (from two into four singlets) giving evidence for the presence of two isomeric tetraligated complexes as shown in Fig. 25.

Presumably, the low ligation numbers found for these first row transition metal complexes is a reflection of small cation size and inherent kinetic lability. To the extent that this is true, it was thought that switching the focus of the investigation to the second and third row transition metals would overcome the problems. Preliminary investigations had indicated that a rhodium dicarbonyl complex was formed from the reaction of sapphyrin with $[RhCl(CO)_2]_2$ [158]. However, no structural information was available to confirm the nature of the rhodium complex obtained. More recent work by Sessler and coworkers has provided this structural information and helped to define in part the coordination abilities of the sapphyrin macrocycle [159]. These workers found that the carbonyl complexes $[RhCl(CO)_2]_2$, and $IrCl(Py)(CO)_2$ react with the free-base sapphyrin macrocycle. However, no evidence for inplane coordination products were obtained (Scheme 38) [159]. Rather, it was found that addition of 0.5 eqv. of $[RhCl(CO)_2]_2$ or 1 eqv. of $IrCl(CO)_2(Py)$ to a solution of free-base sapphyrin **204** afforded $[Rh(CO)_2(H_2Sap)]$ **(217)** and $[Ir(CO)_2(H_2Sap)]$ **(218)**, respectively. The complexes **217** and **218** were found to undergo protonation to form the cationic complexes $[Rh(CO)_2(H_3Sap)]^+$ **(219)** and $[Ir(CO)_2(H_3Sap)]^+$ **(220)**. The structure of $[Ir(CO)_2(H_3Sap)Cl]$ **(220)** was confirmed by a single crystal X-ray diffraction analysis shown in Fig. 26 [157, 159]. Treatment of **217** with an additional molar equivalent of $[RhCl(CO)_2]_2$, of reaction of **204** with an excess of $[RhCl(CO)_2]_2$, gives $[[RhCl(CO)_2]_2(HSap)]$ **(221)**. Similar bimetallic complexes and hetero-bimetallic complexes, namely $[[Ir(CO)_2]_2(HSap)]$ **(222)**, and $[[Ir(CO)_2][Rh(CO)_2](HSap)]$ **(223)**, and the propyl-substituted system **224** can also be formed from analogous reactions. The structures of complexes

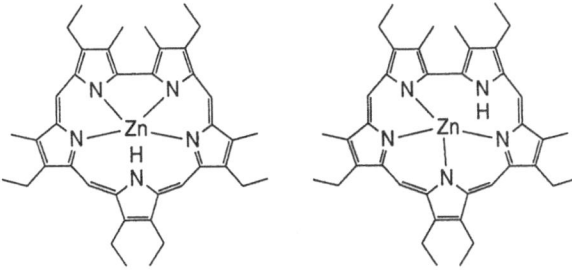

Fig. 25

221 and **224** (Scheme 38) were confirmed by single crystal X-ray structure determinations (Fig. 27) [157, 159]. The structures of both of these complexes was found to be similar to that of $[Rh(CO)_2]_2$(octaethylporphyrin) [160] and the earlier-reported N-methylcorrole-bis[dicarbonylrhodium(I)] complexes [161] in that each metal center is bridging between an imine and an amine nitrogen atom.

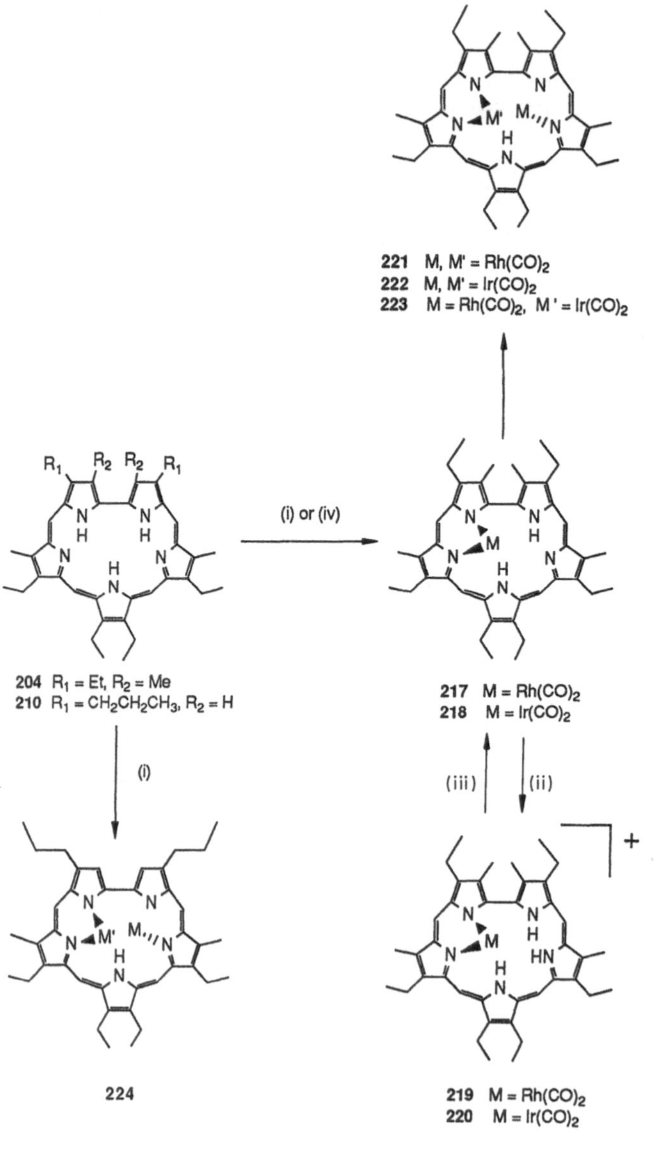

221 M, M' = Rh(CO)$_2$
222 M, M' = Ir(CO)$_2$
223 M = Rh(CO)$_2$, M' = Ir(CO)$_2$

204 R$_1$ = Et, R$_2$ = Me
210 R$_1$ = CH$_2$CH$_2$CH$_3$, R$_2$ = H

(i) or (iv)

217 M = Rh(CO)$_2$
218 M = Ir(CO)$_2$

224

219 M = Rh(CO)$_2$
220 M = Ir(CO)$_2$

(i) [RhCl(CO)$_2$]$_2$, NEt$_3$; (ii) silica gel ; (iii) NEt$_3$; (iv) IrCl(CO)$_2$(py), NEt$_3$

Scheme 38

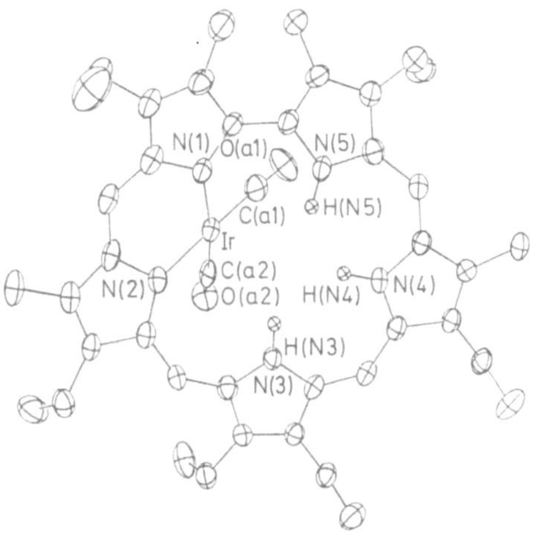

Fig. 26

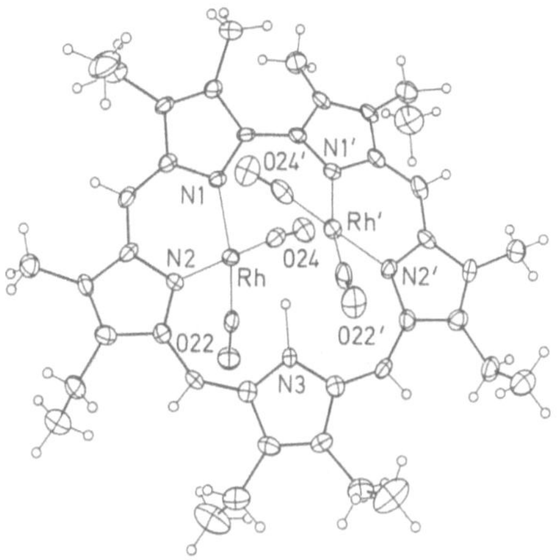

Fig. 27

8.3 Anion Binding Properties of Protonated Sapphyrins

As mentioned above, the sapphyrin core is basic with the two pyridine-like pyrroles being very easily protonated to give a full pentameric NH-containing core. This core is quite unique in terms of containing five potential hydrogen bonding donors within a relatively large and near circular planar array. As a consequence, diprotonated sapphyrins are endowed with unique and interesting anion binding capabilities.

The first diprotonated sapphyrin-derived anion complex, a fluoride-containing chelate [15], was obtained in quite an unusual fashion. During an attempt to obtain an X-ray diffraction quality crystal of the diprotonated sapphyrin 216,

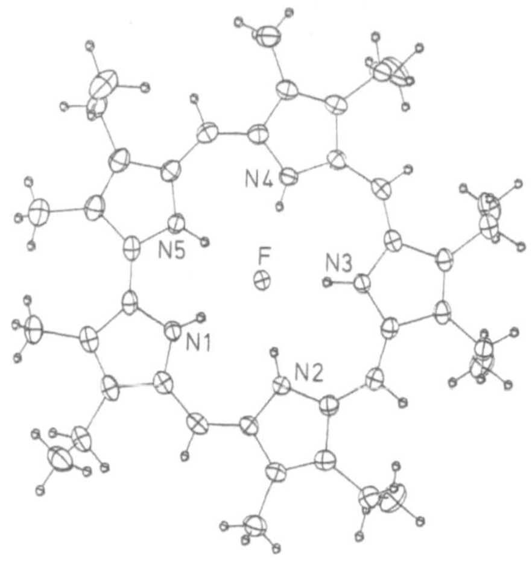

Fig. 28

a simple counter ion exchange reaction (PF_6^- for Cl^-), that was meant to give better crystallization properties, led serendipitously to the stabilization of a centrally bound fluoride anion complex (Fig. 28) within the ca. 5.5 Å diameter sapphyrin core. This stabilization in the solid state, which was most unexpected at the time, is apparently made possible by the five N–H–F hydrogen bonds radiating out from the proton-bearing pyrrolic nitrogens. This solid state structural result led to the consideration that sapphyrins may act as fluoride anion binding agents in solution. Verification of this hypothesis was obtained *via* various spectroscopic means, including UV-visible, ^{19}F NMR, and ^1H NMR [155, 157]. Here UV-visible spectroscopic studies proved particularly informative. Upon treatment with increasing aliquots of fluoride, the Soret-like absorption band of diprotonated sapphyrin in the optical spectrum shifts to the blue, with the maximum hypsochromic shift being ca. 10 nm relative to the starting dihydrochloride salt [157].

The above results served to demonstrate the validity of considering the protonated sapphyrins as both specific fluoride binding agents [162–166] and as novel members of an increasingly large class of anion receptors [167–172]. This is of considerable interest since protonated porphyrins, with a core diameter of approximately 4 Å, are too small to bind anions centrally, let alone accommodate the needed two "extra" protons within the macrocyclic core [173–176].

At present some preliminary evidence has been obtained which suggests that sapphyrins can bind anions other than fluoride. In fact, an X-ray diffraction study shows that N_3^- binds to the monoprotonated form of sapphyrin [157]. Specifically, as shown in Fig. 29, the monoprotonated sapphyrin, $H_4Sap \cdot N_3$, does not complex azide anion in an in-plane fashion but in an end-on manner, with the terminal azide nitrogen atom being 1.13 Å above the sapphyrin plane [157]. Nonetheless, this atom is still within typical hydrogen bonding distance (2.8 to 3.0 Å) of at least four of the five pyrrolic nitrogens [59, 173–176].

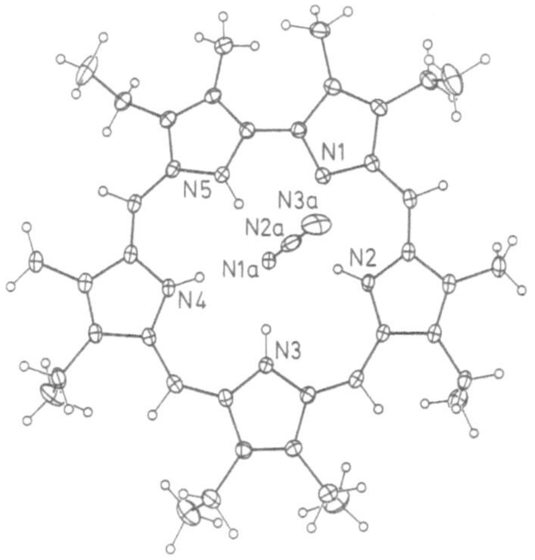

Fig. 29

8.4 Biological Applications

Sapphyrins would appear to be particularly attractive targets in terms of the potential photodynamic therapeutic (PDT) and photodynamic inactivation (PDI) applications discussed in Sect. 12.1. This is because the sapphyrins absorb strongly in the 680–710 nm spectral region. These transitions, which are red-shifted by approximately 50 nm relative to the corresponding transitions in porphyrins, fall within a physiological "window of transparency" [177]. This, along with recent findings indicating that the free-base form of sapphyrin acts as an effective producer of singlet oxygen [178], makes sapphyrin and its derivatives attractive candidates as potential PDT and/or PDI photosensitizers [76–80]. Although complicated by dimerization processes [178, 179], recent experiments have served to indicate that sapphyrin **204** acts as an effective photosensitizer for the photodynamic inactivation of HSV-1 [180]. In fact, this compound displays an efficacy on a *per macrocycle* basis comparable to that obtained with the better studied dihemotoporphyrin ether (DHE) mixture. However, when account is made of the different light absorbing capabilities of these two materials (λ_{max} = 690 nm vs. 630 nm), what these results mean is that sapphyrin **204** is considerably more effective than DHE on a *per photon* basis [180]. Subsequently, investigations into the inactivation of cell-free HIV-1 using sapphyrin **204** were undertaken, with a preliminary report on the startling effectiveness of this macrocycle for *anti*-HIV-1 PDI recently appearing in the literature [181].

Not surprisingly, with encouraging results such as the above now emerging, this aspect of expanded porphyrin-related research is receiving considerable attention. It is, therefore, discussed in greater length in Sect. 12.

9 Smaragdyrins (Nor-sapphyrins)

Just as the removal of one of the four carbon bridges from the 18 π-electron porphyrin system gives corrole, a new 18 π-electron system, removal of one the four methine bridges from the sapphyrin results in a new structure (Fig. 30). This new nor-sapphyrin structure, which was given the trivial name "smaragdyrin" from the Greek, smaragdos, meaning emerald, retains the conjugated 22 π-electron system of sapphyrin but contains two direct links between the pyrroles.

The synthesis of this expanded porphyrin was successfully achieved using [66] a "2 + 3" MacDonald coupling similar to the one used to obtain sapphyrin [26, 66] and pentaphyrin [182, 183]. In this case, a diformyl bipyrrole 194 and a pyrrolyldipyrromethane dicarboxylic acid 225 were condensed in the presence of HBr to give the new macrocycle 226 (Scheme 39) [67]. Here, the requisite pyrrolydipyrromethanes were generated by HBr catalyzed condensations of 2-formylpyrroles with 5,5'-disubstituted bipyrroles [184]. A related smaragdyrin-like system was produced by a similar procedure. In this case, a pyrrolydipyrro-methane, 227 or 228, was condensed with 5,5'-diformylbifuran 192, to give the dioxasmaragdyrins 229 or 230, respectively (Scheme 39) [26, 66].

The electronic absorption spectrum of hexamethylsmaragdyrin 226 displays a series of broad bands in the 700–725 nm spectral region. As with the sapphyrins there is a strong Soret-like absorption at approximately 450 nm. Interestingly, however, the dioxasmaragdyrin 230 displays two strong Soret-like absorptions, at 448 and 459 nm, when treated with acid [26]. Nonetheless, the NMR spectrum reported for dioxasmaragdyrin 230 [26] confirms the presence of a considerable diamagnetic ring current. The internal NH proton resonates at δ −4.85 ppm and the two types of *meso* protons appear as separate singlets at δ 10.52 and 10.06 ppm, respectively.

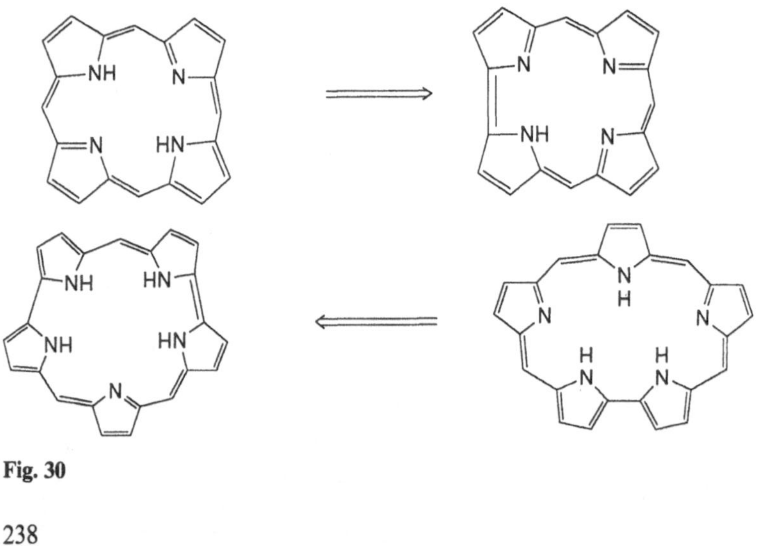

Fig. 30

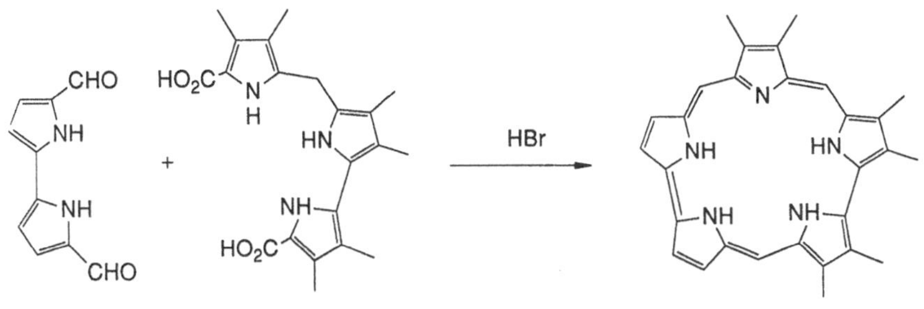

194   225   226

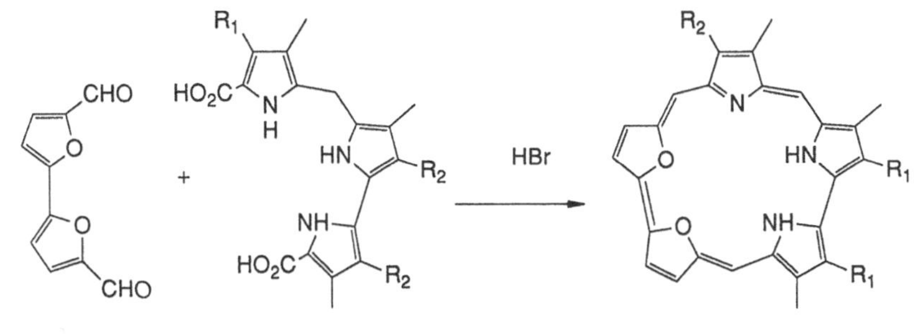

192   227 R$_1$ = R$_2$ = CH$_2$CH$_3$.   229 R$_1$ = R$_2$ = CH$_2$CH$_3$.
               228 R$_1$ = H, R$_2$ = CH$_3$.   230 R$_1$ = H, R$_2$ = CH$_3$.

Scheme 39

Although representing a considerable synthetic achievement, the smaragdyrins **226, 229,** and **230** proved to be extremely sensitive toward acids and light. In fact, all reported attempts to form metal complexes have so far resulted in decomposition of the macrocycle.

To date, the poor stability of the smaragdyrins **226, 229,** and **230** has unfortunately precluded a complete study of their chemistry. However, the unique nature of these unusual aromatic materials suggests that further efforts may be warranted. Just as a reinvestigation of the sapphyrin system has provided a wealth of new chemistry [157], a similar study of the synthesis and properties of smaragdyrin and related compounds could uncover further interesting observations including, perhaps, stabilized forms of this class of expanded porphyrins.

10 Pentaphyrins and Hexaphyrins

10.1 Pentaphyrins

With the original reports of the successful syntheses of the sapphyrins [26, 66, 152] and uranyl superphthalocyanine [112, 118, 119], interest in other expanded porphyrin systems, was kindled. The next logical step (after sapphyrin), in the expanding series of all-pyrrole systems, was the pentaphyrin macrocycle **231** which contains five pyrroles and five *meso*-like methine bridges. In 1983 Gossauer *et al.* reported the synthesis of the first prototypical member **231** of this macrocyclic family [158, 182, 183, 185–187]. This first synthesis was achieved by a "2 + 3" MacDonald-type condensation between an α-free dipyrromethane **233** and a tripyrrane dialdehyde **236**. More recently, the synthesis of pentaphyrin **231** has been achieved by using a dipyrromethane 5,5'-dicarboxylic acid **235** in place of an α-free dipyrromethane [21]. Here, as is the case in many of these kind of reactions [21, 26, 27, 66, 155], decarboxylation occurs under the reaction conditions to produce the corresponding α-free species **233** *in situ*. (Scheme 40) [21].

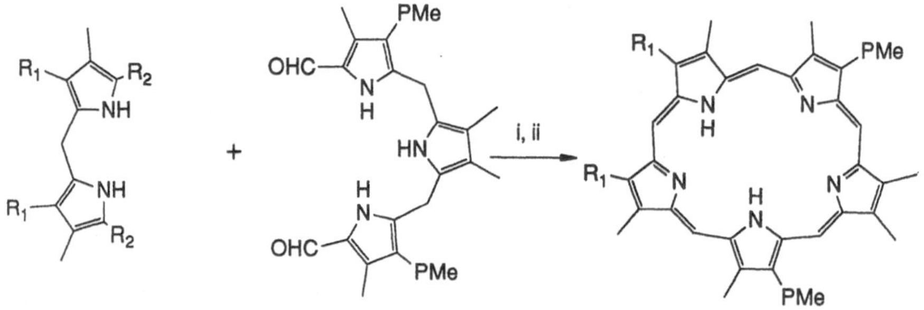

233 PMe = CH₂CH₂CO₂Me, R₂ = H   **236**
234 R₁ = Me, R₂ = H
235 R₁ = PMe; R₂ = CO₂H

231 PMe = CH₂CH₂CO₂Me, R₂ = H
232 R₁ = Me, R₂ = H

i, HBr-AcOH; ii, chloranil

Scheme 40

In contrast to the deep blue-green color of the sapphyrins and emerald green color of the smaragdyrins [26, 66], solid free-base pentaphyrins are orange [66]. Methanolic solutions of pentaphyrins, however, are green, whereas those in dichloromethane are yellowish-green. Addition of acid protonates the pyridine-like pyrrolic nitrogens to form the fully protonated tricationic compound, which is deep green in solution.

Both the free-base pentaphyrin and the triprotonated form display exceptionally intense UV-visible absorption bands. The free-base pentaphyrin, **231** for instance, exhibits three bands in the 330 to 750 nm spectral region. The triply protonated trications, on the other hand, display five bands within the same spectral region.

In addition, as is true for the other reported polypyrrolic aromatic macrocycles, the pentaphyrins display a particularly intense Soret-like absorption in the 450–500 nm spectral region.

The ^{1}H NMR spectra of these expanded porphyrins are consistent with the given aromatic assignment. For instance, in the free-base system **231**, the chemical shifts of the internal NH and methine bridges are roughly -5 ppm and 12.5 ppm, respectively. These shifts agree with the postulation of a strong diamagnetic ring current within a delocalized 22 π-electron perimeter system.

Initial reports served to indicate that the pentaphyrin macrocycle is capable of complexing Zn^{2+}, Co^{3+}, and Hg^{2+} [158]. The exact nature of this coordination, however, was not determined. Nonetheless, it was speculated to be *via* ligation to only two of the five pyrrolic nitrogen centers (c.f. **238–240** in Scheme 41). Also, a complex **237** was found to form between what is formally the doubly deprotonated pentaphyrinato dianion of **232** and uranyl cation (Scheme 41). The uranyl center may be readily displaced by treatment with acid [158, 187].

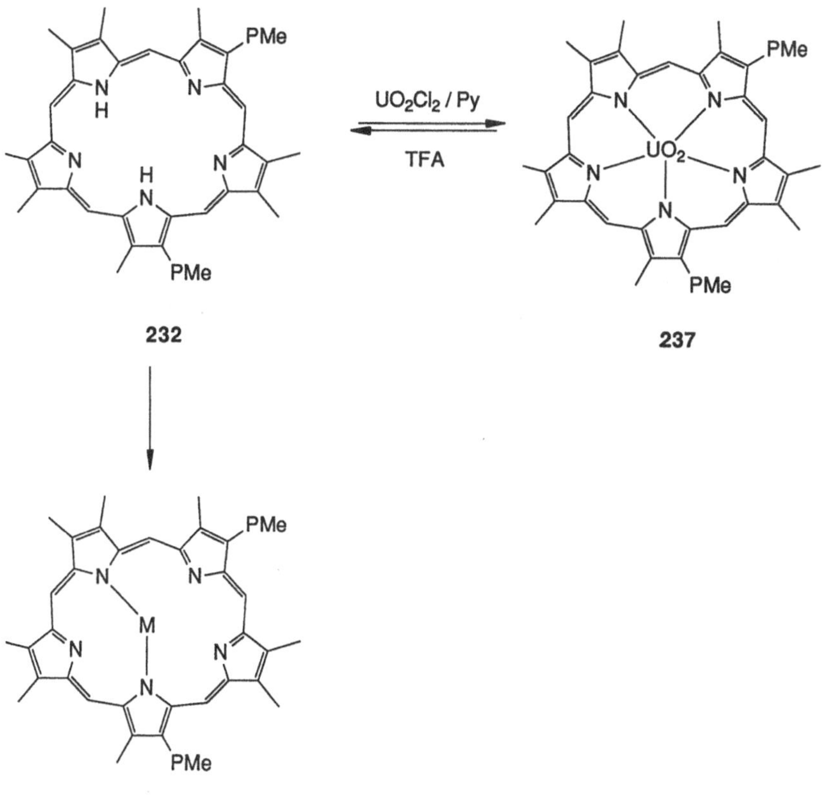

	UO_2Cl_2 / Py	
	TFA	

232   **237**

238 M = Zn
239 M = Co-CN
240 M = Hg

Scheme 41

241

A synthetic approach to a structural isomer of a pentaphyrin-type macrocycle was recently reported by Franck and coworkers [188]. They described the synthesis of an inverted porphyrinoid **241** in which the five nitrogen atoms are on the periphery of the macrocycle instead on in the interior (Scheme 42). Apparently, this is the first report of such an inverted porphyrinoid. It is of special interest that in the biomimetic condensation used to produce **241**, the pyrrole rings undergo inversion and produce the cyclic pentapyrrolic product instead of the corresponding tetrapyrrolic analogue. Such a preference has never been observed in similar condensations involving the related porphobilinogen [189]. Unfortunately, attempts to oxidize this inverted porphyrinoid gave only decomposition products and did not yield any quantities of the aromatic pentaphyrin isomer **242** [188]. The reason given by the authors for such sensitivity involves the destabilizing accumulation of charge that would result upon oxidation of **241**.

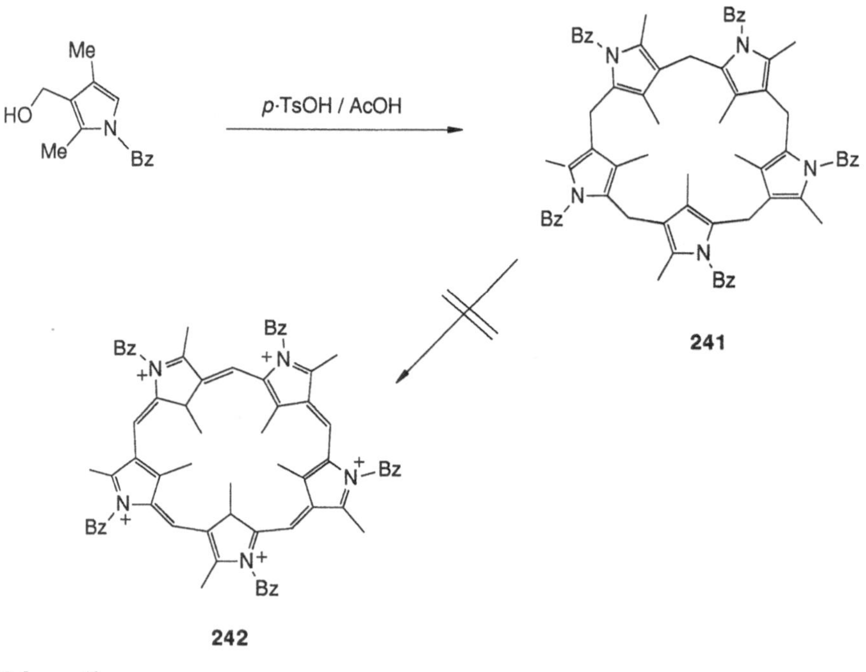

Scheme 42

10.2 Hexaphyrins

Stimulated by the successful synthesis of pentaphyrin, Gossauer, sought to extend the range of expanded porphyrins by condensing the bis- α-free tripyrranes **243–246** with the tripyrrane dialdehydes **236, 247–249** [183]. After oxidation with p-benzo-quinone, several violet products, namely **250–253**, were isolated (Scheme 43). The analytical data obtained are consistent with these being hexapyrrolic macrocycles containing six *meso*-like methine bridges. Molecular models show that the "hexaphyrin" **251** can only be planar when two opposite exocyclic double bonds have an *E* configuration. Due to the substitution pattern on the tripyrrane precursors, **243, 245, 246, 236, 248** and **249**, two isomeric structures **A** and **B** are possible for the products **250, 252** and **253**. NMR evidence (*vide infra*) indicated that both isomers were present in relatively equal amounts.

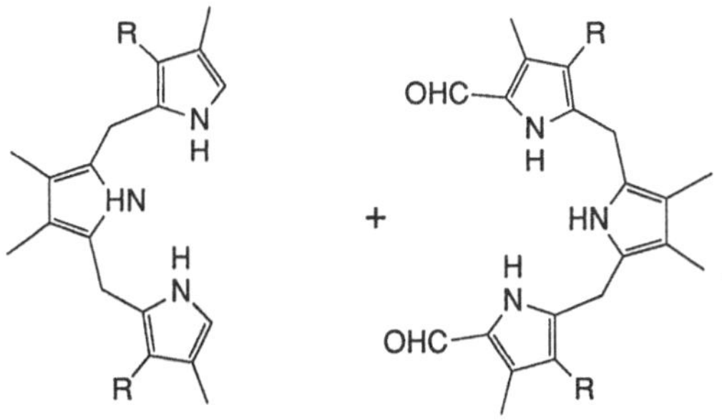

243 R = PMe = CH$_2$CH$_2$CO$_2$Me	**236** R = PMe = CH$_2$CH$_2$CO$_2$Me
244 R = Me	**247** R = Me
245 R = C$_8$H$_{17}$	**248** R = C$_8$H$_{17}$
246 R = CH$_2$CO$_2$Me	**249** R = CH$_2$CO$_2$Me

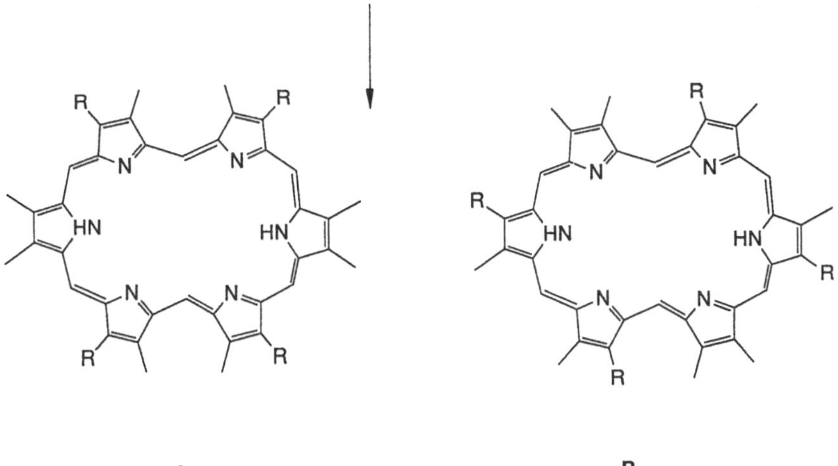

A   B

250 R = PMe = CH$_2$CH$_2$CO$_2$Me
251 R = Me
252 R = C$_8$H$_{17}$
253 R = CH$_2$CO$_2$Me

Scheme 43

The ^{1}H NMR spectrum of the hexapyrrolic macrocycle **250A** and **250B** reveals three signals associated with the protons of the Z-configuration methine bridges. These appear as singlets at approximately δ 12 ppm with relative integrated intensities of $2:1:1$. The singlet at δ 12.42 ppm was assigned as deriving from the four homotopic protons of isomer **250A** which has D_{2h} symmetry. Likewise, the signals at δ 12.33 and 12.19 ppm were assigned to the two pairs of homotopic methine protons thought to belong to isomer **250B** of C_{2h} symmetry. The protons of the E configuration methines appear as two signals at δ −7.40 and −7.54 ppm. These correspond to isomers **250A** and **250B**, respectively. These assignments are confirmed by the ^{1}H NMR of the dodecamethylhexaphyrin **251**, which can exist as only one isomer. Here, only two singlets, at δ 12.5 and −7.3 ppm, are seen for the methine protons. In this case, the peripheral methyl groups give rise to two signals at δ 4.55 and 4.60 ppm, respectively.

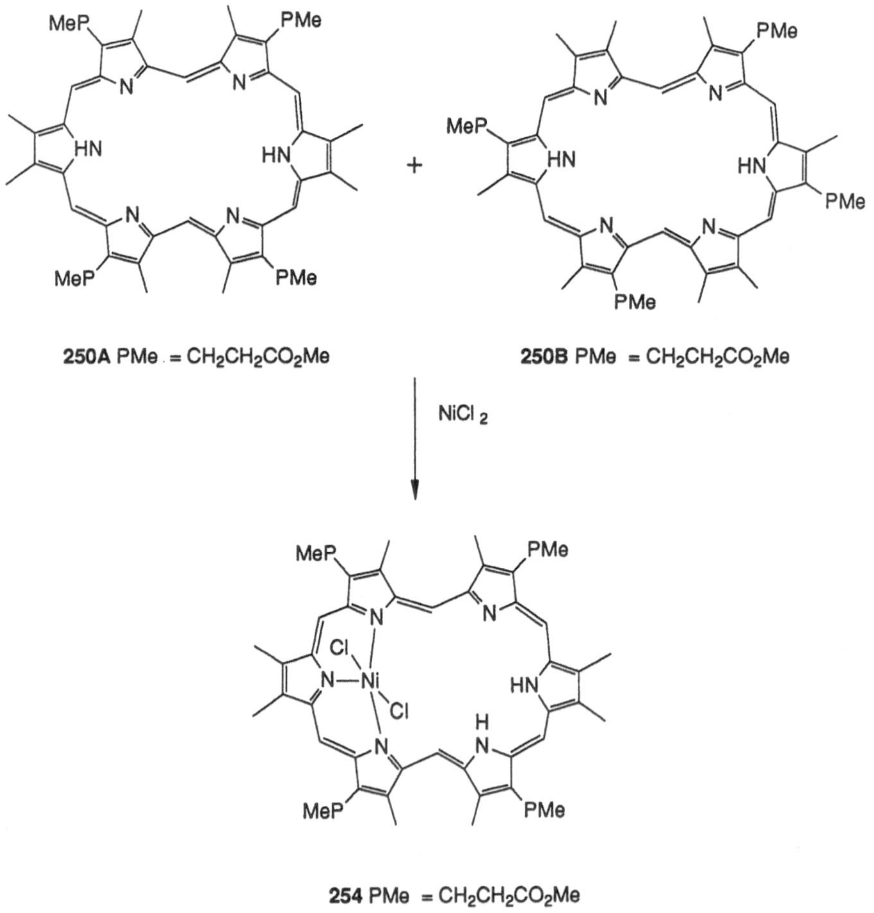

250A PMe = CH$_2$CH$_2$CO$_2$Me   250B PMe = CH$_2$CH$_2$CO$_2$Me

254 PMe = CH$_2$CH$_2$CO$_2$Me

Scheme 44

The coordination chemistry of the hexaphyrin system has provided a number of surprises [190]. The addition of nickel chloride to the isomeric mixture of hexaphyrins **250** leads to a single product. The proposed structure of the resulting nickel complex is shown as compound **254** in Scheme 44. Other metals were also found to react with this and other isomeric mixtures of hexaphyrins to give rise to only one isomeric product. For instance, treatment of either **250** or **252** with zinc chloride gives the bimetallic complexes **255** or **256** as one isomer, respectively (Scheme 45). In contrast, the reaction of **253** under identical conditions gives rise to a similar complex **257**, wherein, the positions of the ester substituents are different from those seen in the previous examples (Scheme 46). The reactions of hexaphyrins with zinc would appear to indicate that the two isomers normally

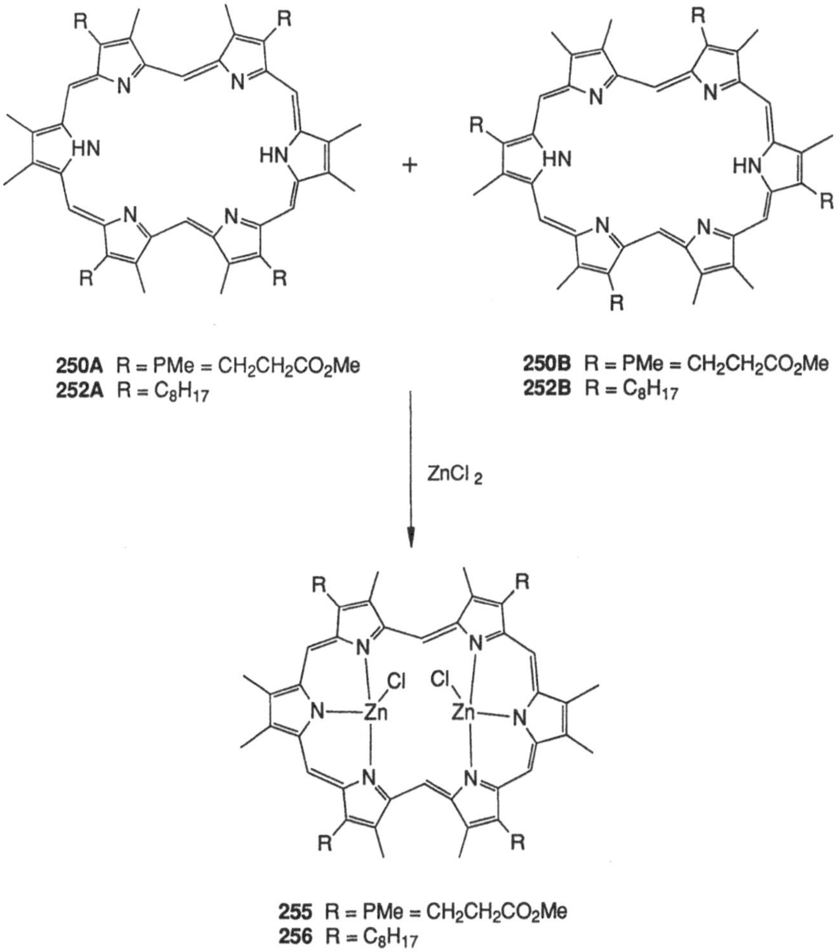

250A R = PMe = CH$_2$CH$_2$CO$_2$Me
252A R = C$_8$H$_{17}$

250B R = PMe = CH$_2$CH$_2$CO$_2$Me
252B R = C$_8$H$_{17}$

ZnCl$_2$

255 R = PMe = CH$_2$CH$_2$CO$_2$Me
256 R = C$_8$H$_{17}$

Scheme 45

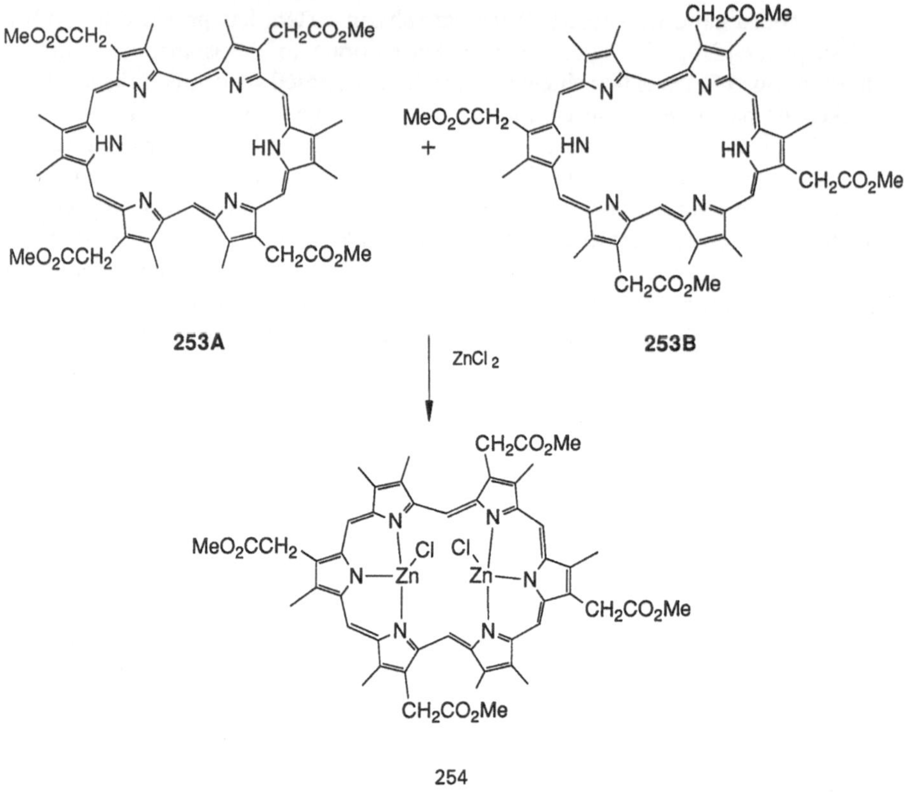

Scheme 46

observed in solution can be interconverted in the presence of certain Lewis acidic metal centers. In addition, the flexibility of this macrocycle is so great that metal complexes may be prepared wherein the metal center is located on the periphery of the macrocycle. For instance, reaction of palladium amine dichloride with the usual hexaphyrin mixture gives rise to an unusual complex, in which two of the pyrrole units are rotated to the outside of the macrocycle and coordinated to the palladium as in **258** or **259** (Scheme 47). Once formed, the ammonia groups on the palladium, in **258**, may be displaced by better ligands such as pyridine to give **260**, as shown in Scheme 48. The geometries of all of the metal complexes described above were derived from NOE difference spectral studies [190]. In no case have any of the postulated structures been confirmed by X-ray diffraction analyses. Given the range and diversity of these structures (and the apparent steric requirements of the ligands) such confirmatory studies might perhaps prove most informative. They could also set the stage for a more complete investigation of the coordination properties of this new class of ligands.

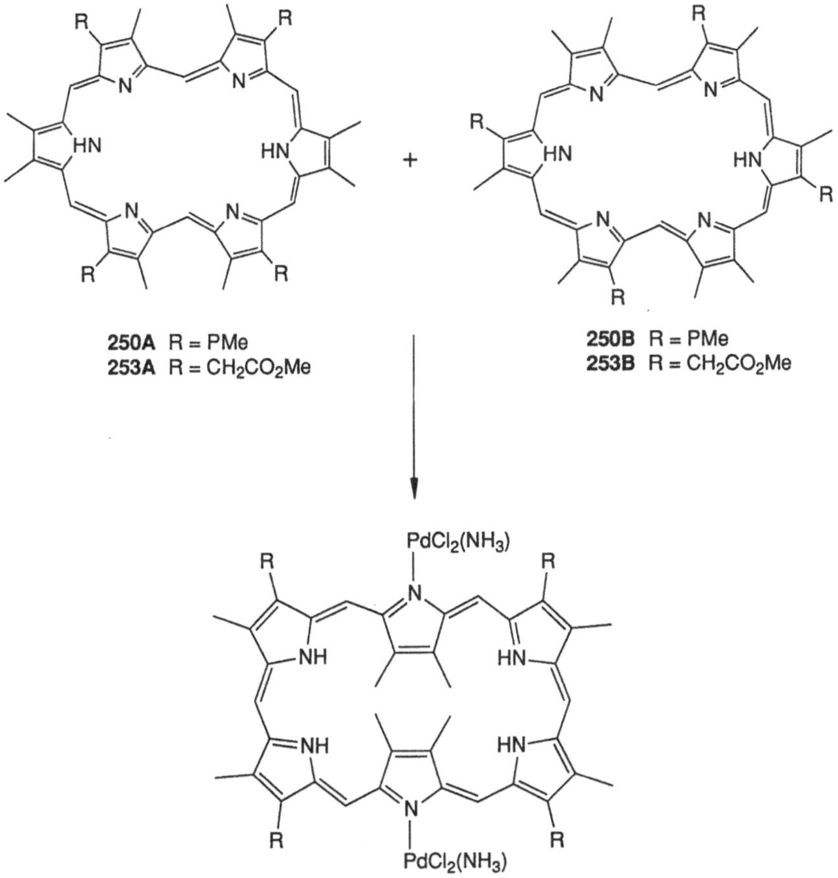

250A R = PMe
253A R = CH₂CO₂Me

+

250B R = PMe
253B R = CH₂CO₂Me

258 R = PMe
259 R = CH₂CO₂Me

Scheme 47

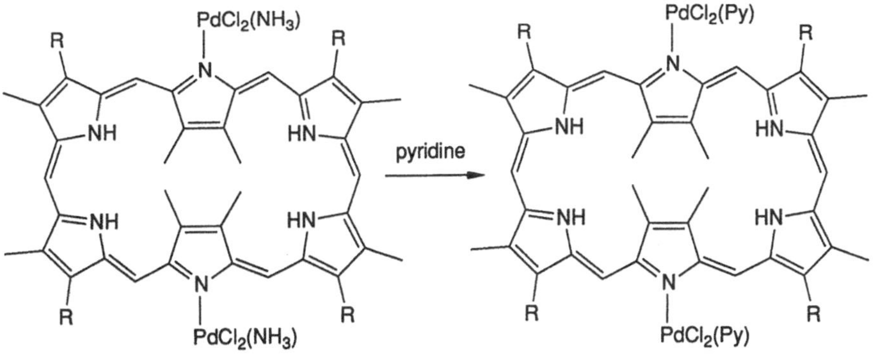

pyridine

258 R = PMe

260 R = PMe

Scheme 48

10.3 Rubyrins

One other hexapyrrolic macrocycle, which bears a close resemblance to the hexaphyrins, has been recently reported. This macrocycle, compound **262**, shown in Scheme 49, contains two bipyrrole fragments linked to two pyrroles *via* methine bridges [61]. This system, which can be considered as being a bisnor-hexaphyrin, and which was assigned the trivial name rubyrin (from the Latin rubews) in light of the bright red color of its diprotonated salt, has much less structural

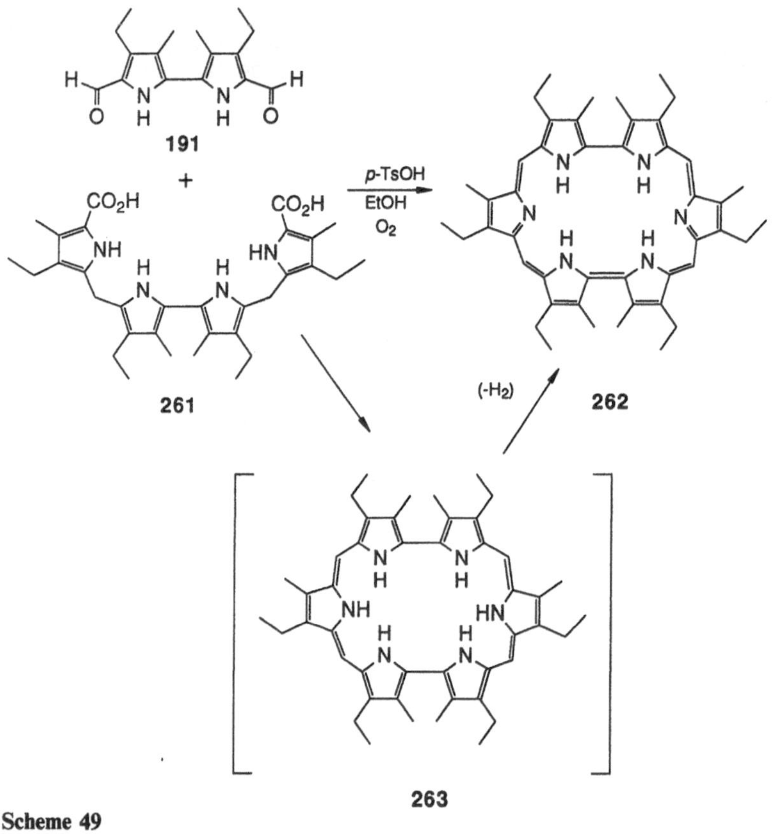

Scheme 49

flexibility than the corresponding parent hexaphyrin. However, rubyrin due to its lower symmetry, could possibly exist in stable form at several different oxidation levels. For instance, stable formulations as either 28 π- or a 26 π-electron annulenes are conceivable (c.f. structures **262** and **263** in Scheme 49). However, results to date suggest it is the diprotonated, rubyrin, form of the 26 π-electron macrocycle **262** which is formed in greatest yield under the synthetic "4 + 2" condensation conditions and which is the most stable form in solutions prepared from proton-containing solvents. The presence of a strong diamagnetic ring current in the macrocycle (as indicated by ^{1}H NMR spectroscopy), strong absorption bands

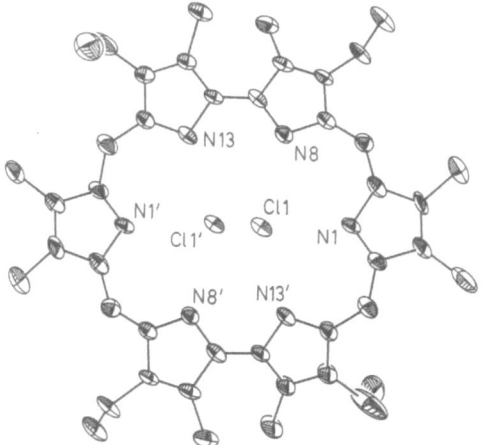

Fig. 31

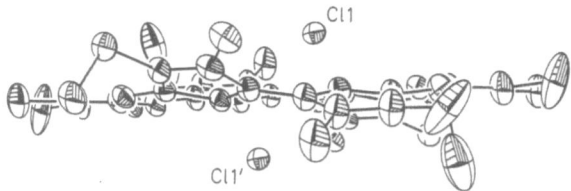

Fig. 32

at 500 and 872 nm in the optical spectrum, and a single crystal X-ray examination of the dihydrochloride salt of **262** (Fig. 31 and 32) are consistent with this assessment. The dihydrochloride salt of **262** has an essentially planar structure consistent with a delocalized aromatic system. At present, this macrocycle is the subject of further ongoing investigations [61].

10.4 Other Related Systems

An attempt to prepare yet an another macrocycle containing six five-membered nitrogen heterocycles, was made by Lind and LeGoff (Scheme 50) [191]. This macrocycle, in analogy to hexaphyrin, was to have been formed by the condensation of two tripyrrane-like segments **264** and **265**. In this work, however, the central pyrrole was have to been replaced by a pyrazole. The condensation between the appropriate pyrazole-containing fragments, in fact, proceeds smoothly but gives an inseparable mixture consisting of two out of the four possible isomers **266–269** (Scheme 50). Unfortunately, the exact nature of these isomers has not been determined. In addition, efforts to remove the benzyl protecting group and effect oxidation to the corresponding aromatic expanded porphyrin species did not prove successful.

249

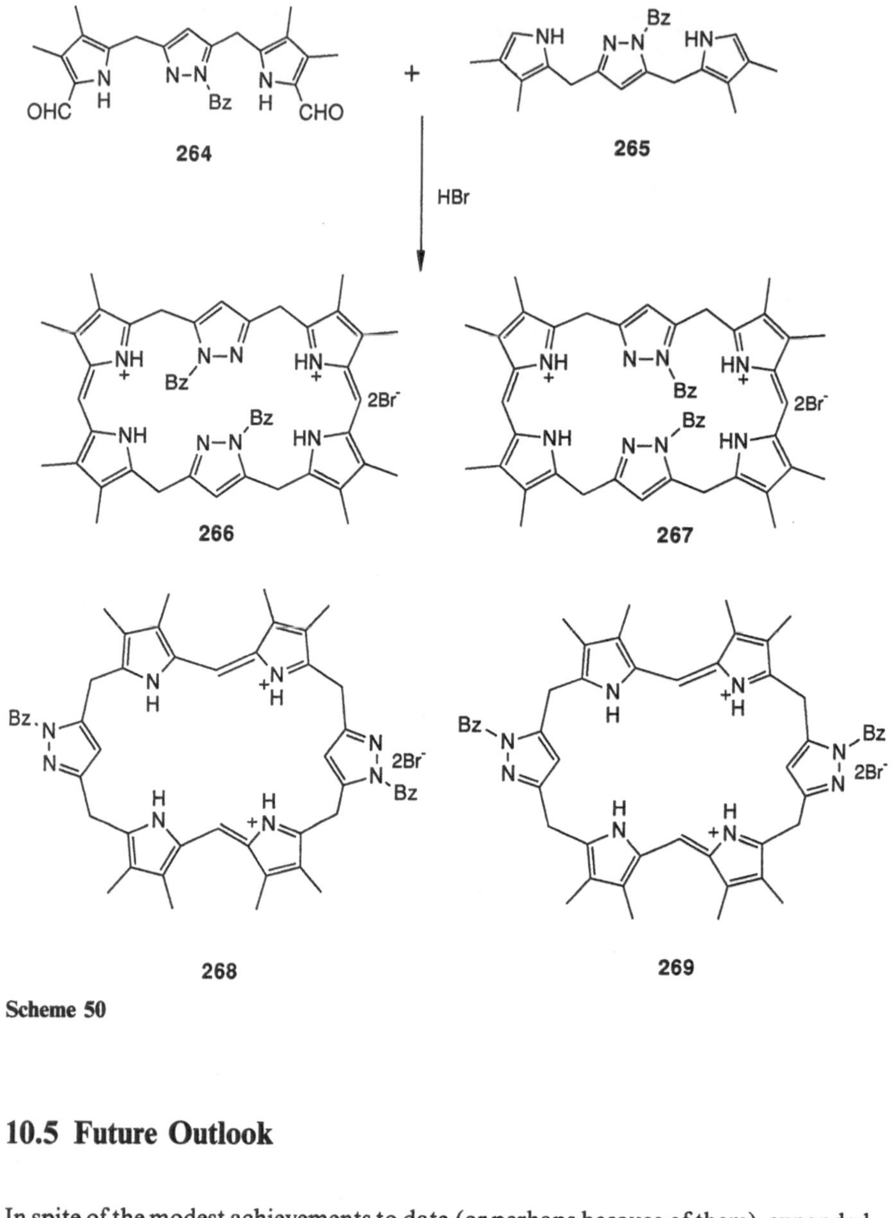

Scheme 50

10.5 Future Outlook

In spite of the modest achievements to date (or perhaps because of them), expanded porphyrins containing five or six (or even more) pyrrole-like heterocyclic subunits appear to constitute a very interesting but relatively unexplored set of systems. Many of these compounds, including the pentaphyrins and hexaphyrins, can be made from readily obtainable starting materials and have already demonstrated preliminary hints of what promises to be an unusual coordination chemistry. Further investigation of these macrocycles and their metal chelating properties, therefore, appears strongly warranted.

11 Vinylogous Porphyrins

11.1 Bisvinylogous Porphyrins

Extending the conjugated bridges between the pyrrole groups in the porphyrin structure is perhaps the most obvious way of forming expanded porphyrins. However, such a strategy was not successfully implemented prior to 1978 [192].

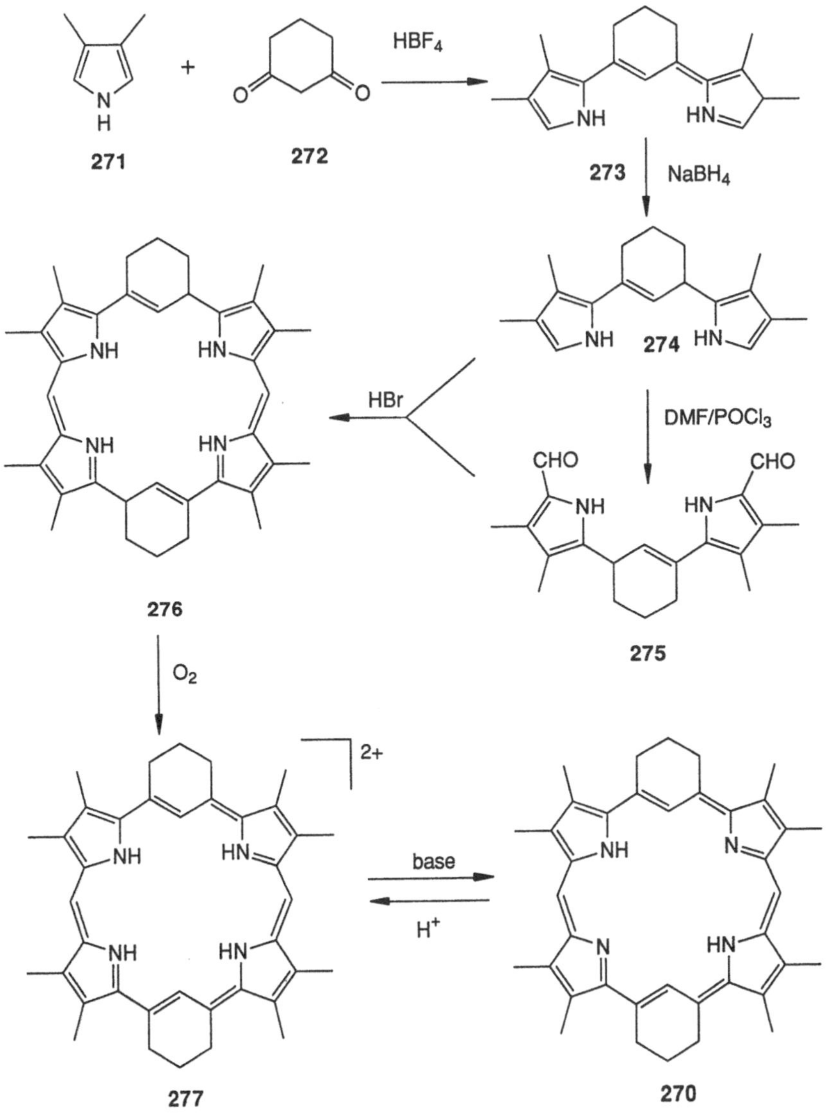

Scheme 51

251

At that time, the first report of a "vinylogous porphyrin" **270** appeared from the group of LeGoff. It was synthesized by the condensation of **274** and **275**, to give the macrocycle **276** in which the dipyrromethene subunits are "stretched apart" or "extended" by a single π-bond (Scheme 51) [192]. This new class of compounds was given the generic name "platyrins" (from the Greek word "platus" meaning wide). The macrocycle **276** was not isolated but was thought to be an intermediate in the formation of the fully oxidized diprotonated platyrin **277**. Deprotonation of the dication **277** then gave the free-base bisvinylogous porphyrin **270**. Both **277** and **270** display an intense Soret-like absorption at 477 nm (ε = 398,000 cm · mol^{-1}). The ^{1}H NMR spectrum of **270** has not been reported. The dicationic platyrin **277** displays a single *meso*-like methine resonance at δ 11.64 ppm. In addition, for this salt, the internal vinyl signals occur at δ −8.97 ppm in the ^{1}H NMR spectrum. These spectral and magnetic resonance features are consistent with a fully conjugated macrocycle. The platyrin **270** was reported to form highly insoluble complexes with metals [192]. However, the nature and composition of these were apparently never determined.

The next macrocycle in the series of bisvinylogous porphyrins **278** to be prepared by the LeGoff group effectively "added" yet another π-bond between the pyrroles [193]. Unfortunately, both **278** and its diprotonated salt, proved to be unstable decomposing within hours even in the solid state. This high reactivity is consistent with the supposition that there is little or no resonance stabilization in this platyrin **278**. However, there is a substantial diamagnetic ring current present

278

in this expanded porphyrin. The internal CH protons, as they appear in the ^{1}H NMR, are shifted to extremely high field (at δ −14.26 ppm), with the proton of the single *meso*-like methine having a low field resonance occurring at δ 11.75 ppm. The visible spectrum of **278** has a strong absorptions at 495 nm (ε = 123,000 cm · mol^{-1}) and 536 nm (ε = 144,000 cm · mol^{-1}) and other absorptions at 705 nm (ε = 12,300 cm · mol^{-1}), 718 nm (ε = 13,500 cm · mol^{-1}),

780 nm (ε = 9,400 cm · mol^{-1}). The dicationic salt of **278** displays a similar spectrum [193].

Subsequent to the initial work of LeGoff, a bisvinylogous porphyrin, which is formally the parent form of the heterocycle **270**, was reported by Franck and coworkers. This newer system was obtained from the reaction of a dipyrromethane **279** with an appropriate vinylaldehyde substituted dipyrromethane **280** (Scheme 52) [194]. The resulting macrocycle **281** was oxidized with bromine to give

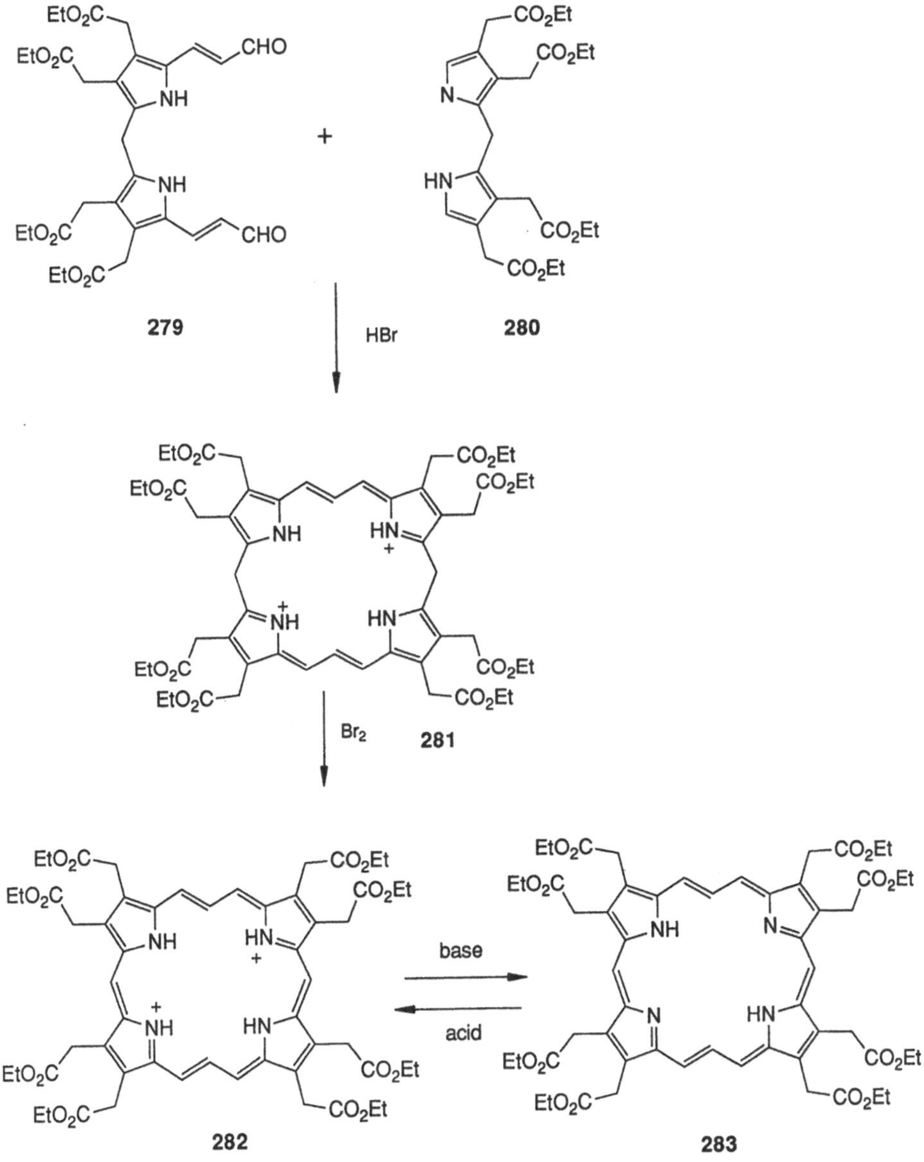

Scheme 52

253

the aromatic expanded porphyrin **282**. Neutralization then enabled the isolation of the free-base form **283** as a green solid. The macrocycle **283** is fairly stable and has a UV/visible spectrum which is dominated by an intense Soret-like band at 469 nm. Protonation of **283**, to give **282**, increases the intensity of the Soret-like band by an order of magnitude, but does not significantly change the position of the absorptions [194]. The ^{1}H NMR spectrum of **283** reveals a strong diamagnetic ring current effect. Thus, the doublet derived from the outer protons in the trimethylene bridge appears at $\delta = 11.91$ ppm, and the triplet derived from the inner trimethylene protons is observed at $\delta = -8.19$ ppm. The signal for the proton of the *meso*-like monomethine bridge is strongly shifted towards lower field and, in fact, is observed at $\delta = 10.59$ ppm.

The octaethyl analogue of **283** was recently synthesized by Franck and co-workers [195]. Condensation of **284** with the α-free dipyrromethane **285** gave the biladiene **286** which, when reacted with formaldehyde and subsequently oxidized with DDQ, gave the vinylogous porphyrin **287** (Scheme 53) [195]. This

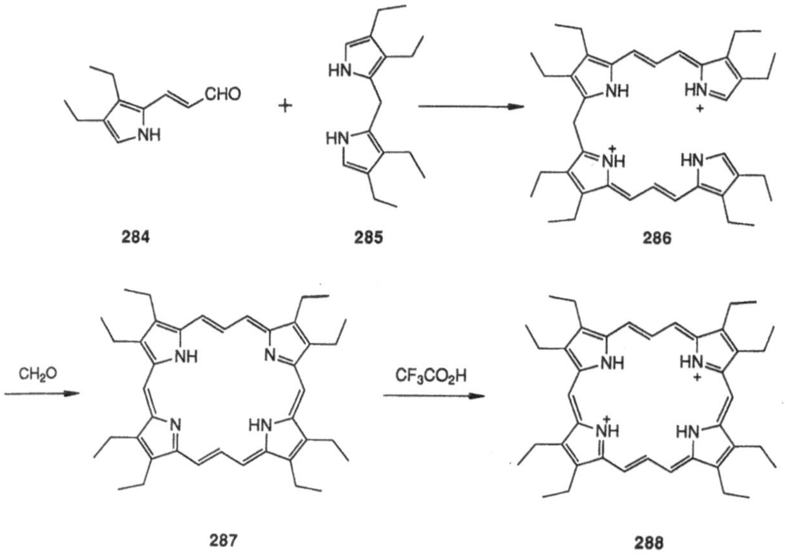

Scheme 53

new bisvinylogoues porphyrin **287** is extremely stable and displays a marked aromaticity. A single crystal X-ray diffraction analysis of the bistrifluoroacetate salt of the dication **288** (Figures 33 and 34) confirmed the planar nature of this macrocycle [195]. The methine protons of **287** display different reactivities depending upon their relative orientation. Thus, the external methine protons undergo exchange with deuterium when **287** is treated with D_2SO_4/D_2O, while the internal methine protons remained unaffected [195].

Currently the singlet oxygen producing capability and general photosensitizing properties, of the macrocycle **287** along with its efficacity *vis a vis* the photodynamic

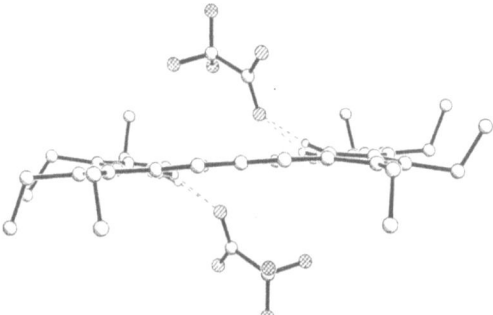

Fig. 33

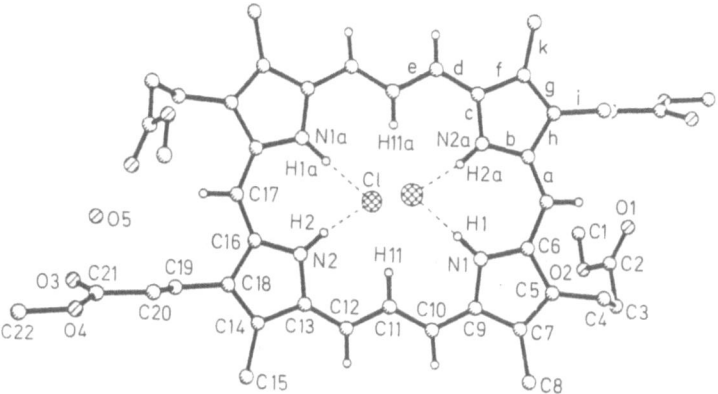

Fig. 34

Fig. 35

inactivation of human pathogenic viruses are being examined extensively [194]. In accordance with this general objective, Franck and co-workers have also produced a vinylogous porphyrin which has functionality similar to that of hematoporphyrin (Scheme 54) [196]. Here a single crystal diffraction X-ray study

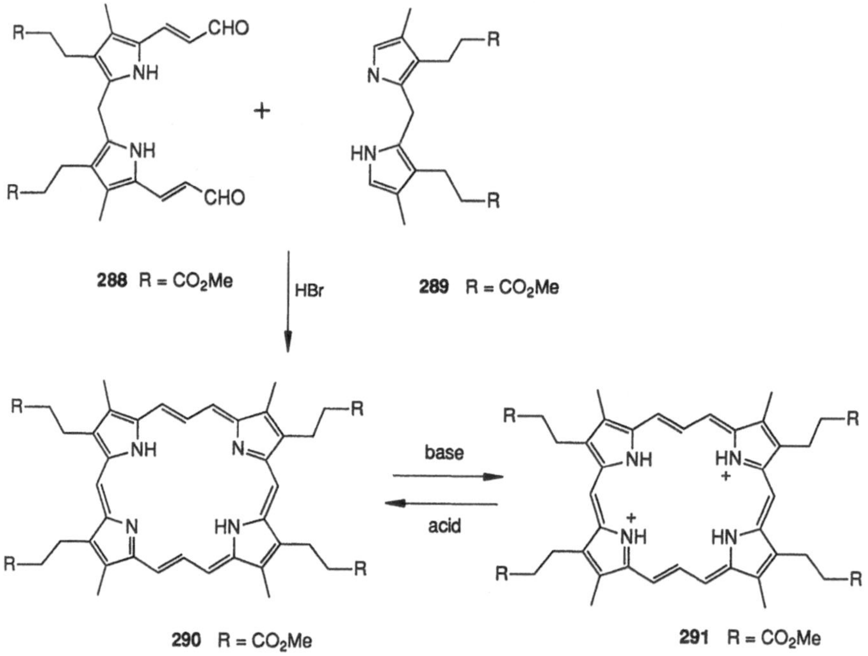

288 R = CO₂Me   HBr   **289** R = CO₂Me

290 R = CO₂Me   **291** R = CO₂Me

Scheme 54

(of **291** as the dichloride salt) again served to confirm the planar nature of this set of vinylogous porphyrins (Figs. 35 and 36) [196]. The photosensitizing ability of **290** was determined chemically. The $^1O_2$ formation of **290** exceeds that of isohematoporphyrin by a factor of 1.7 [196].

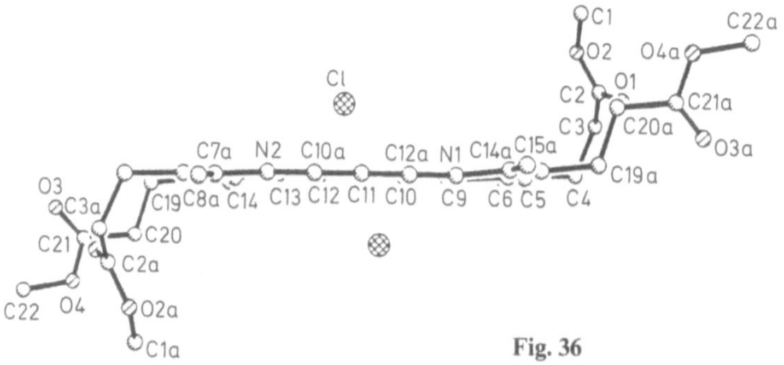

Fig. 36

11.2 Tetravinylogous Porphyrins

The natural progression from bisvinylogous expanded porphyrins is to systems in which all four of the normally one atom *meso* bridges are enlarged. Compounds of this class were first reported by Franck and Gosmann in 1986 [197]. These workers used an acid catalyzed self-condensation of the N-protected, pyrrole substituted allyl alcohol **292** to obtain macrocycle **293** (Scheme 55). Oxidation of **293** with bromine then gave the "tetravinylogous" expanded porphyrin **294**. As with all of these "stretched" compounds, the ^{1}H NMR spectrum of **294** confirms the presence of a strong diamagnetic ring current in the macrocycle. The internal methine proton signal is found at $\delta = -11.46$ ppm and the external methine resonances are found at $\delta = 13.67$ ppm. The protons of the N-methyl blocking groups are also shifted to very high field ($\delta = -9.09$ ppm). In addition, a very intense Soret-like band is observed at 547 nm ($\varepsilon = 909{,}600$ cm $\cdot$ mol^{-1}) in the UV/visible spectrum with no other absorption features being reported [197].

Franck and co-workers also prepared the next higher homolog in the series by using pyrrylpentadienol instead of **292** in the initial condensation procedure

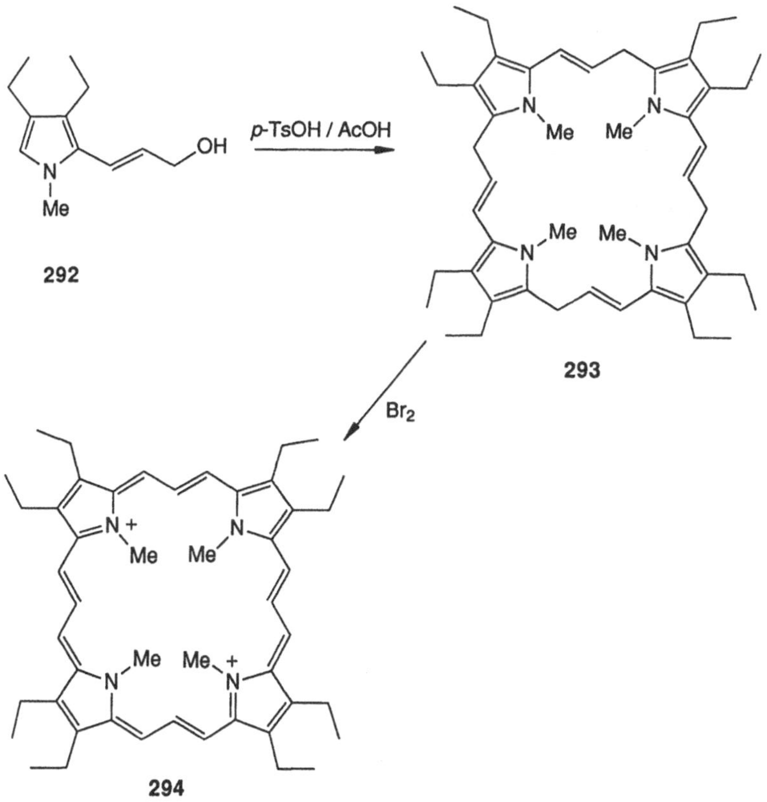

Scheme 55

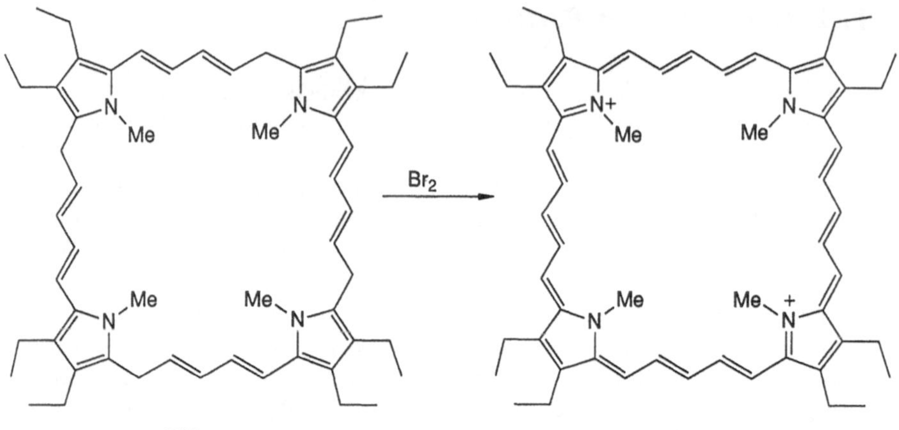

295   **296**

Scheme 56

(to give compounds **295** and **296**; Scheme 56) [198]. Interestingly, however, in addition to the tetrapyrrolic macrocycle **295**, the pentapyrrolic macrocycle **297** is also formed in small amounts during the condensation procedure. Oxidation of **295** with bromine gives the fully conjugated expanded porphyrin **296**. The macrocycle **296** forms deep blue solutions and shows evidence for an extraordinarily large ring current in the ^{1}H NMR spectrum. The external protons of the methine bridges now resonate at $\delta = 16.18$ and 17.19 ppm while the internal methine resonate at $\delta = -14.27$ ppm. In addition, the protons of the nitrogen methyl substituents are now found at even higher field ($\delta -11.44$ ppm) than those of **294**

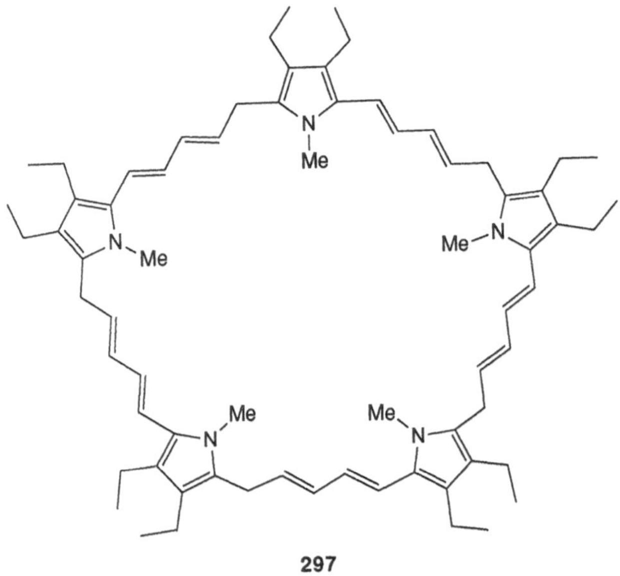

297

(c.f. Scheme 55). Two interesting and very intense absorption features at 663 nm ($\varepsilon = 370{,}000$ cm · mol^{-1}) and 997 nm (2,400 cm · mol^{-1}) were also reported for this system [198].

11.3 Bisvinylogous Porphycenes

A new series of "bisvinylogous porphycene" compounds have been reported by Vogel and co-workers [199]. These compounds are "extended" forms of porphycene, an isomer of porphine [4, 200]. The self-condensation of the bis-pyrroleacetylene derivative **298** was used to obtain the expanded porphyrin **300**, presumably *via* the intermediate **299** (Scheme 57). The ^{1}H NMR spectrum of **300** reveals a single resonance for the methine proton at $\delta = 9.99$ ppm and a signal for the NH protons at δ 2.28 ppm. The UV/visible spectrum of **300** is considerably red shifted in comparison to that of porphycene [4]. An intense Soret-like absorption

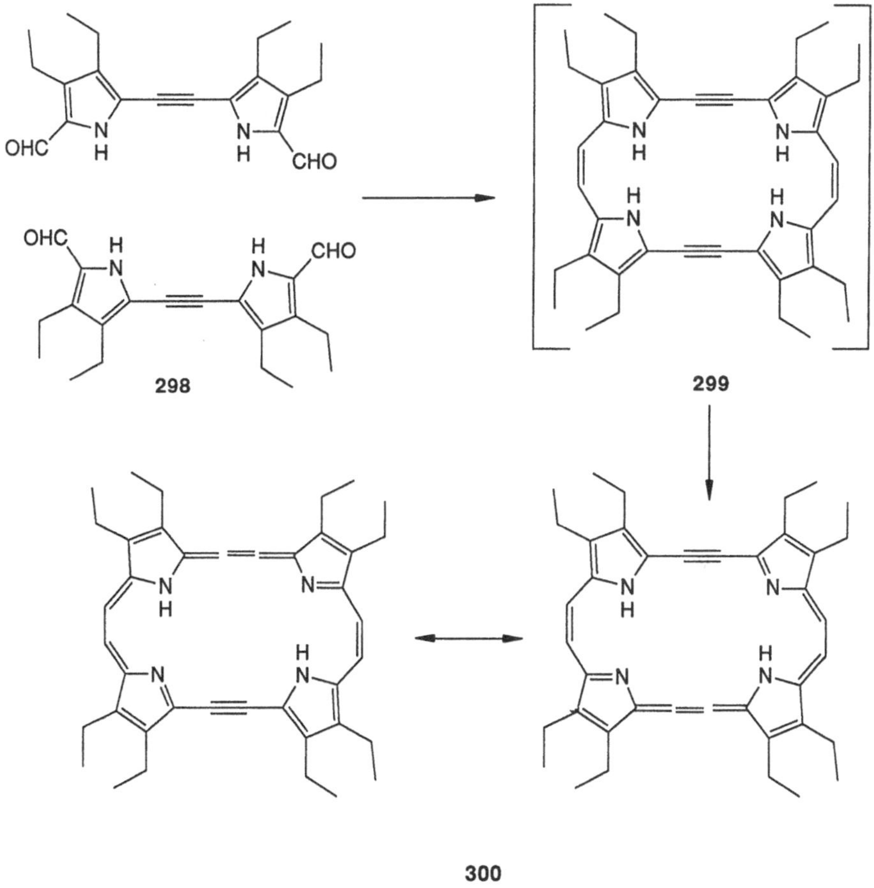

300

Scheme 57

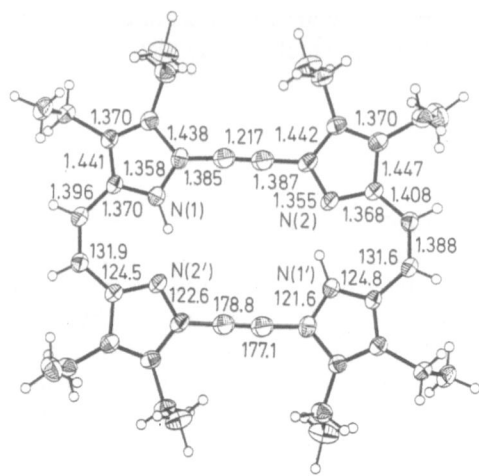

Fig. 37

at 405 nm ($\varepsilon = 188{,}700$ cm · mol^{-1}) is observed with other relatively strong absorption features also being found in the 651–766 nm spectral region. A single X-ray crystal diffraction study of **300** confirmed the planarity of the macrocycle (Figs. 37 and 38), and the expected linear nature of the acetylene and cumulene bridges [199].

Another bisvinylogous porphycene compound was also generated by the reduction of **300** (Scheme 58) [201]. In this latter expanded porphyrin (i.e. **301**) the acetylene and cumulene bridges of **300** have been converted into two sets of two methines. An alternative synthesis of **301** is also shown in Scheme 58. The structure of **301**, as suggested by ^{1}H NMR spectroscopic studies, consists of alternating *cis* and *trans* olefinic bridges between the pyrrolic subunits. In this compound, the internal protons of the *trans* bridge occur at $\delta = -7.5$ ppm while those on the exterior resonate at $\delta = 11.70$ ppm. In contrast, the protons of the *cis* bridge occur at $\delta = 9.83$ and 9.88 ppm, respectively. The visible spectrum of **301** is characterized by strong absorption features at 270 nm ($\varepsilon = 14{,}400$ cm mol^{-1}), 440 nm ($\varepsilon = 207{,}900$ cm · mol^{-1}), 464 nm ($\varepsilon = 104{,}400$ cm · mol^{-1}), 672 nm ($\varepsilon = 38{,}900$ cm · mol^{-1}), 726 nm ($\varepsilon = 17{,}100$ cm · mol^{-1}) and 790 nm

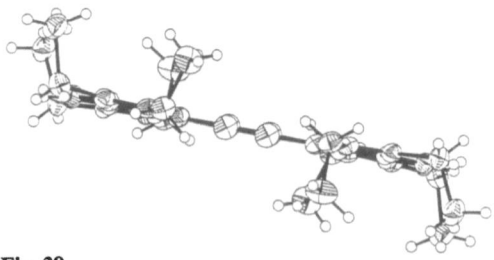

Fig. 38

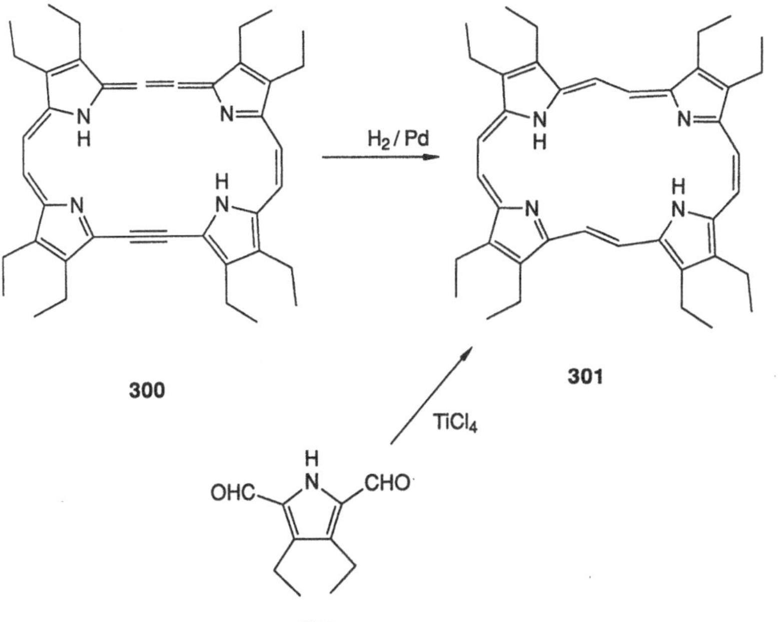

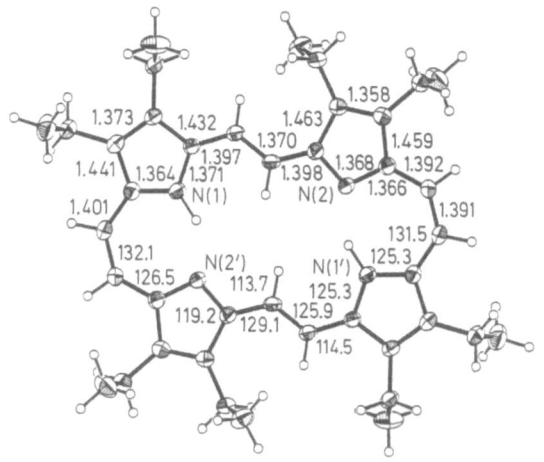

Scheme 58

$(\varepsilon = 57{,}200 \text{ cm} \cdot \text{mol}^{-1})$. The single crystal X-ray structure of **301** (Figs. 39 and 40) served to confirm the postulated alternating *cis-trans-cis-trans* nature of the bridging ethylenes. It also established unequivocally the planar nature of the macrocycle which, of course, was to be expected for such an aromatic system [201]. As a result of its unusual light absorption properties (*viz.* absorptions in

Fig. 39

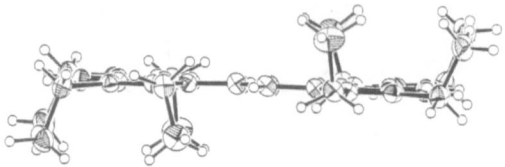

Fig. 40

the 650–800 nm spectral region) future investigations into the possible use of **301** as a sensitizer for photodynamic therapy are sure to follow.

In terms of both size and spectral properties, this class of expanded porphyrins is among the most interesting. The observation of extreme diamagnetic NMR ring currents and strong low-energy absorption features are particularly noteworthy. With regards to the latter, we note that, at present, only two preliminary reports related to the study of vinylogous porphyrins as potential photosensitizers have appeared [196, 202]. Thus, it is clear that further study will be needed to establish the extent to which this class of molecules will be of use in photodynamic applications.

12 Potential Applications and Future Directions

The major emphasis of the research carried out to date in the expanded porphyrin area has been concerned with the synthesis and characterization of new systems. However, significant effort has been devoted recently to exploring the use of these macrocycles as sensitizers for photodynamic therapy (PDT) and as magnetic resonance imaging (MRI) contrast agents. These two applications serve as the most visible of several potential areas where expanded porphyrins could provide an improvement over existing technology. In this section, therefore, we review the work to date relative to both PDT and MRI and also comment briefly about the other areas of potential utility.

12.1 Photodynamic Therapy (PDT)

At present, diamagnetic porphyrins and their derivatives are the dyes of choice for PDT. For a considerable time it has been known that porphyrins localize selectively in rapidly growing tissues such as sarcomas and carcinomas [203–205], although the reasons for this selectivity remain unknown. Most attention to date has focused on the hematoporphyrin derivative (HPD) [203–215] an incompletely characterized mixture of monomeric and oligomeric porphyrins produced by treating hematoporphyrin dihydrochloride with acetic acid-sulfuric acid followed by dilute base [83, 216–220]. The mechanism of the action is thought to be largely, if not entirely, due to the photoproduction of singlet oxygen ($O_2(^1\Delta g)$), although alternative mechanisms of action, including those involving superoxide anion or hydroxyl and/or porphyrin-based radicals cannot be entirely ruled out [221–224].

Singlet oxygen is also believed to be the critical toxic species operative in experimental photosensitized blood purification procedures [81, 225–230]. This new application of photodynamic therapy (termed PDI for photodynamic inactivation) is of tremendous potential importance. It shows promise, as yet not realized clinically, of providing a safe and effective means of removing enveloped viruses such as HIV-1, cytomegalovirus (CMV), various forms of hepatitis, as well as other opportunistic blood-borne infections, such as *Trypanosoma cruzi* (the causative agent of Chagas' disease) and malaria plasmodium, from transfused blood. Given that blood products are major vectors for the transmission of a number of diseases, the potential of such a blood purification procedure is clear.

As critical as the potential PDI and PDT applications, which are currently being explored using the hematoporphyrin derivative (HPD) and its partially purified active fractions dihematoporphyrin ether (DHE), it is important to realize that these "first generation" photosensitizers are not ideal. They contain a range of chemical species, they are neither catabolized nor excreted rapidly from the body, and they absorb poorly in the red part of the spectrum where blood and other bodily tissues are most transparent [90, 205, 231–234]. This latter deficiency is particularly relevant to expanded porphyrin research. This is because the longest wavelength absorption maximum of HPD (and DHE) falls at 630 nm. As a result, most of the incipient energy is either dispersed or attenuated before reaching a blood-borne pathogen and/or the center of a deep-seated tumor. Less of the initial light is, therefore, available for singlet oxygen production and photodynamic action [90, 205, 231–234]. It is thus apparent that far more effective photosensitizers could be developed if dyes could be prepared which absorb in the ca. 690 nm region, provided, of course, that they retain the desirable features of HPD and DHE (e.g. selective localization on pathogens, low dark toxicity, and efficient photosensitization). This, of course, is where an expanded porphyrin approach could provide an advantage.

Initial studies carried out to date suggest that certain expanded porphyrins such as the sapphyrins [178, 180, 181], texaphyrins [94], and vinylogous porphyrins [202] possess unique long-wavelength light absorbing and singlet oxygen producing properties that make them attractive as potential photosensitizers for use in anti-viral blood purification and/or tumor phototherapy. For instance, a decaalkyl sapphyrin derivative has been shown to be effective for the PDI eradication of cell-free HIV-1 and a cadmium(II) texaphyrin complex has been found active for the *in vitro* inactivation of human leukemic cells [94, 180, 181]. Given these promising results, it is likely that, with further research, these and other expanded porphyrins could emerge to play a significant role in this field.

12.2 Magnetic Resonance Imaging (MRI) Contrast Agents

New techniques that may allow neoplastic tissue to be observed and recognized at the early stages of development are currently attracting considerable interest. One such promising technique is magnetic resonance imaging (MRI) [96–101,

235–237]. Although new, this noninvasive, non-ionizing, apparently innocuous method, is now firmly entrenched as a diagnostic tool. Unfortunately, however, the degree of signal enhancement for diseased *vs.* normal tissues is often insufficient to allow this approach to be used in many clinical situations. To overcome this problem (lack of signal intensity enhancement) considerable effort is currently being devoted to the preparation of MRI contrast reagents. Here, highly paramagnetic metal complexes, such as those derived from gadolinium(III) (which has seven unpaired electrons), have proved particularly efficient in clinical use and/or preclinical tests.

In most cases reported to date the coordination of gadolinium in MRI contrast agents has been achieved using carboxylate-type ligands [96–101]. As a result the systems under current consideration are all of high thermodynamic stability but nonetheless high intrinsic lability. Expanded porphyrins on the other hand offer the possibility of binding Gd(III) (which is too large to fit into a normal porphyrin) in a stable in-plane porphyrin-like manner. As a result, they could provide an improved approach to MRI contrast agent development. Recent work by Sessler and co-workers has shown that at least one expanded porphyrin forms extremely stable Gd(III) complexes *in vitro* [95]. Here, as expected, the large macrocyclic core of texaphyrin provides a stable coordination environment for the ca. 1.0 Å ionic radius Gd(III) cation. Investigations into other expanded porphyrin systems, which have suitable core sizes, may give complexes capable of acting as viable MRI agents while meeting all the necessary accompanying biological requirements, such as low toxicity, good tissue localization and high *in vivo* relaxivity.

12.3 Other Potential Applications and Future Directions

Describing possible applications and areas of future research in which the expanded porphyrins may play a role is somewhat difficult task. This is not because this class of compounds is lacking in potential. Rather, on the contrary, it is because of the large array of possible uses to which the expanded porphyrins are suited, that it is difficult at present not to sound over-enthusiastic. One has only to consider the amount of literature devoted to the properties of the porphyrins, to begin to appreciate what could be a bright future for the expanded porphyrins. While initial investigations have, for the most part, focused on topics of purely academic interest, more and more information concerning the possible practical applications of expanded porphyrins is becoming available. Here, we wish to present a few of the, as yet unrealized, potential areas where expanded porphyrins may be useful. This subsection is thus designed to complement the above discussion of PDT and MRI and also set the stage for possible future work in other areas.

First, one can envision the expanded porphyrins being useful in a number of applications involving metal chelation: The increased core size of these macrocycle could prove particularly useful for the coordination of large cations. This would suggest their utility as detoxification chelators in medical situations. One could also conceive that the expanded porphyrins could prove of particular use in the removal of large cations, such as those of the lanthanides and actinide series, from radioactive or non-radioactive waste water.

Second, in addition to the above, the fact that many expanded porphyrins are highly colored makes their use as dyes an obvious possibility. Here their planar nature makes them particularly attractive as chromophores for use in liquid crystals and optical data storage applications. Also, these properties could make them of interest as photo-sensors in various clinical or pseudo-clinical situations. For instance, the high affinity by certain sapphyrins for enveloped viruses and cholesterol rich liposomes suggests that expanded porphyrins could be used to detect and/or destroy a variety of unwanted biological targets, including arterial sclerotic plaque.

Finally, quite apart from any practical applications, several features of the expanded porphyrins could make them of considerable academic interest. For instance synthetic porphyrins have proved extremely useful in determining the biosynthetic pathways and reaction mechanisms of a variety of metalloproteins and enzymes which incorporate porphyrins or related tetrapyrrolic complexes as the active prosthetic groups. It may be of interest to examine these systems (and the associated chemistry) using expanded porphyrins. The different ligation and optical properties offered by the expanded porphyrins could give rise to interesting new behavior and provide further insight into even the best studied of these systems. In particular, the optical properties of certain expanded porphyrins could make them of interest as models for the red-absorbing chlorophyll and bacteriochlorophyll centers of various photosynthetic systems. Here, one idea might be to explore the use of various expanded porphyrins as novel photo-donors in synthetic charge separating systems.

With almost all of the conceivable coordination chemistry of the expanded porphyrins still left to be explored, it cannot be over-stressed that the potential for new chemistry is enormous. This is particularly true when account is made of the fact that the chemistry of the metalloporphyrins has played a dominant role in modern inorganic chemistry. What with the possibility to enhance the stability of unusual coordination geometries (and, perhaps oxidations states) and the ability to form stable coordination complexes with a variety of unusual cations including those of the lanthanide and actinide series, the potential for new inorganic and organometallic discoveries are almost unlimited. For instance, as with the porphyrins, one may envision linear arrays of stacked expanded porphyrin macrocycles which may have unique conducting properties and/or which could display beneficial super- or semiconducting capabilities. Here, of course, the ability to coordinate not only to cations but also to anions could prove to be of tremendous utility.

To conclude, the proposals mentioned here by no means represent the limits of the possible, but merely the most obvious next directions that we the authors conceive for the future development of the field. Certainly, with the range and scope of discoveries already recorded in the relatively short time that this class of compounds has been known, we feel confident in predicting that these macrocycles will continue to remain among the most intriguing of those macrocycle chemical entities currently available.

13 Acknowledgments

The authors would like to express gratitude to Dr. Vincent Lynch and Mr. Thaddeus George for their help in redrawing the X-ray structures. Also, we wish to thank Professors Vogel, LeGoff, Smith, and Gossauer for providing us with material prior to publication. We would also like to acknowledge research support from the National Institute of Health (AI 28845). J.L.S. is a Presidential Young Investigator (1986), a Camille and Henry Dreyfus Foundation Teacher-Scholar (1988–1992) and a Sloan Foundation Fellow (1989–1991).

14 References

1. Dolphin D (ed) (1978) The Porphyrins, Academic, New York
2. Sessler JL, Cyr M, Murai T (1988) Comments Inorg Chem 7: 333
3. Grigg R (1978) In: Dolphin D (ed) The Porphyrins; Academic, New York, vol II, p 351
4. Vogel E, Köcher M, Schmickler H, Lex J (1986) Angew Chem Int (ed) Engl 25: 257
5. Callot HJ, Tschamber T (1974) Tetrahedron Lett 36: 3155
6. Callot HJ, Tschamber T (1975) J Am Chem Soc 97: 6175
7. Chevrier B, Weiss R (1974) J Chem Soc, Chem Commun 884
8. Chevrier B, Weiss R (1975) J Am Chem Soc 97: 1416
9. Callot HJ, Tschamber T (1974) Tetrahedron Lett 36: 3159
10. Callot HJ, Tschamber T, Schaeffer E (1975) J Am Chem Soc 97: 6178
11. Callot HJ, Tschamber T, Schaeffer E (1975) Tetrahedron Lett 33: 2919
12. Callot HJ, Tschamber T, Schaeffer E (1977) J Org Chem 42: 1567
13. Chevrier B, Weiss R (1976) Inorg Chem 4: 770
14. Callot HJ, Schaeffer E (1978) Tetrahedron 34: 2295
15. Louati A, Schaeffer E, Callot HJ, Gross M (1978) Nouv J Chem 2: 163
16. Callot HJ, Schaeffer E (1978) J Chem Research (S), 51
17. Callot HJ, Schaeffer E (1978) J Chem Research (M), 0690
18. Louati A, Schaeffer E, Callot HJ, Gross M (1979) Nouv J Chem 3: 191
19. Liddell PA, Olmstead MM, Smith KM (1990) J Am Chem Soc 112: 2038
20. Swanson KL, Snow KM, Jeyakumar D, Smith KM (1991) Tetrahedron Symposium in Print (in press)
21. Shiau F-Y, Liddell PA, Vicente GH, Ramana NV, Ramachandran K, Lee S-J, Pandey RK, Dougherty TJ, Smith KM (1989) SPIE Future directions and applications in photodynamic therapy, IS 6: 71
22. Grigg R (1967) J Chem Soc, Chem Commun 1238
23. Grigg R (1971) J Chem Soc, (C) 3664
24. Broadhurst MJ, Grigg R, Johnson AW (1969) J Chem Soc, Chem Commun 23
25. Broadhurst MJ, Grigg R, Johnson AW (1970) J Chem Soc, Chem Commun 807
26. Broadhurst MJ, Grigg R, Johnson AW (1972) J Chem Soc, Perkin Trans 1 2111
27. Broadhurst MJ, Grigg R, Johnson AW (1969) J Chem Soc, Chem Commun 1480
28. Busch DH (1978) Acc Chem Res 11: 392
29. Busch DH (1967) Helv Chim Acta 5: 174
30. Busch DH, Farmery K, Goedken V, Katovic V, Melnyk AC, Sperati, Tokel N (1971) Adv Chem Ser 100: 44
31. Nelson SM (1980) Pure and Appl Chem 52: 4261
32. Fenton DE (1986) Pure and Appl Chem 58: 1437
33. Lindoy LF and Busch DH in: Jolly W (ed) Preparative inorganic reactions. Interscience, New York, vol 6

34. Curtis NF (1968) Coord Chem Rev 3: 3
35. Cook DH, Fenton DE, Drew MGB, McFall SG, Nelson SM (1977) J Chem Soc, Dalton Trans 446
36. Drew MGB, Rodgers A, McCann M, Nelson SM (1978) J Chem Soc, Chem Commun 415
37. Nelson SM, Knox CV, McCann M, Drew MGB (1981) J Chem Soc, Dalton Trans 1669
38. McKee V, Shepard WB (1985) J Chem Soc, Chem Commun 158
39. Brooker S, McKee V, Shepard WB, Pannell LK (1987) J Chem Soc, Dalton Trans 2555
40. Drew MGB, Esho FS, Nelson SM (1982) J Chem Soc, Chem Commun 1347
41. Drew MGB, Esho FS, Nelson SM (1983) J Chem Soc, Dalton Trans 1 653
42. Nelson SM, Esho F, Lavery A, Drew MGB (1983) J Am Chem Soc, 105: 5693
43. Nelson SM, Esho FS, Drew MGB (1981) J Chem Soc, Chem Commun 388
44. Drew MGB, Yates PC, Trocha-Grimshaw J, McKillop KP, Nelson SM (1985) J Chem Soc, Chem Commun 262
45. Kauffmann T, Albrecht J, Berger D, Legler J (1967) Angew Chem Int (ed) Engl 6: 633
46. Cotton FA, Wilkinson G (eds) (1980) Advanced inorganic chemistry. 4th edn, Wiley, New York p 804
47. Abid KK, Fenton DE (1984) Inorg Chim Acta 82: 223
48. Drew MGB, Yates PC (1987) J Chem Soc, Dalton Trans 2563
49. Fenton DE, Murphy BP, Leong AJ, Lindoy LF, Bashall A, McPartlin M (1987) J Chem Soc, Dalton Trans 2543
50. Fenton DE, Moody R (1987) J Chem Soc, Dalton Trans 219
51. Adams H, Bailey NA, Fenton DE, Good RJ, Moody R, Rodriguez de Barbarin CO (1987) J Chem Soc, Dalton Trans 207
52. Bailey NA, Eddy MM, Fenton DE, Jones G, Moss S, Mukhopadhyay A (1981) J Chem Soc, Chem Commun 628
53. Bailey NA, Eddy MM, Fenton DE, Jones G, Moss S, Mukhopadhyay A, Jones G (1984) J Chem Soc, Dalton Trans 2281
54. Miller R, Olsson K (1981) Acta Chem Scand Ser B 35: 303
55. Adams H, Bailey NA, Fenton DE, Moss S (1984) Inorg Chim Acta 83: L 79
56. Adams H, Bailey NA, Fenton DE, Moss S, Rodriguez de Barbarin CO, Jones G (1986) J Chem Soc, Dalton Trans 693
57. Acholla FV, Mertes KB (1984) Tetrahedron Lett 25: 3269
58. Acholla FV, Takusagawa F, Mertes KB (1985) J Am Chem Soc 107: 6902
59. Sessler JL, Johnson MR, Lynch V (1987) J Org Chem 52: 4394
60. Sessler JL, Murai T, Lynch V, Cyr M (1988) J Am Chem Soc 110: 5586
61. Sessler JL, Morishima T, Mody TD, Hemmi G (1991) Abstract of 201st National Am Chem Soc Meeting, Inorganic Division
62. Sessler JL, Johnson MR, Lynch V, Murai T (1988) J Coord Chem 18: 99
63. Chem Eng News (1988) (Aug 8) p 26
64. Whitlock HW Jr, Buchanan DH (1969) Tetrahedron Lett 42: 3711
65. Sessler JL, Murai T, Lynch V (1989) Inorg Chem 28: 1333
66. Bauer VJ, Clive DLJ, Dolphin D, Paine III JB, Harris FL, King MM, Loder J, Wang S-WC, Woodward RB (1983) J Am Chem Soc 105: 6429
67. Gouterman M (1978) In: Dolphin D (ed) The Porphyrins, Academic Press, New York, vol III, chapt 1
68. Becker RS, Allison JB (1963) J Phys Chem 67: 2669
69. Hoard JL (1976) in: Smith KM (ed) Porphyrins and metalloporphyrins, Elsevier, Amsterdam, chapt 8
70. Kennedy MA, Sessler JL, Murai T, Ellis PD (1990) Inorg Chem 29: 1050
71. Regev A, Berman A, Levanon H, Murai T, Sessler JL (1989) Chem Phys Lett 160: 401
72. Regev A, Levanon H, Murai T, Sessler JL (1990) J Chem Phys 92: 4718
73. Michl J, (1978) J Am Chem Soc 100: 6801
74. Waluk J, Michl J (1991) J Org Chem 56: 2729
75. Waluk J, Hemmi G, Sessler JL Michl J (1991) J Org Chem 56: 2735
76. Gomer CJ (1987) Photochem Photobiol 46: 561

77. Dahlman A, Wile AG, Burns RG, Mason GR, Johnson FM, Berns MW (1983) Cancer Res 43: 430
78. Dougherty TJ (1985) in: Kessel D (ed) Methods in porphyrin photosensitization. Plenum Press, New York, p 313
79. Dougherty TJ (1987) Photochem Photobiol 45: 879
80. Gomer CJ (1989) Seminars in hematology 26: 27
81. Matthews JL, Newsam JT, Sogandares-Bernal F, Judy ML, Skiles H, Levenson JE, Marengo-Rowe AJ, Chanh TC (1988) Transfusion 28: 81
82. Detty MR, Merkel PB, Powers SK (1988) J Am Chem Soc 110: 5920
83. Bonnett R, McGarvey DJ, Harriman A, Land EJ, Truscott TG, Winfield U-J (1988) Photochem Photobiol 48: 271
84. Bonnett R, Ioannou S, White RD, Winfield U-J, Berenaum MC (1987) Photochem Photobiolphys 45
85. Scourides PA, Bohmer RM, Kaye AH, Morstyn G (1987) Cancer Res 47: 3439
86. Berenbaum MC, Akande SL, Bonnett R, Kaur H, Ioannou S, White RD, Winfield U-J (1986) Br J Cancer 54: 717
87. Spikes JD (1986) Photochem Photobiol 43: 691
88. Kessel D, Dutton CJ (1984) Photochem Photobiol 40: 403
89. Firey PA, Rodgers MAJ (1987) Photochem Photobiol 45: 535
90. Wan S, Parrish JA, Anderson RR, Madden M (1981) Photochem Photobiol 34: 679
91. Harriman A, Maiya BG, Murai T, Hemmi G, Sessler JL, Mallouk TE (1989) J Chem Soc Chem Commun 314
92. Harriman A, Maiya BG, Murai T, Hemmi G, Sessler JL, Mallouk TE (1990) J Phys. Chem 93: 8111
93. Maiya BG, Mallouk TE, Hemmi G, Sessler JL (1990) Inorg Chem 29: 3738
94. Sessler JL, Hemmi G, Maiya BG, Harriman A, Judy ML, Boriak R, Matthews JL, Ehrenberg B, Malik Z, Nitzan Y, Ruck A (1991) SPIE Soc 1426: 318
95. Sessler JL, Murai T, Hemmi G (1989) Inorg Chem 28: 3390
96. Lauffer RB (1987) Chem Rev 87: 901
97. Kornguth SE, Turski PA, Perman WH, Schultz R, Kalinke T, Reale R, Raybaud F (1987) J Neursurg 66: 898
98. Koenig SH, Spiller M, Brown RD, Wolf GL (1986) Invest Radiol 21: 697
99. Cacheris WP, Nickel SK, Sherry AD (1987) Inorg Chem 26: 958
100. Loncin MF, Desreux JF, Merciny E (1986) Inorg Chem 25: 2646
101. Chang CA, Sekhar VC (1987) Inorg Chem 26: 1981
102. Horrocks WD, Hove EG (1978) J Am Chem Soc 100: 4386
103. Lyon RC, Faustino PJ, Cohen JS, Katz A, Mornex F, Colcher D, Baglin C, Koenig SH, Hambright P (1987) Magn Reson Med 4: 24
104. Jelicks LA, Gupta RK (1989) J Biol Chem 264: 15230
105. Szklaruk J, Marecek JF, Springer AL, Springer CS Jr (1990) Inorg Chem 29: 660–667
106. Yarmush D (1983) NMR Studies of detergent-induced transmembrance cation transport, Ph.D. Dissertation State University New York, Stony Brook, NY
107. Sherry AD, Mallory CR, Jeffrey FMH, Cacheris WP, Geraldes CFGC (1988) J Magn Reson 76: 528
108. Balschi JA, Bittl JA, Springer CS, Ingwall JS (1991) NMR Biomed in press
109. Gupta RK, Gupta P (1982) J Magn Reson 47: 344
110. Mody TD, Sessler JL, Sherry AD, Ramasamy R (1991) New J Chem in press
111. Sink RM, Buster DC, Sherry AD (1990) Inorg Chem 29: 3645
112. Day VW, Marks TJ, Wachter WA (1975) J Am Chem Soc 97: 4519
113. Frigerio NA, Coley RF (1963) J Inorg Nucl Chem 25: 1111
114. Kobyschev GK, Lyalin GN, Terenin AN (1963) Dokl Akad Nauk SSSR 148: 1053, 1294
115. Lyalin GN, Kobyshev GK (1963) Opt i spektroskopiya 15: 253
116. Bloor JE, Schlabitz J, Walden CC, Demerdache A (1964) Can J Chem 42: 2201
117. Lux F (1973) Proceedings, Tenth rare earth research conference, Carefree, Ariz. p 871
118. Marks TJ, Stojakovic DR (1978) J Am Chem Soc 100: 1695
119. Cuellar EA, Stojakovic DR, Marks T (1980) Inorg Synth 20: 97

120. Keen IM (1964) Platinum Met Rev 8: 143
121. Keen IM, Malerbi BW (1965) J Inorg Nucl Chem 27: 1311
122. Robertson JM (1936) J Chem Soc 1195
123. Hoskins BF, Mason SA, White JCB (1969) J Chem Soc, Chem Commun 554
124. Brown CJ (1968) J Chem Soc, A 2488
125. Robertson JM, Woodward I (1937) J Chem Soc, 219
126. Brown CJ (1968) J Chem Soc, A 2494
127. Cuellar EA, Marks TJ (1981) Inorg Chem 20: 3766
128. Clavijo RE, Aroca R, Kovacs GJ, Jennings CA, Duff J, Loutfy RO (1989) J Raman Spect 20: 461
129. Lever ABP (1965) Adv Inorg Chem Radiochem 7: 27
130. Moser FA, Thomas AL (eds) (1963) Phthalocyanine complexes, Reinhold, New York, NY
131. Smith KM (ed) (1975) Porphyrins and metalloporphyrins, Elsevier, Amsterdam
132. Marks TJ, Stojakovic DR (1975) J Chem Soc Chem Commun 28
133. Silver J, Jassim QAA (1988) Inorg Chem Acta 144: 281
134. Hückel E (1931) Z Phys Chem 70: 204
135. Hückel E (1931) Z Phys Chem 72: 310
136. Hückel E (1932) Z Phys Chem 76: 628
137. Ginsberg D (1959) Non-benzenoid aromatic compounds, Interscience Publishers, Inc New York, NY
138. Loyd D (1966) Carbocyclic Non-benzenoid aromatic compounds, Elsevier Publishing Co New York, NY
139. Garrat PJ, Sargent MV (eds) (1975) Advances in organic chemistry, methods and results. Interscience Publishers, Inc New York, NY vol. 6
140. Sondheimer F (1972) Accts Chem Res 5: 81
141. Sondheimer F (1963) Pure Appl Chem 7: 363
142. Longuett-Higgins HC, Salem L (1959) Proc Roy Soc (London) 251a: 172
143. Dewar MJS, Gleicher GJ (1965) J Am Chem Soc 87: 685
144. Vollhardt KP (1975) Synthesis 765
145. Elix JA (1969) Aust J Chem 22: 1951
146. Elix JA, Sargent MV (1968) J Am Chem Soc 90: 1631
147. Saikachi H, Ogawa H, Sato K (1971) Chem Pharm Bull (Tokyo) 19: 97
148. Cresp TM, Sargent MV (1973) J Chem Soc, Perkin Trans 1: 1786
149. Cresp TM, Sargent MV (1974) J Chem Soc, Chem Commun 101
150. Cresp TM, Sargent MV (1974) J Chem Soc, Perkin Trans 1: 2145
151. LeGoff E, Leung W-Y (1988) Organic division, 196th Am Chem Soc Meeting Los Angeles, Organic division, Paper 126
152. First reported by RB Woodward at the aromaticity conference, Sheffield UK 1966
153. King MM (1970) PhD Dissertation, Harvard University, Cambridge MA
154. Johnson AW (1976) In: Smith KM (ed) Porphyrins and metalloporphyrins, Elsevier, Amsterdam, p 750
155. Sessler JL, Cyr MJ, Lynch V, McGhee E, Ibers JA (1990) J Amer Chem Soc 112: 2810
156. Buchler JW (1975) In: Smith KM (ed) Porphyrins and metalloporphyrins, Elsevier, Amsterdam, p 177
157. Sessler JL, Cyr M, Burrell AK (1991) Synlett 2: 127
158. Rexhausen H (1984) PhD Dissertation, University of Berlin, Federal Republic of Germany
159. Burrell AK, Sessler JL, Cyr M, McGhee E, Ibers JA (1991) Angew Chem, Int (ed) Engl 30: 91
160. Takenaka A, Sasada Y, Ogoshi H, Omura T, Yoshida Z-I (1975) Acta Cryst B 31: 1
161. Abeysekera AM, Grigg R, Trocha-Grimshaw J, Viswanatha V, King TJ, (1979) J Chem Soc Perkins 1: 2184
162. Graf E, Lehn J-M (1976) J Amer Chem Soc 98: 6403
163. Suet E, Handel H (1984) Tetrahedron Lett 25: 645
164. Dietrich B, Guilhem J, Lehn J-M, Pascard C, Sonveaux E (1984) Helv Chim Acta 67: 91

165. Newcomb M, Blanda MT (1988) Tetrahedron Lett 29: 4261
166. Dietrich B, Lehn J-M, Guilhem J, Pascard C (1989) Tetrahedron Lett 30: 4125
167. Lindoy LF (1990) The Chemistry of macrocyclic ligands, Cambridge University Press: Cambridge, Chapter 5.
168. Lehn J-M (1988) Angew Chem. Int (ed) Engl 27: 89
169. Potvin PG, Lehn J-M (1987) In: Izatt RM, Christensen JJ (ed) Synthesis of macrocycles (Progress in macrocyclic chemistry, vol 3) Wiley, New York, p 167
170. Vögtle F, Sieger H, Müller WM (1981) Top Curr Chem 98: 107
171. Schmidtchen FP (1986) Top Curr Chem 132: 101
172. Schulthess P, Ammann D, Simon W, Caderas C, Stepánek R, Kräutler B (1984) Helv Chim Acta 67: 1026
173. Hrung CP, Tsutsui M, Cullen DL, Meyer EF Jr (1976) J Amer Chem Soc 98: 7878
174. Fleischer EB (1970) Acc Chem Res 3: 105
175. Scheidt WR (1978) In: Dolphin D (ed) The Porphyrins, Academic Press, New York, vol III, p 463
176. Hirayama N, Takenaka A, Sasada Y, Watanabe E-I, Ogoshi H, Yoshida Z-I (1974) J Chem Soc, Chem Commun 330
177. Kreimer-Birnbaum M (1989) Seminars in hematology 26: 157
178. Maiya BG, Cyr M, Harriman A, Sessler JL (1990) J Phys Chem 94: 3597
179. Levanon H, Regev A, Michaeli S, Galili T, Cyr M, Sessler JL (1990) Chem Phys Lett 174: 235
180. Judy ML, Matthews JL, Newman JT, Skiles H, Boriack R, Cyr M, Maiya BG, Sessler JL (1991) Photochem Photobiol 53: 101
181. Sessler JL, Cyr M, Maiya BG, Judy ML, Newman JT, Skiles H, Boriack R, Matthews JL, Chanh TC (1990) (Photodynamic therapy: Mechanisms II) Proc SPIE Int Opt Eng 1203: 233
182. Rexhausen H, Gossauer A (1983) J Chem Soc, Chem Commun 275
183. Gossauer A (1983) Bull Soc Chim Belg 92: 793
184. Bullock E, Grigg R, Johnson AW, Wasley JWF (1963) J Chem Soc 2326
185. Gossauer A (1984) Chimia 37: 341
186. Gossauer A (1984) Chimia 38: 45
187. Czurylowski M (1987) Thesis University de Fribourg (Suisse)
188. Schumacher K-H, Franck BL (1989) Angew Chem Int (ed) Engl 28: 1243
189. Bogorad L (1979) in Dolphin D (ed) The Porphyrins, vol 6, Academic Press, New York, NY pp 125–178
190. Charriere R (1987) Thesis University de Fribourg (Suisse)
191. Lind E (1987) MSc Thesis, Michigan State University
192. Berger RA, LeGoff E (1978) Tetrahedron Lett 44: 4225
193. LeGoff E, Weaver OG (1987) J Org Chem 52: 710
194. Gosmann M, Vogt A, Franck B (1990) Liebigs Ann Chem 163
195. Konig H, Eickemeier C, Möller M, Rodewald U, Franck B (1990) Angew Chem Int (ed) Engl 29: 1393
196. Beckmann S, Wessel T, Franck B, Hönle W, Borrmann H, Schnering H-G (1990) Angew Chem Int (ed) Engl 29: 1395
197. Gosmann M, Franck B (1986) Angew Chem Int (ed) Engl 25: 1100
198. Knübel G, Franck B (1988) Angew Chem Int (ed) Engl 27: 1170
199. Jux N, Koch P, Schmickler H, Lex J, Vogel E (1990) Angew Chem Int (ed) Engl 29: 1385
200. Vogel E (1990) Pure Appl Chem 62: 557
201. Vogel E, Jux N, Rodriquez-Val E, Lex J, Schmicker H (1990) Angew Chem Int (ed) Engl 29: 1387
202. Franck B, Fulling M, Gosmann G, Mertes H, Schroder D (1988) Proc SPIE Int Soc Opt Eng Ser 5, 997: 107
203. Figge FHJ, Weiland GS (1948) Anat Rec 100: 659
204. Rasmussen-Taxdal DS, Ward GE, Figge FH (1955) Cancer (Phila) 8: 78
205. Berenbaum MC, Bonnett R, Scourides PA (1982) Br J Cancer 47: 571

206. Berns W, Dahlman A, Johnson FM, Burns R, Sperling D, Guiltinan M, Siemens A, Walter R, Wright W, Hammer-Wilson M, Wile A (1982) Cancer Res 42: 2326
207. Evensen JF, Sommer S, Moan J, Chistensen T (1984) Cancer Res 44: 482
208. Gibson SL, Hilf R (1985) Photochem Photobiol 42: 367
209. Herra-Ornelas L, Petrelli NJ, Mittleman A, Dougherty TJ, Boyle DG (1986) Cancer 57: 677
210. Kessel D (1986) Photochem Photobiol 44: 489
211. Kessel D (1986) Int J Radiat Biol 49: 901
212. Klaunig JE, Selman SH, Shulok JR, Schaefer PJ, Britton SL, Goldblatt PJ (1985) Am J Path 119: 230
213. Moan J, Sommer S (1987) Cancer Lett 21: 167
214. Singh G, Jeeves WP, Wilson BC, Jang D (1987) Photochem Photobiol 46: 645
215. Bonnett R, Ridge RJ, Scourides PA (1981) J Chem Soc, Perkin Trans I 3135
216. Chang CK, Takamura S, Musselman BD, Kessel D (1986) ACS Adv Chem Ser 321: 347
217. Dougherty TJ (1987) Photochem Photobiol 46: 569
218. Kessel D (1986) Photochem Photobiol 44: 193
219. Moan J, Christensen T, Somer S (1982) Cancer Lett 15: 161
220. Blum A, Grossweiner LI (1985) Photochem Photobiol 41: 27
221. Henderson BW, Miller AC (1986) Radiat Res 108: 196
222. Keene JP, Kessel D, Land EJ, Redmond RW, Truscott TG (1986) Photochem Photobiol 43: 117
223. Parker JG (1986) Lasers Surg Med 6: 258
224. Tanielian C, Heinrich G, Entezami A (1988) J Chem Soc, Chem Commun 1197
225. Matthews JL, Sogandares-Bernal F, Judy ML, Marengo-Rowe A, Skiles H, Leveson J, Chanh T, Newman J (1989) Transfusion (Suppl 6 S), 28: S 110
226. Newman JT, Matthews JL, Songandares-Bernal F, Judy ML, Skiles H, Leveson J, Marengo-Rowe A, Chanh TC, Dreesman G (1988) Proc Baylor Univ Med Centr and references therein 1: 3
227. Gulliya KS, Matthews JL, Fay JW, Dowben RM (1988) Life Sciences and references therein 42: 2651
228. Skiles H, Sogandares-Bernal F, Judy ML, Matthews JL, Newman JT (1987) Abstracts of 6th southern biomedical engineering conference, 83
229. Chanh TC, Allan JS, Matthews JL, Songandares-Bernal F, Judy ML, Skiles H, Leveson J, Marengo-Rowe A, Newman JT (1989) J Virol Meth 26: 125
230. Lewin AA, Schnipper LE, Crumpacker CS (1980) Proc Soc Exptl Biol Med 163: 81
231. Profio AE, Doiron DR (1987) Photochem Photobiol 46: 591
232. Eichler J, Knop J, Lenz H (1977) Rad Environ Biophys 14: 239
233. Gemert MJC, Welch AJ, Amin AP (1986) Lasers Surg Med 6: 76
234. Welch AJ, Yoon G, van Gemert MJC (1986) Lasers Surg Med 6: 488
235. Morris PG (1986) Nuclear magnetic resonance imaging in medicine and biology, Claredon, Oxford
236. MacKenzie NE, Gooley PR (1988) Med Res Rev 8: 57
237. Tweedle MF, Brittain HG, Eckelman WC, Gaughan GT, Hagan JJ, Wedeking PW, Runge VM (1988) In: Partain CL (ed) Magnetic resonance imaging, 2nd (ed) Philadelphia, vol I, pp 793-809

Addendum

In the period between the submission of this manuscript and it's going to press there have been several significant additions to the chemistry of the expanded porphyrins. While these results will not be described in detail they are nonetheless important with respect to this review.

As we describe in section 8 Woodward and coworkers reported that the sapphyrin macrocycle failed to give any complexes with the uranyl cation [66]. However, it has recently been discovered that the sapphyrin **204** does indeed form a coordination complex with UO_2 [238]. This complex, which has been characterized structurally, is not formally a sapphyrin. In fact, the sapphyrin macrocycle has been attacked, at a *meso*-position by methoxide to give a new and non-aromatic expanded porphyrin complex. Sapphyrins have also attracted interest as anion transport agents. Particularly in the binding and transport of phosphate anion. Effective transport of nucleotides and analogues, AMP and GMP, through a dichloromethane membrane was achieved with the protonated sapphyrin **216** [239]. In other work the use of rubyrin **262** for similar transport was investigated and a specific structural effect has been noted [239] with the protonated form of **262** appearing relatively more effective for the transport of diphosphorylated species such as GDP.

Another expanded porphyrin previously known to form a complex with the uranyl cation was the pentaphyrin **232** [158, 187]. An improved synthesis of a new pentaphyrin derivative and its corresponding, structurally characterized, uranyl complex was recently reported [240]. This new uranyl pentaphyrin, has a very distorted solid state structure reminiscent of the closely related uranyl superphthalocyanine complex **160** [112] (Figures 22 and 23).

Two new expanded porphyrins, that would perhaps best be classified as vinylogous porphyrins, have also been reported in the last several months. Corriu *et al.* have developed a facile synthesis of new tetrapyrrolic macrocyclic derivatives [241]. A notable difference between these new vinylogous porphyrins and those described above (Section 11) is the incorporation of pyridine, in one case, as part of the macrocycle. While these new expanded porphyrins are not aromatic, they may have great potential as ligands. Indeed, the formation of a bimetallic palladium complex was described.

A variation of the general theme of pyrrole containing vinylogous porphyrins has been attempted by Street. In this new system imidazoles replace the pyrroles to give a new coronand class [242].

Lastly a very large expanded porphyrin has been prepared by Bell and coworkers [243]. This new macrocycle has both pyrrole and pyridine heterocycles contained within a large octaaza-macrocycle.

238. Sessler JL, Burrell AK, Cyr MJ, Ford D, Hemmi G, Maiya BG, Murai T, Mody TD, Morishima T, Schreder K (1991) Abstract of Fourth Chemical Congress of North America; Organic division
239. Furuta H, Cyr MJ, Sessler JL (1991) J Am Chem Soc 113: 6677
240. Burrell AK, Hemmi G, Lynch V, Sessler JL (1991) J Am Chem Soc 113: 4690
241. Corriu PJP, Bolin G, Moreau JJE, Vernhet C (1991) J Chem Soc Chem Commun 211
242. Street JP (1991) Abstract of Fourth Chemical Congress of North America; Organic Division
243. Papoulis A, Bell TW (1991) Abstract of Fourth Chemical Congress of North America; Organic Division

Author Index Volumes 151–161

Author Index Vols. 26–50 see Vol. 50
Author Index Vols. 50–100 see Vol. 100
Author Index Vols. 101–150 see Vol. 150

The volume numbers are printed in italics